EMS Textbooks in Mathematics

EMS Textbooks in Mathematics is a series of books aimed at students or professional mathematicians seeking an introduction into a particular field. The individual volumes are intended not only to provide relevant techniques, results, and applications, but also to afford insight into the motivations and ideas behind the theory. Suitably designed exercises help to master the subject and prepare the reader for the study of more advanced and specialized literature.

Jørn Justesen and Tom Høholdt, *A Course In Error-Correcting Codes*
Markus Stroppel, *Locally Compact Groups*
Peter Kunkel and Volker Mehrmann, *Differential-Algebraic Equations*
Dorothee D. Haroske and Hans Triebel, *Distributions, Sobolev Spaces, Elliptic Equations*
Thomas Timmermann, *An Invitation to Quantum Groups and Duality*
Oleg Bogopolski, *Introduction to Group Theory*
Marek Jarnicki und Peter Pflug, *First Steps in Several Complex Variables: Reinhardt Domains*
Tammo tom Dieck, *Algebraic Topology*
Mauro C. Beltrametti et al., *Lectures on Curves, Surfaces and Projective Varieties*
Wolfgang Woess, *Denumerable Markov Chains*

Eduard Zehnder

Lectures on Dynamical Systems

Hamiltonian Vector Fields and Symplectic Capacities

European Mathematical Society

Author:

Eduard Zehnder
Departement Mathematik
ETH Zürich, LEO D 3
Leonhardstrasse 27
8092 Zürich
Switzerland

E-mail: zehnder@math.ethz.ch

2010 Mathematical Subject Classification (primary; secondary): 37-01; 37A25, 37B25, 37C10, 37C20, 37C27, 37C29,37C50, 37C57, 37D05, 37J05, 37J35, 37J45, 37N05, 34C25, 34C28, 34C37, 34C45, 34D20, 70H15,70H05, 70H25, 53D35, 53C24, 53D05, 58E05, 58E30

Key words: Dynamical systems, ergodicity, transitivity, structural stability, stable and unstable manifolds, hyperbolic sets, shadowing lemma, homoclinic and heteroclinic orbits, Bernoulli systems, chaos, limit sets, Lyapunov function, gradient systems, Morse inequalities, symplectic maps, Hamiltonian vector fields, Hamiltonian formalism, integrable systems, symplectic capacities, Hofer–Zehnder capacity, action functional, mini-max principles, periodic motions on prescribed energy surfaces

ISBN 978-3-03719-081-4

The Swiss National Library lists this publication in The Swiss Book, the Swiss national bibliography, and the detailed bibliographic data are available on the Internet at http://www.helveticat.ch.

Contact address:

European Mathematical Society Publishing House
Seminar for Applied Mathematics
ETH-Zentrum FLI C4
CH-8092 Zürich
Switzerland

Phone: +41 (0)44 632 34 36
Email: info@ems-ph.org
Homepage: www.ems-ph.org

Typeset using the author's TEX files: I. Zimmermann, Freiburg
Printed in Germany

9 8 7 6 5 4 3 2 1

Preface

The field of dynamical systems originated in the difficult mathematical questions related to movements of the planets and the moon, questions like: Are there periodic orbits? Will the solar system keep its present beautiful form, also in the distant future, or could it happen that one of the planets, Jupiter for instance, leaves the system? Or could it come to a collision between planets, leading to a dramatic change of the solar system?

The mathematical theory of dynamical systems provides concepts, ideas and tools, in order to analyze and model dynamical processes in all fields of natural sciences, making use of nearly all branches of mathematics. On the other hand, already in the past, questions of dynamical systems in the real world have triggered new mathematical developments and led to whole new branches of mathematics. Here, typical questions would be: Knowing its present state, how will a dynamical system develop in the long run? Will it, for example, tend to an equilibrium state or will it come back to itself? What will happen to the long-time behavior if we change the initial conditions a little bit? And what will happen to the whole orbit structure of a system if we perturb the system itself?

The book addresses readers familiar with standard undergraduate mathematics. It is not a systematic monograph, but rather the lecture notes of an introductory course in the field of dynamical systems given in the academic year 2004/2005 at the ETH in Zürich for third year students in mathematics and physics. I selected relatively few topics, tried to keep the requirements of mathematical techniques minimal and provided detailed (sometimes excruciatingly detailed) proofs.

The introductory chapter discusses simple models of discrete dynamical systems, in which the dynamics is determined by the iteration of a map. There are examples for minimal, transitive, structurally stable and ergodic systems. Mappings that preserve the measure of a finite measure space have strong recurrence properties in view of a classical result due to H. Poincaré. In order to describe the statistical distribution of their orbits, the ergodic theorem of G. Birkhoff is proved.

Chapters II and III are devoted to unstable phenomena caused by a hyperbolic fixed point of a diffeomorphism. Such a point gives rise to two global invariant sets, the so-called stable, respectively unstable, invariant manifolds issuing from fixed point. These consist of points which tend to the fixed point under the iteration of the map and under the iteration of the inverse map, respectively. The transversal intersection of the stable and unstable manifolds in the so-called homoclinic points is one of the roads to chaos. The existence of homoclinic points, discovered by H. Poincaré in the 3-body problem of celestial mechanics, complicates the orbit structure considerably and gives rise to invariant hyperbolic sets. The chaotic structure of the orbits near such sets is analyzed by means of the Shadowing Lemma, which is also used to demonstrate S. Smale's theorem about the embeddings of Bernoulli systems near homoclinic orbits. The interpretation of unpredictable orbits, which are determined by random sequences, will be demonstrated in the simple system of a periodically perturbed mathematical pendulum.

In Chapter IV we deal with smooth flows generated by vector fields and with continuous flows on metric spaces. The concepts of limit set, attractor and Lyapunov function are introduced. The bounded solutions of a gradient-like flow tend to rest points in forward and in backward time. The rest points are found by mini-max principles. The intimate and fruitful relation between the dynamics of gradient flows and the topology of the underlying compact manifold is described by the Morse inequalities. The Morse theory is sketched at the end of the chapter.

Chapter V introduces the special class of Hamiltonian vector fields that are determined by a single function and defined on symplectic manifolds. These manifolds are even-dimensional and carry a symplectic structure. A symplectic structure is a 2-form that is closed and nondegenerate. In contrast to Riemannian structures which do exist on every manifold, not every even-dimensional manifold admits a symplectic structure. Symplectic manifolds of the same dimension are locally indistinguishable (Darboux). There are no local symplectic invariants. The Hamiltonian formalism is developed in the convenient language of the exterior calculus which will be briefly introduced. The very special integrable Hamiltonian systems are characterized by the property that they possess sufficiently many integrals of motion, so that the task of solving the Hamiltonian equations for all time becomes almost trivial. This will follow from the existence of action- and angle-variables established by V. Arnold and R. Jost.

Chapter VI motivates the study of global symplectic invariants different from the volume which will be introduced in Chapter VII. They are called symplectic capacities and go back to I. Ekeland and H. Hofer. In view of their monotonicity properties they represent, in particular, obstructions to symplectic embeddings. The Gromov non-squeezing phenomenon is an immediate consequence. A symplectic capacity of dynamical nature (Hofer–Zehnder capacity) measures the minimal oscillations of Hamiltonian functions needed to conclude the existence of a fast periodic solution of the associated Hamiltonian vector field. Its construction is based on a variational principle for the action functional of classical mechanics. The tools from the calculus of variations are developed from scratch.

Chapter VIII deals with applications of dynamical symplectic capacity to Hamiltonian systems. It turns out that a compact and regular energy surface gives rise to an abundance of periodic orbits nearby, if a neighborhood of the surface possesses a finite Hofer–Zehnder capacity. The existence of a periodic solution on the given energy surface (and not only nearby) necessarily requires additional properties of the surface. The contact type property, for example, immediately leads to the solution of the Weinstein conjecture due to C. Viterbo which generalizes the pioneering results of P. Rabinowitz and A. Weinstein. Finally, a classical result of H. Poincaré shows that a periodic solution of a Hamiltonian system is, in general, not isolated but belongs to a smooth family of periodic solutions parametrized by the energy and having similar periods.

The chapters devoted to Hamiltonian systems and their global periodic orbits related to the symplectic capacities rely heavily on the book *Symplectic Invariants and Hamiltonian Dynamics* [52] by H. Hofer and E. Zehnder and on the *Notes in Dynamical Systems* [74] by J. Moser and E. Zehnder.

Each chapter begins with a short survey of its contents and ends with a brief selection of references to literature giving an alternative view on the subjects or describing related and more advanced topics.

I would like to thank Marcel Nutz for setting the German text into LaTeX, for many improvements and for producing the beautiful figures. I also would like to thank Manfred Karbe and Irene Zimmermann for their encouragement, their patience, their careful editing and for many helpful suggestions. For his generous help in preparing and improving the text of these lectures, I am very grateful to Felix Schlenk. I thank Beata and Kris Wysocki for their patient assistance with technical problems. For her translation of the German text into English, I am very grateful to Jeannette Zehnder.

Zürich, March, 2010

Contents

Chapter I
Introduction

The first chapter is devoted to simple and explicit examples of dynamical systems that illustrate some concepts and help to ask the appropriate questions. For simplicity, the systems under consideration are discrete and hence given by mappings acting on sets. The aim is to study the behavior of points under all iterates of a map (orbits of the points) and also to see what happens under perturbation of a map. A dynamical system consisting of a continuous map acting on a topological space is called transitive, if it possesses a dense orbit. The transitivity of a system will be guaranteed by the criterion of G. Birkhoff. An example of such a system is the rigid irrational rotation of the unit circle where every orbit of the system is dense on the circle. This example will lead us to the equidistribution (mod 1) theorem of H. Weyl. In sharp contrast to the stable systems of rigid rotations, a simple expansive map on the circle shows already a quite chaotic behavior described by the shift map in a sequence space. In this example, orbits of completely different behavior over a long-time interval (many iterates) coexist side by side. In the language of physics, the system shows a sensitive dependence on the initial conditions. It is a typical phenomenon that such an unstable behavior survives under a perturbation of the system, as will be demonstrated by a special case of the so-called structural stability theorem. Measure preserving mappings acting on a measure space will show strong recurrence properties and the question arises, how an orbit of such a system is distributed statistically in the space. As an answer we shall prove the individual ergodic theorem of G. Birkhoff following the strategy designed by A. M. Garsia.

The origin of the field of dynamical systems lies in the deep mathematical problems of celestial mechanics. That is why we shall first recall the N-body system whose dynamics is determined by the Newton equations.

I.1 N-body problem of celestial mechanics

In the N-body problem of celestial mechanics one studies N points $x_k \in \mathbb{R}^3$ in the 3-dimensional Euclidean space having masses $m_k > 0$. The *evolution in time* of these mass points,

$$x_k(t), \quad 1 \leq k \leq N,$$

is determined by the Newton equations

$$m_k \ddot{x}_k = \sum_{j \neq k} m_k m_j \frac{x_j - x_k}{|x_j - x_k|^3}, \quad 1 \leq k \leq N,$$

according to which every mass point x_k is attracted by every other mass point, which gives rise to an extremely complicated dynamics. The equation is not defined at *the collisions* $x_j = x_k$ for $j \neq k$. Introducing the collision sets $\Delta_{ij} = \{x \in \mathbb{R}^{3N} \mid x_i = x_j\}$ and

$$\Delta = \bigcup_{i<j} \Delta_{ij},$$

the Newton equations are defined on the *configuration space*

$$\mathbb{R}^{3N} \setminus \Delta \subset \mathbb{R}^{3N}$$

which is the set of points $x = (x_1, \dots, x_N)$ without collisions. We reformulate the Newton equations as a system of first-order differential equations in the form of a *vector field* as

$$\begin{aligned} \dot{x}_k &= y_k \in \mathbb{R}^3, \\ \dot{y}_k &= \sum_{j \neq k} m_j \frac{x_j - x_k}{|x_j - x_k|^3} \in \mathbb{R}^3. \end{aligned}$$

The *phase space* of the N-body system is the set

$$\Omega = (\mathbb{R}^{3N} \setminus \Delta) \times \mathbb{R}^{3N} \subset \mathbb{R}^{6N}.$$

Denoting the points in the phase space by $z = (x, y) = (x_1, \dots, x_N, y_1, \dots, y_N)$, we write the equations in short form as

$$\dot{z} = V(z), \quad z \in \Omega \subset \mathbb{R}^{6N}.$$

The vector field $V : \Omega \subset \mathbb{R}^{6N} \to \mathbb{R}^{6N}$ is continuously differentiable and consequently, in particular, *locally Lipschitz-continuous*. The development of the system in time is described by a solution, which is a *continuously differentiable curve*

$$t \mapsto z(t) \in \Omega,$$

solving the equation

$$\dot{z}(t) = V(z(t))$$

for the time t in an open interval. Prescribing the point

$$z(0) = (x(0), y(0)) \in \Omega,$$

we are confronted with the Cauchy initial value problem. In view of the classical Cauchy–Lipschitz–Picard theorem in ordinary differential equations, there exists for every initial condition $z(0)$ exactly one solution $z(t)$ satisfying $z(t) = z(0)$ at the time $t = 0$ and this solution exists on an open interval

$$t \in I = I(z(0), V).$$

Locally the Cauchy initial value problem is not of great interest for us. We are interested in a long-term prediction of the future of the system as well as in a reconstruction of its past. Over long-time intervals solutions can behave very differently, even if their initial conditions are very close. Concerning the long-time behavior of the N-body-problem the following natural questions arise.

1. Are there solutions without collisions, which do exist for all times $t \in \mathbb{R}$, hence satisfying the estimate
$$0 < |x_j(t) - x_k(t)| < \infty, \quad j \neq k, \ t \in \mathbb{R}?$$

2. Are there singularities that are not collisions? Are there solutions that explode in finite time $t^* \in \mathbb{R}$ without collisions, so that
$$\operatorname{diam}\{x_1(t), \dots, x_N(t)\} \to \infty, \quad t \to t^* \in \mathbb{R}?$$
This question of P. Painlevé was answered only in 1988 by J. Xia who succeeded in finding such solutions in the $N = 5$-body-problem. His construction relies on the analysis of the 3-body-collision by R. McGehee. The history of the problem is described in [95] by D. Saari and J. Xia.

3. Does our solar system (in which one of the masses is much bigger than the others) keep its present nice shape for all future times or does, in the distant future, one of the planets, e.g. Jupiter, escape, or will a collision provoke a dramatic change of the situation?

The planetary system is an example of a classical dynamical system, defined by an ordinary differential equation on a manifold.

We now turn to discrete dynamical systems which are defined by a mapping acting on a set. The number of iterations of the map plays the role of the time.

I.2 Mappings as dynamical systems

Let X be a set, $x \in X$ a point in the set and let $\varphi\colon X \to X$ be a mapping of the set into itself. Iterating the map φ we obtain the following sequence of points:

$$\begin{aligned}
x_0 &= x = \varphi^0(x),\\
x_1 &= \varphi(x),\\
x_2 &= \varphi(x_1) = \varphi \circ \varphi(x) = \varphi^2(x),\\
&\vdots\\
x_j &= \varphi \circ \cdots \circ \varphi(x) = \varphi^j(x), \quad j \geq 1.
\end{aligned}$$

Definition. Given $x \in X$, the *sequence* $(\varphi^j(x))_{j\geq 0}$ is called the *parameterized orbit* of φ through the point x. The *set*

$$\mathcal{O}^+(x) := \bigcup_{j\geq 0}\{\varphi^j(x)\} \subset X$$

is called the *unparametrized orbit* (or simply *orbit*) of x under the mapping φ.

A point $x \in X$ is called a *periodic point* of φ, if there exists an integer $N \geq 1$ satisfying

$$\varphi^N(x) = x.$$

The corresponding (unparametrized) orbit consists of the *finitely many* points

$$x,\ \varphi(x),\ \varphi^2(x), \dots, \varphi^N(x) = x$$

and satisfies

$$\varphi^{N+1}(x) = \varphi(\varphi^N(x)) = \varphi(x), \quad \varphi^{N+2}(x) = \varphi^2(x), \quad \dots.$$

The sequence $(\varphi^j(x))_{j\geq 0}$ through a periodic point is periodic and the orbit is the finite set

$$\mathcal{O}^+(x) = \bigcup_{0\leq j\leq N-1}\{\varphi^j(x)\} \quad \text{and} \quad |\mathcal{O}^+(x)| \leq N.$$

The cardinality of the set is equal to $|\mathcal{O}^+(x)| = N$ if the period is *minimal*, i.e., if $\varphi^j(x) \neq x$ for $1 \leq j \leq N-1$ and $\varphi^N(x) = x$.

We are interested in the *asymptotic* behavior of the orbits. How do the sequences $(\varphi^j(x))_{j\geq 0}$ behave in the limit as $j \to \infty$?

Example (Contraction principle of Banach). We require additional conditions on the set and the mapping under consideration and assume (X, d) to be a complete metric space and the mapping $\varphi\colon X \to X$ to be a *contraction*. Hence, there exists a constant $0 \leq \lambda < 1$, satisfying

$$d(\varphi(x), \varphi(y)) \leq \lambda\, d(x, y) \quad \text{for all } x, y \in X.$$

It is well known that under these assumptions, there exists precisely one *fixed point* $x^* \in X$ satisfying $\varphi(x^*) = x^*$. In addition, for every point $x \in X$ the orbit $(\varphi^j(x))_{j\geq 0}$ converges to the fixed point,

$$\lim_{j\to\infty} \varphi^j(x) = x^*.$$

In this example the asymptotic behavior is easy to describe. *Every* orbit eventually comes to a standstill at the fixed point x^*. The set $\{x^*\}$ is a *global attractor* of the mapping φ. In addition, the system is *stable under perturbation*!

If the mapping $\varphi\colon X \to X$ is *bijective*, we can iterate also the inverse mapping $\varphi^{-1}\colon X \to X$ and study the (two-sided) *sequences*

$$(\varphi^j(x))_{j\in\mathbb{Z}}.$$

Accordingly, we distinguish between the positive orbit of a point, the negative orbit and the (full) orbit for which we introduce the following notation:

$$\mathcal{O}^+(x) = \bigcup_{j\geq 0} \varphi^j(x), \quad \mathcal{O}^-(x) = \bigcup_{j\geq 0} \varphi^{-j}(x), \quad \mathcal{O}(x) = \bigcup_{j\in\mathbb{Z}} \varphi^j(x).$$

For a given point $x \in X$ we now try to determine *the future as well as to reconstruct the past*!

Example (Rigid rotations). We look at the unit circle $S^1 \subset \mathbb{C}$, defined by

$$S^1 = \{z \in \mathbb{C} \mid |z| = 1\} = \{z = e^{2\pi i x} \mid x \in \mathbb{R}\}$$

and study the rotation φ of the circle around the angle $2\pi\alpha$ for a real number $\alpha \in \mathbb{R}$, in complex notation given by the map

$$\varphi\colon S^1 \to S^1, \quad z \mapsto \theta z, \quad \text{where } \theta = e^{2\pi i\alpha},$$

so that

$$\varphi(e^{2\pi i x}) = e^{2\pi i\alpha} e^{2\pi i x} = e^{2\pi i(x+\alpha)}.$$

In the covering space $\mathbb{R}$ of S^1 the map is the *translation* $\Phi\colon \mathbb{R} \to \mathbb{R}$, $x \mapsto x + \alpha$.

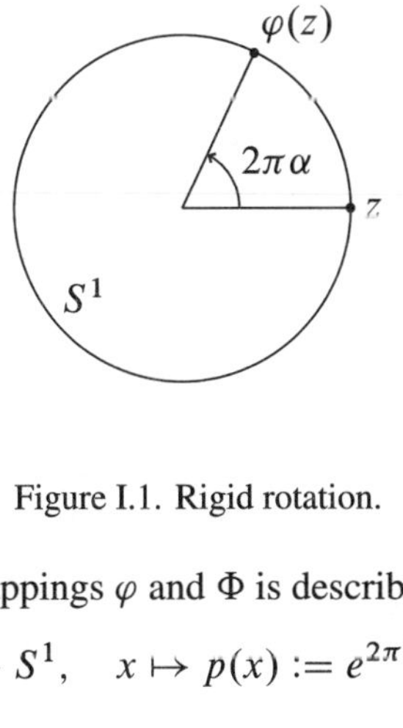

Figure I.1. Rigid rotation.

The relation between the mappings φ and Φ is described by the covering map

$$p\colon \mathbb{R} \to S^1, \quad x \mapsto p(x) := e^{2\pi i x} \in S^1$$

and one reads off that

$$\begin{aligned} p(x + j) &= p(x), \quad j \in \mathbb{Z}, \\ \Phi(x + j) &= \Phi(x) + j, \quad j \in \mathbb{Z} \end{aligned}$$

and

$$\varphi(p(x)) = p(\Phi(x)).$$

The last equation expresses the commutativity of the diagram

$$\begin{array}{ccc} S^1 & \xrightarrow{\varphi} & S^1 \\ {\scriptstyle p}\uparrow & & \uparrow{\scriptstyle p} \\ \mathbb{R} & \xrightarrow{\Phi} & \mathbb{R}. \end{array}$$

On the quotient space $\mathbb{R}/\mathbb{Z}$ the map Φ induces the map $\widehat{\Phi}$ on the equivalence classes defined by

$$\widehat{\Phi}\colon \mathbb{R}/\mathbb{Z} \to \mathbb{R}/\mathbb{Z} \cong S^1, \quad [x] \mapsto [x+\alpha],$$

or, in short notation, by $x \mapsto x + \alpha \bmod 1$. Rigid rotations are *isometric*, if we choose as *metric* the smallest arc length between two points (or the smallest angle),

$$d(e^{2\pi i x}, e^{2\pi i y}) = \min_{j\in\mathbb{Z}} |x - y - j|, \quad x, y \in \mathbb{R}.$$

Then $0 \le d \le 1/2$. The metric on $\mathbb{R}/\mathbb{Z}$ is defined by the same formula,

$$d([x],[y]) = \min_{j\in\mathbb{Z}} |x - y - j|.$$

Proposition I.1. *If $\varphi\colon S^1 \to S^1$ is the rigid rotation $\varphi(z) = e^{2\pi i\alpha} z$, the following holds true.*

(i) *If $\alpha = p/q$ is **rational** and the integers $p, q \in \mathbb{Z}$ relatively prime, then every orbit is **periodic** with the same minimal period q,*

$$\varphi^q(z) = z \quad \text{for all } z \in S^1.$$

(ii) *If $\alpha \in \mathbb{R}\setminus\mathbb{Q}$ is **irrational**, then every orbit is **dense** in S^1, i.e., the closure of an orbit is the circle,*

$$\overline{\mathcal{O}^+(z)} = \overline{\bigcup_{j\ge 0} \varphi^j(z)} = S^1.$$

For a fixed real number α *all* solutions behave in the same way, independent of their initial condition z. We have stability under the perturbation of the initial conditions, because φ is isometric.

In case (i) all solutions are periodic of the same period so that the orbits are finite sets. In case (ii) every orbit is dense and consists of infinitely many points.

Proof. (i) If $z \in S^1$ and $\alpha = p/q$ is a rational number, then $\varphi^j(z) = e^{2\pi i \frac{p}{q} j} z = z$ for $j = q$. The integer q is a minimal period, since p, q are relatively prime and

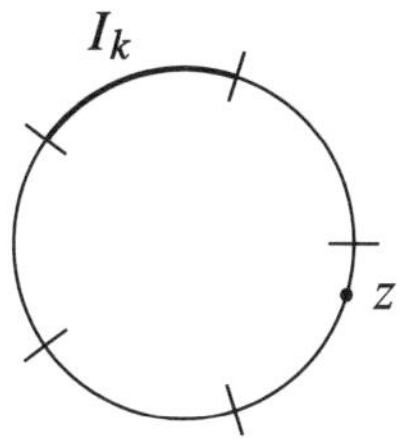

Figure I.2. Dirichlet's pigeon hole principle.

the orbit is the finite set $\mathcal{O}^+(z) = \{z, \varphi(z), \ldots, \varphi^{q-1}(z)\}$. In the covering space $\mathbb{R}$ we have $\Phi^j(x) = x + j\alpha$.

(ii) If $\alpha \in \mathbb{R}\backslash\mathbb{Q}$ is an irrational number and $z \in S^1$, then $\varphi^j(z) \neq \varphi^k(z)$ for $j \neq k$. Indeed, if $\varphi^j(z) = \varphi^k(z)$ and hence $e^{2\pi i(x+j\alpha)} = e^{2\pi i(x+k\alpha)}$, then it follows that $x + j\alpha = x + k\alpha + l$ for an integer $l \in \mathbb{Z}$, so that $\alpha = l/(j-k) \in \mathbb{Q}$ is a rational number, contradicting the assumption on the number.

Next, we apply the so-called *pigeon hole principle of Dirichlet*. We fix z and divide S^1 into N disjoint intervals $I_k = \{e^{2\pi i\beta} \mid \frac{k-1}{N} < \beta \leq \frac{k}{N}\}$ of length $1/N$ for $1 \leq k \leq N$. The $N+1$ points $z = \varphi^0(z), \ldots, \varphi^N(z)$ are different from each other on S^1. Therefore, there exists an interval that contains at least two of the points, let us say $\varphi^j(z)$ and $\varphi^k(z)$, where $0 \leq j < k \leq N$. By construction, $d(\varphi^j(z), \varphi^k(z)) < 1/N$ and since φ is isometric, it follows that $d(\varphi^{k-j}(z), z) < 1/N$, and

$$d(\varphi^{(k-j)(n+1)}(z), \varphi^{(k-j)n}(z)) = d(\varphi^{k-j}(z), z) < 1/N, \quad n \geq 0,$$

so that the points $\varphi^{(k-j)n}(z)$ for $n \geq 0$ traverse the circle in equidistant steps of length $< 1/N$. For every $w \in S^1$ there exists an integer $n \geq 0$ satisfying $d(\varphi^{(k-j)n}(z), w) < 1/N$. This is true for every integer $N > 0$, so that the set $\mathcal{O}^+(z)$ is dense in S^1. □

How is the orbit statistically distributed on the circle S^1? If $I \subset S^1$ is an interval we can ask, how often the orbit $(\varphi^j(z))_{j\geq 0}$ visits the interval I on the average. To be precise, we introduce the function

$$H(z, n, I) = \frac{1}{n}\#\{0 \leq j \leq n-1 \mid \varphi^j(z) \in I\}$$

and investigate the convergence as $n \to \infty$. We shall see for α irrational that

$$H(z, n, I) \to \frac{1}{2\pi}|I| = \frac{|I|}{|S^1|}$$

as $n \to \infty$, where the measure is defined by the arc length. In order to reformulate the problem, we introduce the characteristic function χ_I of the interval I,

$$\chi_I(z) = \begin{cases} 1, & z \in I, \\ 0, & z \notin I. \end{cases}$$

Then

$$H(z,n,I) = \frac{1}{n}\sum_{j=0}^{n-1} \chi_I(\varphi^j(z)).$$

More generally, instead of χ_I we can take any Riemann integrable (short: R-integrable) function $f : S^1 \to \mathbb{C}$ and prove the following classical result.

Theorem I.2 (Equidistribution (mod 1) by H. Weyl). *Let $\varphi : S^1 \to S^1$ be the rigid rotation $\varphi(e^{2\pi i x}) = e^{2\pi i(x+\alpha)}$ where $\alpha \in \mathbb{R} \setminus \mathbb{Q}$ is an irrational number. Then, for every Riemann integrable function $f : S^1 \to \mathbb{C}$,*

$$\lim_{n\to\infty} \frac{1}{n}\sum_{j=0}^{n-1} f(\varphi^j(z)) = \frac{1}{2\pi}\int_{S^1} f := \frac{1}{2\pi}\int_0^{2\pi} f(e^{it})\,dt$$

*for **every point** $z \in S^1$.*

Remark. (i) We point out that the limit exists for *every* $z \in S^1$, if f is Riemann integrable. Later on, we shall show that the limit exists for *almost every* $z \in S^1$, if f is merely Lebesgue integrable.

(ii) If for any function $f : S^1 \to \mathbb{C}$ the limit on the left-hand side exists,

$$\lim_{n\to\infty} \frac{1}{n}\sum_{j=0}^{n-1} f(\varphi^j(z)) =: f^*(z) \in \mathbb{C},$$

then the limit is called the *mean value of the function f over the orbit $\mathcal{O}^+(z)$*, or mean value in time. The number $f^*(z)$ can depend on the orbit.

(iii) If $f : S^1 \to \mathbb{C}$ is a (Lebesgue) integrable function, the number

$$\bar{f} := \frac{1}{2\pi}\int_{S^1} f = \frac{1}{2\pi}\int_0^{2\pi} f(e^{it})\,dt = \int_0^1 f(e^{2\pi i t})\,dt$$

is called the *mean value of f over the space S^1*.

Considering the function $f : S^1 \to \mathbb{C}$ on the *covering space* $\mathbb{R}$ of S^1 we define the 1-periodic function $F : \mathbb{R} \to \mathbb{C}$ by $F(x) = f(e^{2\pi i x})$ and Theorem I.2 becomes

$$\lim_{n\to\infty} \frac{1}{n}\sum_{j=0}^{n-1} F(x+\alpha j) = \int_0^1 F(y)\,dy$$

for every $x \in \mathbb{R}$ and every 1-periodic, locally Riemann integrable function $F : \mathbb{R} \to \mathbb{C}$.

Corollary I.3. *If $I \subset S^1$ is an* ***interval****, then*

$$\frac{1}{n}\sum_{j=0}^{n-1}\chi_I(\varphi^j(z)) \longrightarrow \frac{1}{2\pi}\int_0^{2\pi}\chi_I(e^{it})\,dt = \frac{1}{2\pi}|I| = \frac{|I|}{|S^1|}$$

for every $z \in S^1$.

We see that statistically the points of an orbit $\mathcal{O}^+(z)$ are equidistributed on S^1. The mechanism to generate the orbit is *deterministic* and not stochastic!

Theorem I.2 implies Proposition I.1.

Corollary I.4. *If α is irrational, then $\overline{\mathcal{O}^+(z)} = S^1$ for every $z \in S^1$.*

Proof. If I is an interval, then

$$\lim_{n\to\infty}\frac{1}{n}\sum_{j=0}^{n-1}\chi_I(\varphi^j(z)) = \frac{|I|}{|S^1|} \neq 0.$$

Hence, the interval I is *visited infinitely often*. This holds true for every interval, so that $\mathcal{O}^+(z)$ is indeed dense in S^1. □

Proof of Theorem I.2 [Definition of R-integrable, Weierstrass]. The proof is carried out in four steps.

(1) We first take the *trigonometric monomial* $f(z) = z^p$ where $p \in \mathbb{Z}$ and $z = e^{2\pi i x}$ for $x \in \mathbb{R}$. Abbreviating $\theta = e^{2\pi i\alpha}$, we have $f(\varphi^j(z)) = (\theta^p)^j z^p$ and therefore,

$$\frac{1}{n}\sum_{j=0}^{n-1} f(\varphi^j(z)) = \frac{1}{n}z^p\sum_{j=0}^{n-1}(\theta^p)^j = \begin{cases} 1, & p = 0,\\ \frac{1}{n}z^p\frac{\theta^{np}-1}{\theta^p-1}, & p \neq 0.\end{cases}$$

Because of $|\theta^{np}-1| \leq 2$ and $\theta^p \neq 1$ for $p \neq 0$ (since α is irrational), it follows that

$$\begin{aligned}\lim_{n\to\infty}\frac{1}{n}\sum_{j=0}^{n-1} f(\varphi^j(z)) &= \begin{cases} 1, & p = 0\\ 0, & p \neq 0\end{cases}\\ &= \int_0^1 f(e^{2\pi i t})\,dt.\end{aligned}$$

(2) Now, we take *linear combinations* and consider the trigonometric polynomial

$$P(z) = \sum_{k=-N}^{N} a_k z^k, \quad z \in S^1,\ a_k \in \mathbb{C}.$$

It follows from step (1) that $P^*(z) = a_0 = \bar{P}$.

(3) Next we *approximate* the R-integrable function $f\colon S^1 \to \mathbb{R}$ (in case of $\mathbb{C}$ we split the function into its real and its imaginary part). We claim that for $\varepsilon > 0$ there exist two trigonometric polynomials P_ε^-, P_ε^+, satisfying

$$(*)\quad \begin{cases} P_\varepsilon^-(z) \le f(z) \le P_\varepsilon^+(z), & z \in S^1, \\ \int_{S^1} P_\varepsilon^+ - \int_{S^1} P_\varepsilon^- < \varepsilon. \end{cases}$$

This can be seen as follows. Since the function f is *R-integrable*, there exist according to a classical theorem by Darboux two-step functions (belonging to lower and upper sums of f), for which $(*)$ holds true with $\varepsilon/4$. Moving these step functions down, respectively up, we approximate them by continuous functions, satisfying $(*)$ with $\varepsilon/2$. Since every continuous, periodic function can be *uniformly* approximated by trigonometric polynomials (K. Weierstrass), the claim follows with ε.

(4) Finally, *integrating* $(*)$, we obtain the estimates

$$-\varepsilon + \int f \le \int P_\varepsilon^- \le \int f \le \int P_\varepsilon^+ \le \int f + \varepsilon.$$

For a function $g\colon S^1 \to \mathbb{R}$ we abbreviate

$$S_n(g, z) = \frac{1}{n} \sum_{j=0}^{n-1} g(\varphi^j(z)).$$

With this abbreviation we can estimate

$$-\varepsilon + \int f \le \int P_\varepsilon^- \overset{(2)}{=} \lim_{n\to\infty} S_n(P_\varepsilon^-, z) = \underline{\lim}\, S_n(P_\varepsilon^-, z) \overset{(*)}{\le} \underline{\lim}\, S_n(f, z)$$

and

$$\overline{\lim}\, S_n(f, z) \overset{(*)}{\le} \overline{\lim}\, S_n(P_\varepsilon^+, z) \overset{(2)}{=} \lim_{n\to\infty} S_n(P_\varepsilon^+, z) \overset{(2)}{=} \int P_\varepsilon^+ \le \int f + \varepsilon,$$

so that altogether

$$-\varepsilon + \int f \le \underline{\lim}\, S_n(f, z) \le \overline{\lim}\, S_n(f, z) \le \int f + \varepsilon.$$

This holds true for every $\varepsilon > 0$ and therefore,

$$\underline{\lim}\, S_n(f, z) = \overline{\lim}\, S_n(f, z) = \lim_{n\to\infty} S_n(f, z) = \int f.$$

This is true for every $z \in S^1$ and the equidistribution theorem is proved. □

The equidistribution theorem is not only valid for S^1, but also for the n-torus

$$T^n := S^1 \times S^1 \times \cdots \times S^1 \subset \mathbb{C}^n$$

where $z \in T^n \iff z = (z_1, \ldots, z_n)$, $|z_j| = 1$, and we can write z_j as $z_j = e^{2\pi i x_j}$ with a real number $x_j \in \mathbb{R}$. The *covering map* p is defined by

$$p\colon \mathbb{R}^n \to T^n, \quad x = (x_1, \ldots, x_n) \mapsto (e^{2\pi i x_1}, \ldots, e^{2\pi i x_n}),$$

so that $p(x+j) = p(x)$ for every integer vector $j \in \mathbb{Z}^n$; introducing the frequency vector $\omega = (\omega_1, \ldots, \omega_n) \in \mathbb{R}^n$, we define the mapping $\varphi\colon T^n \to T^n$ of the torus by

$$(z_1, \ldots, z_n) \mapsto (e^{2\pi i \omega_1} z_1, \ldots, e^{2\pi i \omega_n} z_n).$$

On the covering space $\mathbb{R}^n$ of the torus the *translation* $\Phi\colon \mathbb{R}^n \to \mathbb{R}^n$, defined by $\Phi(x) = x + \omega$, satisfies $\Phi(x + j) = \Phi(x) + j$ for $j \in \mathbb{Z}^n$. The induced map on the quotient is denoted by

$$\widehat{\Phi}\colon \mathbb{R}^n/\mathbb{Z}^n \to \mathbb{R}^n/\mathbb{Z}^n \cong T^n, \quad [x] \mapsto \widehat{\Phi}([x]) = [x + \omega].$$

With the projection $\hat{p}\colon \mathbb{R}^n/\mathbb{Z}^n \to T^n$, defined by $\hat{p}([x]) = p(x)$, the diagram

$$\begin{array}{ccc} T^n & \xrightarrow{\ \varphi\ } & T^n \\ \uparrow{\scriptstyle \hat{p}} & & \uparrow{\scriptstyle \hat{p}} \\ \mathbb{R}^n/\mathbb{Z}^n & \xrightarrow{\ \widehat{\Phi}\ } & \mathbb{R}^n/\mathbb{Z}^n \end{array}$$

is commutative, so that $\varphi \circ \hat{p}([x]) = \hat{p} \circ \widehat{\Phi}([x])$.

In order *to visualize* the mapping we choose the representative x of the equivalence class $[x]$ in the fundamental domain $[0,1]^n = [0,1] \times \cdots \times [0,1] \subset \mathbb{R}^n$(with identified sides), then, the map $\widehat{\Phi}$ is a translation in $[0,1]^n$. If a point x is pushed out on one side, it enters again as illustrated in Figure I.3.

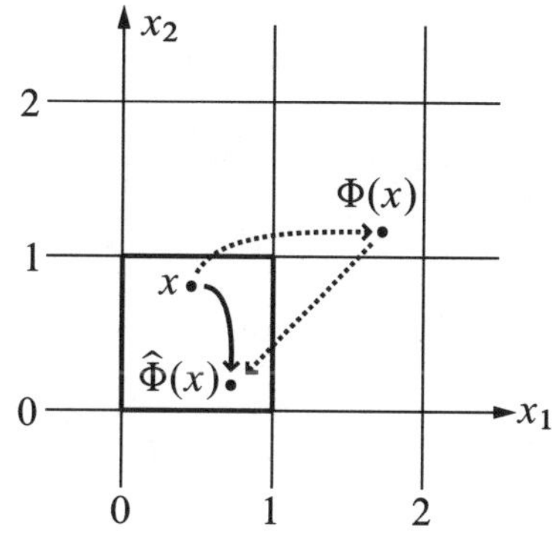

Figure I.3. The mapping $\widehat{\Phi}$ in the fundamental domain $[0,1] \times [0,1]$ of T^2.

We can embed the torus T^2 into $\mathbb{R}^3$ by means of the mapping $\psi\colon \mathbb{R}^2 \to \mathbb{R}^3$, $\psi(x_1, x_2) = (\xi_1, \xi_2, \xi_3)$, defined by

$$\begin{aligned}\xi_1 &= (a + b\cos 2\pi x_1)\cos 2\pi x_2,\\ \xi_2 &= (a + b\cos 2\pi x_1)\sin 2\pi x_2,\\ \xi_3 &= b\sin 2\pi x_1,\end{aligned}$$

where $a > b > 0$, see Figure I.4. The image $\psi(\mathbb{R}^2)$ is an embedded torus and the induced mapping $\hat{\psi}\colon \mathbb{R}^2/\mathbb{Z}^2 \to \mathbb{R}^3$ is bijective onto the torus $\psi(\mathbb{R}^2)$. Introducing the frequencies $\omega = (\omega_1, \omega_2)$ we define the translation φ on the embedded torus by

$$\varphi(\psi(x)) = \psi(x + \omega) = \psi(\Phi(x)).$$

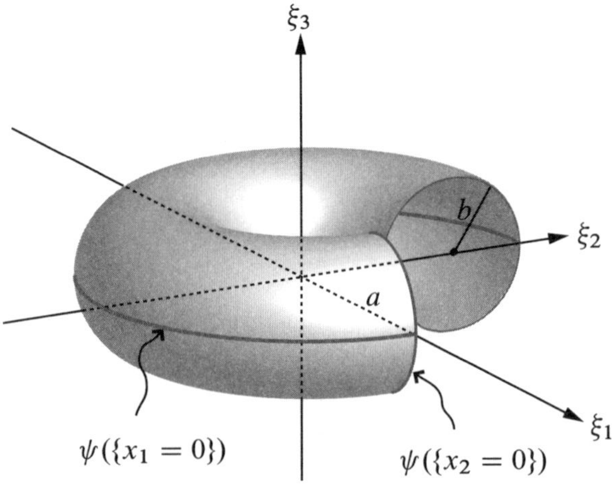

Figure I.4. The embedded torus.

The image of the line $x + t\omega$ on $\mathbb{R}^2$ is the curve $t \mapsto \psi(x + t\omega)$ on the embedded torus. Requiring

$$\langle \omega, j\rangle := \sum_{k=1}^{2} \omega_k j_k \notin \mathbb{Z} \quad \text{for all } j = (j_1, j_2) \in \mathbb{Z}^2 \setminus \{0\},$$

the curve spirals around on the torus without self intersections. Indeed, arguing by contradiction and assuming that $\psi(x + t_1\omega) = \psi(x + t_2\omega)$ for $t_1 \neq t_2$, we obtain $x + t_1\omega = x + t_2\omega \mod \mathbb{Z}^2$, and hence

$$x + t_1\omega = x + t_2\omega + r$$

for an integer vector $0 \neq r = (r_1, r_2) \in \mathbb{Z}^2$. Consequently, $\omega = \tau^{-1} r$, where $\tau := t_1 - t_2$, so that $\langle \omega, j \rangle = 0$ for the integer vector $j := (r_2, -r_1)$ which contradicts our assumption on the frequencies. The next theorem shows that not only the curve $\psi(x + t\omega)$, but already the set of points

$$\varphi^s(\psi(x)) = \psi(x + s\omega), \qquad \text{for all integers } s \in \mathbb{N}_0$$

are dense on the torus $\psi(\mathbb{R}^2)$.

Theorem I.5 (Equidistribution mod $\mathbb{Z}^n$). *If $\varphi\colon T^n \to T^n$ is the translation $\varphi(z_1, \ldots, z_n) = (e^{2\pi i \omega_1} z_1, \ldots, e^{2\pi i \omega_n} z_n)$ and if*

$$\langle \omega, j \rangle = \sum_{k=1}^{n} \omega_k j_k \notin \mathbb{Z} \quad \text{for all } j = (j_1, \ldots, j_n) \in \mathbb{Z}^n \setminus \{0\},$$

then for every R-integrable function $f\colon T^n \to \mathbb{C}$ the equality

$$\lim_{n \to \infty} \frac{1}{n} \sum_{s=0}^{n-1} f(\varphi^s(z)) = \frac{1}{m(T^n)} \int_{T^n} f$$

holds for every point $z \in T^n$, where

$$\begin{aligned} \frac{1}{m(T^n)} \int_{T^n} f &= \int_0^1 \cdots \int_0^1 f(e^{2\pi i x_1}, \ldots, e^{2\pi i x_n})\, dx_1 \ldots dx_n \\ &= \frac{1}{(2\pi)^n} \int_0^{2\pi} \cdots \int_0^{2\pi} f(e^{i x_1}, \ldots, e^{i x_n})\, dx_1 \ldots dx_n. \end{aligned}$$

Proof. Exercise. Hint: the proof is analogous to the proof of Theorem I.2. As for step (1) one takes $f(z_1, \ldots, z_n) = z_1^{p_1} \ldots z_n^{p_n}$ with the integer vector $p = (p_1, \ldots, p_n) \in \mathbb{Z}^n$, so that $f(\varphi^s(z)) = (e^{2\pi i \langle \omega, p \rangle})^s f(z)$. By assumption, $e^{2\pi i \langle \omega, p \rangle} \neq 1$ if $p \neq 0$, while $|e^{2\pi i \langle \omega, p \rangle}| = 1$. The steps (2), (3) and (4) are as before. □

Corollary I.6 (Equidistribution). *If $\langle \omega, j \rangle \notin \mathbb{Z}$ for all $j \in \mathbb{Z}^n \setminus \{0\}$ and if $I \subset T^n$ is an interval satisfying $m(I) > 0$, then*

$$\lim_{n \to \infty} \frac{1}{n} \sum_{j=0}^{n-1} \chi_I(\varphi^j(z)) = \frac{m(I)}{m(T^n)}$$

for every $z \in T^n$.

Corollary I.7 (Kronecker, density). *If $\langle \omega, j \rangle \notin \mathbb{Z}$ for all $j \in \mathbb{Z}^n \setminus \{0\}$, then every orbit φ is dense on the torus:*

$$\overline{\bigcup_{j \geq 0} \varphi^j(z)} = \overline{\mathcal{O}^+(z)} = T^n$$

for every point $z \in T^n$.

Proof. If $I \subset T^n$ is an interval satisfying $m(I) > 0$ we consider the step function $f = \chi_I$ and conclude from Corollary I.6 that every orbit visits the interval I infinitely often. This holds true for every open interval and the proof is complete. □

Corollary I.8 (Kronecker, rational approximation). *Assume that* $\langle \omega, j \rangle \notin \mathbb{Z}$ *for all* $j \in \mathbb{Z}^n \setminus \{0\}$ *and let* $\varepsilon > 0$ *and* $N \geq 1$ *be given. Then there exist an integer* $s \geq N$ *and an integer vector* $j \in \mathbb{Z}^n$ *satisfying*

$$|s\omega - j| < \varepsilon$$

or equivalently

$$|\omega - j/s| < \varepsilon/s.$$

Proof. Exercise. Hint: consider orbits of the translation map $\widehat{\Phi}$ on the quotient $\mathbb{R}^n/\mathbb{Z}^n$ and use the metric $d([x],[y]) = \min_{j \in \mathbb{Z}^n} |x - y - j|$. Then,

$$\widehat{\Phi}^s([x]) = [\Phi^s(x)] = [x + s\omega].$$

The orbit through the point $p(0) \in T^n$ is dense. If $\varepsilon > 0$, there exists an integer $s \geq N$ satisfying $d(\widehat{\Phi}^s([0]),[0]) = \min_{j \in \mathbb{Z}^n} |s\omega - j| < \varepsilon$. □

I.3 Transitive dynamical systems

In the following (X, d) is a metric space and $\varphi \colon X \to X$ a continuous map.

Definition. The dynamical system (X, φ) is called *transitive*, if φ possesses *a dense orbit* i.e., if there exists a point $x \in X$ whose orbit is dense, so that its closure satisfies

$$\overline{\mathcal{O}^+(x)} = \overline{\bigcup_{j \geq 0} \varphi^j(x)} = X.$$

The system (X, φ) is called *minimal*, if *every* orbit of φ is dense.

Example. The irrational rotations of S^1 are *transitive and minimal.*

If the system (X, φ) is transitive, then there exists for every two non-empty open sets $\emptyset \neq U, V \subset X$ an integer $n \geq 0$ satisfying

$$\varphi^n(U) \cap V \neq \emptyset,$$

as illustrated in Figure I.5. Indeed, according to the assumption, there exists a dense orbit $(\varphi^j(x))_{j \geq 0}$ and therefore there are two integers j, k satisfying $\varphi^j(x) \in U$ and $\varphi^k(x) \in V$. Assuming $k \geq j$, we define $y = \varphi^j(x) \in U$ and set $n = k - j$, then $\varphi^n(y) = \varphi^n(\varphi^j(x)) = \varphi^k(x) \in V$, proving the claim.

Under *additional assumptions* the converse also holds true, as the next result will show.

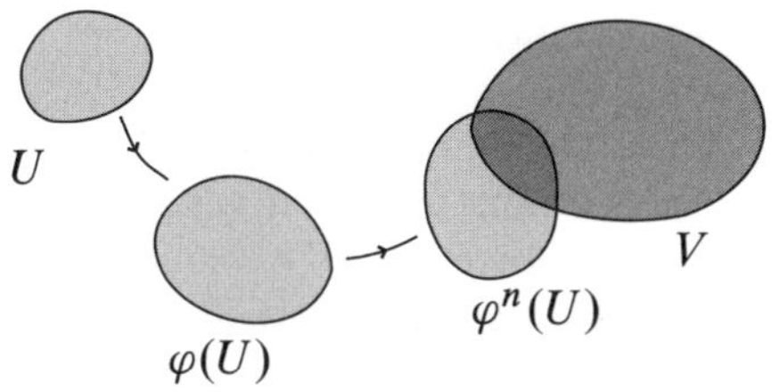

Figure I.5. Necessary condition for transitivity.

Theorem I.9 (Transitivity theorem by G. Birkhoff). *We assume that the metric space (X, d) is complete and that X possesses a countable basis of open sets. Let $\varphi\colon X \to X$ be a **continuous** map.*

If for every pair $\emptyset \neq U, V \subset X$ of open sets there exists an integer $n = n(U, V) \geq 0$ satisfying

$$\varphi^n(U) \cap V \neq \emptyset,$$

*then there exists a **dense** set $R^+ \subset X$, so that for every point $p \in R^+$,*

$$\overline{\mathcal{O}^+(p)} = X.$$

*In addition, R^+ is of **second Baire** category.*

Remark. The assumptions on the metric space (X, d) are fulfilled, in particular in the following cases.

- *(X, d) is a complete, separable, metric space.* Indeed, in a metric space, separability (i.e., the existence of a countable, dense subset) is equivalent to the existence of a countable basis of open sets. To see this, it is sufficient to find a countable system of open sets $(B_i)_{i\geq 1}$, so that for every $x \in X$ and every neighborhood U_x of x there exists an index i satisfying $B_i \subset U_x$. This is easy to accomplish. Indeed, if $(y_n)_{n\geq 1}$ is a dense sequence in X, then there exists, for every $x \in X$ and every $\varepsilon > 0$, a point y_n satisfying $d(x, y_n) < \varepsilon$. Therefore, the system of open balls $\{B(y_n, 1/m) \mid m, n \geq 1\}$ has the desired properties.
- *(X, d) is a compact metric space.* Due to the compactness every sequence has a convergent subsequence. This applies, in particular, to every Cauchy sequence. A Cauchy sequence, however, possessing a convergent subsequence is convergent. This proves the completeness of the metric space. It remains to show that X is *separable*. The open balls $\{B(x^n, 1/n) \mid x^n \in X\}$ are an open covering of X for every $n \in \mathbb{N}$. If we choose the finite subcovers $B(x_1^n, 1/n), \dots, B(x_{m_n}^n, 1/n)$, the set of points $\{x_i^n \mid n \geq 1, 1 \leq i \leq m_n\}$ is countable and dense in X.

- The space X is a closed subset of $\mathbb{R}^n$.

Proof of Theorem I.9 [Completeness, Baire category theorem]. For two open sets $\emptyset \neq U, V \subset X$ there exists by assumption an integer $n \geq 0$ satisfying $\varphi^n(U) \cap V \neq \emptyset$. Hence $U \cap \varphi^{-n}(V) \neq \emptyset$, where $\varphi^{-n}(V) := \{x \in X \mid \varphi^n(x) \in V\}$ is the preimage of V under the iterated map φ^n. Consequently,

$$U \cap \bigcup_{j \geq 0} \varphi^{-j}(V) \neq \emptyset,$$

and in view of the assumption this holds true for *every* open set $U \neq \emptyset$. Therefore, the open set

$$\mathcal{O}^-(V) = \bigcup_{j \geq 0} \varphi^{-j}(V)$$

is dense in X. This holds true for every open set $V \neq \emptyset$. If $(V_j)_{j \geq 1}$ is a *countable basis of open sets* in X, then the sets $\mathcal{O}^-(V_j)$ are open and dense. The countable intersection

$$R^+ := \bigcap_{j \geq 1} \mathcal{O}^-(V_j)$$

is still a dense subset of X in view of the following result.

Lemma I.10 (Baire category theorem). *If (X, d) is a complete metric space and if $(V_j)_{j \geq 1}$ is a* countable *family of* open and dense *subsets of X, then the countable intersection $\bigcap_{j \geq 1} V_j$ is dense in X.*

Postponing the proof of the Baire category theorem, we first complete the proof of Theorem I.9 and choose a point $p \in R^+$. Then

$$p \in \mathcal{O}^-(V_j) = \bigcup_{s \geq 0} \varphi^{-s}(V_j)$$

for every $j \geq 0$. Hence for every integer j there exists an integer $s \geq 0$ satisfying $\varphi^s(p) \in V_j$, so that

$$\mathcal{O}^+(p) \cap V_j \neq \emptyset$$

for every j. Because the family $(V_j)_{j \geq 1}$ is a basis of open sets, every open subset $U \neq \emptyset$ of X contains some subset $V_j \subset U$, so that

$$\mathcal{O}^+(p) \cap U \neq \emptyset.$$

Consequently, the orbit $\mathcal{O}^+(p)$ is dense in X. This holds true for every point $p \in R^+$ and completes the proof of Theorem I.9. □

It remains to prove the lemma.

Proof of Lemma I.10 [Completeness]. We have to show that for a given point $x \in X$ and a given real number $\varepsilon > 0$ there exists a point x^* satisfying

$$x^* \in B(x, \varepsilon) \cap \bigcap_{j \geq 1} V_j$$

where $B(x, \varepsilon)$ is an open ball of radius ε centered at x. Since $B(x, \varepsilon)$ is open and V_1 open and dense, the intersection $B(x, \varepsilon) \cap V_1$ is open and not empty. Therefore, there exists a point $x_1 \in B(x, \varepsilon)$ satisfying

$$K_1 := \overline{B(x_1, r_1)} \subset B(x, \varepsilon) \cap V_1$$

for a radius $0 < r_1 < 1/2$. Since $B(x_1, r_1)$ is an open ball and since V_2 is open and dense, we find a point $x_2 \in B(x_1, r_1)$ satisfying

$$K_2 := \overline{B(x_2, r_2)} \subset B(x_1, r_1) \cap V_2$$

for a radius $0 < r_2 < 1/2^2$. Proceeding inductively, we find for every $j \geq 2$ a point x_j satisfying

$$K_j := \overline{B(x_j, r_j)} \subset B(x_{j-1}, r_{j-1}) \cap V_j$$

for a radius $0 < r_j < 1/2^j$. Then

$$B(x, \varepsilon) \supset K_1 \supset K_2 \supset \cdots$$

and $\operatorname{diam}(K_j) \to 0$. Since K_j is a nested sequence of closed sets, it follows from the completeness of the metric space X that the countable intersection

$$\bigcap_{j \geq 1} K_j = \{x^*\}$$

consists of a single point $x^* \in X$. In view of our construction, $x^* \in K_j \subset B(x, \varepsilon) \cap V_j$ for every $j \geq 1$ and therefore,

$$x^* \in B(x, \varepsilon) \cap \bigcap_{j \geq 1} V_j.$$

This completes the proof of the lemma. □

Definition. A set $A \subset X$ is called *invariant* under the map $\varphi \colon X \to X$ if

$$\varphi^{-1}(A) := \{x \in X \mid \varphi(x) \in A\} = A.$$

We note that the set A is invariant precisely if

$$\varphi(A) \subset A \quad \text{and} \quad \varphi(A^c) \subset A^c$$

where A^c denotes the complement of the set A in X.

Proposition I.11. *If the dynamical system (X, φ) is minimal, then every* ***closed invariant*** *set $A \neq \emptyset$ is already the whole space, $A = X$.*

Proof [Definitions]. Since A is invariant under φ, $\varphi(A) \subset A$. Hence if $x \in A$,then the positive orbit $\mathcal{O}^+(x)$ is contained in A. Since (X, φ) is minimal, it is dense in X and taking its closure, we conclude

$$X = \overline{\mathcal{O}^+(x)} \subset \bar{A} = A \subset X,$$

so that $A = X$, as claimed in the proposition. □

I.4 Structural stability

In order to illustrate the concept of structural stability, we study the special example of an expanding map on the circle defined by

$$\varphi\colon S^1 \to S^1, \quad z \mapsto z^2, \quad z = e^{2\pi i x},\ x \in \mathbb{R},$$

or $\varphi(e^{2\pi i x}) = e^{2\pi i (2x)}$ for $x \in \mathbb{R}$. This map is *not* bijective, but two-to-one. The covering map on $\mathbb{R}$ is the function

$$\Phi\colon \mathbb{R} \to \mathbb{R}, \quad \Phi(x) = 2x$$

satisfying $\Phi(x + j) = \Phi(x) + 2j$ for $j \in \mathbb{Z}$. The projection map $p\colon \mathbb{R} \to S^1$, defined by $p(x) = e^{2\pi i x}$, satisfies $\varphi(p(x)) = p(\Phi(x))$, so that the diagram

$$\begin{array}{ccc} S^1 & \xrightarrow{\varphi} & S^1 \\ {\scriptstyle p}\uparrow & & \uparrow{\scriptstyle p} \\ \mathbb{R} & \xrightarrow{\Phi} & \mathbb{R} \end{array}$$

is commutative. The map Φ induces the map $\widehat{\Phi}\colon \mathbb{R}/\mathbb{Z} \to \mathbb{R}/\mathbb{Z}$, $\widehat{\Phi}([x]) = [\Phi(x)]$, on the quotient space, in short notation, $\widehat{\Phi}(x) = 2x \bmod \mathbb{Z}$. With the *homeomorphism* $\hat{p}\colon \mathbb{R}/\mathbb{Z} \to S^1$, defined by $\hat{p}([x]) := p(x) = e^{2\pi i x}$, we obtain $\varphi \circ \hat{p}([x]) = \hat{p} \circ \widehat{\Phi}([x])$, so that the diagram

$$\begin{array}{ccc} S^1 & \xrightarrow{\varphi} & S^1 \\ {\scriptstyle \hat{p}}\uparrow & & \uparrow{\scriptstyle \hat{p}} \\ \mathbb{R}/\mathbb{Z} & \xrightarrow{\widehat{\Phi}} & \mathbb{R}/\mathbb{Z} \end{array}$$

commutes. The (restricted) projection $p\colon (0, 1] \to S^1$ is *bijective* and, identifying $x \in (0, 1]$ with its equivalence class $[x]$, the map $\widehat{\Phi}$ can be represented in the *fundamental domain* $(0, 1]$, by means of the formula

$$\widehat{\Phi}\colon (0,1] \to (0,1], \quad \widehat{\Phi}(x) = \begin{cases} 2x, & 0 < x \le 1/2, \\ 2x - 1, & 1/2 < x \le 1, \end{cases}$$

illustrated in Figure I.6. We recall that every real number in $0 < x \le 1$ can be represented by a unique dyadic expansion containing infinitely many nonzero digits,

$$x = \frac{x_1}{2} + \frac{x_2}{2^2} + \cdots = \sum_{j \ge 1} \frac{x_j}{2^j}$$

where $x_j \in \{0, 1\}$. The standard notation for the dyadic expansion is the following

$$x = 0.x_1 x_2 x_3 \ldots.$$

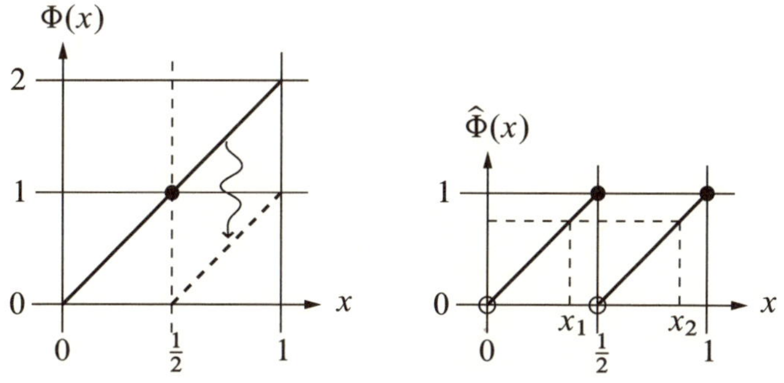

Figure I.6. The maps Φ and $\widehat{\Phi}$ in the fundamental domain $(0, 1]$.

In this notation, the multiplication of x by 2 corresponds to the shift of the point, namely

$$\begin{aligned} \Phi(x) &= 2x \\ &= x_1 + \frac{x_2}{2} + \frac{x_3}{2^2} + \cdots \\ &= x_1.x_2x_3\ldots \\ &= 0.x_2x_3x_4\ldots \quad \bmod 1. \end{aligned}$$

Consequently,

$$\widehat{\Phi}(x) = \widehat{\Phi}(0.x_1x_2x_3\ldots) = 0.x_2x_3x_4\ldots,$$

and we see that the mapping $\widehat{\Phi}$ is a *shift map*. This shift map simply forgets the first entry in a sequence and shifts all other entries one place to the left. Of course, $\widehat{\Phi}$ is not bijective, since every image point $y = \widehat{\Phi}(x)$ has two preimages, namely $0.1y_1y_2\ldots$ and $0.0y_1y_2\ldots$.

We now introduce the space

$$\mathcal{F} = \{(x_j)_{j\geq 1} \mid x_j \in \{0,1\}\}$$

of *one-sided sequences of the two symbols* 0 and 1, and consider the subset

$$\mathcal{F}_0 = \{(x_j)_{j\geq 1} \mid x_j \neq 0 \text{ for infinitely many } j\}.$$

There is a *bijective* mapping

$$\psi\colon (0,1] \to \mathcal{F}_0,$$

defined by the dyadic expansion containing infinitely many nonzero digits as follows:

$$\psi(x) = (x_j)_{j\geq 1} \text{ if } x = 0.x_1x_2\cdots \in (0,1].$$

Denoting by $\sigma\colon \mathcal{F} \to \mathcal{F}$ the shift $(x_1, x_2, x_3, \dots) \mapsto (x_2, x_3, x_4, \dots)$, in formulas defined as $\sigma((x_j)_{j\geq 1}) = (x_{j+1})_{j\geq 1}$, we obtain the commutative diagram

$$\begin{array}{ccc} (0,1] & \xrightarrow{\ \widehat{\Phi}\ } & (0,1] \\ {\scriptstyle\psi}\big\downarrow & & \big\downarrow{\scriptstyle\psi} \\ \mathcal{F}_0 & \xrightarrow{\ \sigma\ } & \mathcal{F}_0 \end{array}$$

illustrating the equation

$$\psi \circ \widehat{\Phi}(x) = \sigma \circ \psi(x).$$

This way, we have represented the *analytical* mapping $\varphi\colon z \mapsto z^2$ on the circle S^1 by the *shift map* σ in the *sequence space* $\mathcal{F}_0$. We will show that there coexist orbits of completely different long-time behavior.

Proposition I.12. *If $\varphi\colon S^1 \to S^1$ is the mapping $z \mapsto z^2$ and m the Lebesgue measure on S^1, the following holds true.*

(i) *The set of periodic points of φ is countable and dense.*

(ii) *The set $R^+ := \{z \in S^1 \mid \overline{\mathcal{O}^+(z)} = S^1\}$ of initial points whose orbits are dense, is a* **dense set** *in S^1.*

(iii) *$m(R^+) = m(S^1)$ and R^+ is of second Baire category.*

In particular, the mapping φ is transitive, but not minimal.

In contrast to the rigid rotations of S^1 this dynamical system possesses, in every open set, points whose orbits behave asymptotically *completely differently*. The distant future is *no longer predictable*, if not every digit in the dyadic expansion of the initial condition z for the orbit $\mathcal{O}^+(z)$ is known. For example, if $x = 0.x_1x_2\dots x_N * * * \dots$, then $\widehat{\Phi}^N(x) = 0. * * * \dots$, where the stars stand for the unknown digits.

In physics one talks about the *sensitive dependence* of the orbits on the initial conditions. It is *hopeless* to gain an insight into the orbit structure of all solutions over infinitely long times by *solving the Cauchy initial value problem*!

Proof of Proposition I.12. In order to prove the statement (i), we assume $n \geq 1$. From $\varphi^n(z) = z \Longleftrightarrow z^{2^n} = z \Longleftrightarrow z^{2^n-1} = 1$ it follows that every complex *root of unity* of order $2^n - 1$ is a periodic point of the period n, and vice versa. There exist exactly $2^n - 1$ such roots of unity, and they are *equidistantly* distributed on S^1, so that the periodic points are countable and dense.

(ii) It is sufficient to verify the assumptions of the transitivity theorem (Theorem I.9). Since S^1 is compact, we have to show that, for every pair $\emptyset \neq U, V \subset S^1$ of open sets, there exists an integer n satisfying $\varphi^n(U) \cap V \neq \emptyset$. An open set $U \subset S^1$ contains the image of a *binary interval* I_n^k under the projection map $p|_{(0,1]}$, where

$$I_n^k := \left(\frac{k}{2^n}, \frac{k+1}{2^n}\right] \subset (0,1], \quad k = 0, 1, 2, \ldots, 2^n - 1,$$

the integer n being sufficiently large. Applying the map $\Phi(x) = 2x$ we obtain

$$\begin{aligned} \Phi^n(I_n^k) &= 2^n\left(\frac{k}{2^n}, \frac{k+1}{2^n}\right] \\ &= (k, k+1] \\ &= (0,1] \mod 1. \end{aligned}$$

In order to see this in terms of dyadic expansions, we take the real number $x \in I_n^k$, represented as

$$\begin{aligned} x &= 0.x_1x_2x_3 \ldots x_nx_{n+1} \ldots \\ &= \frac{x_1}{2} + \frac{x_2}{2^2} + \cdots + \frac{x_n}{2^n} + \frac{x_{n+1}}{2^{n+1}} + \cdots \\ &= \frac{k}{2^n} + \frac{x_{n+1}}{2^{n+1}} + \cdots . \end{aligned}$$

Then

$$\begin{aligned} \Phi^n(x) &= 2^n x \\ &= k + \frac{x_{n+1}}{2} + \frac{x_{n+2}}{2^2} + \cdots \\ &= k.x_{n+1}x_{n+2} \ldots \\ &= 0.x_{n+1}x_{n+2} \ldots \mod 1 \end{aligned}$$

and therefore $\widehat{\Phi}^n(I_n^k) = (0,1]$. In view of $p(I_n^k) \subset U$, we find

$$S^1 = p((k, k+1]) = p \circ \Phi^n(I_n^k) = \varphi^n \circ p(I_n^k) \subset \varphi^n(U).$$

Having verified that the assumptions of Theorem I.9 are met, the statement (ii) follows.

(iii) The statement (iii) will be proved later in Section I.5. □

Considering the mapping $\varphi_0 \colon S^1 \to S^1$ defined by $\varphi_0(z) = z^2$, we shall study what happens to its complex orbit structure under a perturbation. It turns out that the complex structure is stable under perturbations, as will be proved in the following statement.

- *For every continuously differentiable mapping $\varphi \colon S^1 \to S^1$ in a sufficiently small C^1-neighborhood of φ_0, there exists a unique homeomorphism $h \colon S^1 \to S^1$, so that the diagram*

$$\begin{array}{ccc} S^1 & \xrightarrow{\varphi_0} & S^1 \\ \downarrow{\scriptstyle h} & & \downarrow{\scriptstyle h} \\ S^1 & \xrightarrow{\varphi} & S^1 \end{array}$$

commutes, i.e.,

$$\varphi \circ h = h \circ \varphi_0.$$

For the iterates, we then have $\varphi^j(z) = h \circ \varphi_0^j \circ h^{-1}(z)$. The mappings φ and φ_0 are called *topologically conjugated*, and the mapping φ_0 is called structurally stable.

Definition. A C^1-mapping φ_0 is called (C^1-)*structurally stable*, if every C^1-mapping φ in a sufficiently small C^1-neighborhood of φ_0 is topologically conjugated to φ_0.

It is useful to describe the mappings in the *covering space* $\mathbb{R}$ of S^1. The unperturbed mapping is equal to $\varphi_0(e^{2\pi i x}) = e^{2\pi i \Phi(x)}$ where $\Phi(x) = 2x$, and the perturbed mapping is represented by

$$\varphi(e^{2\pi i x}) = e^{2\pi i \psi(x)}$$

where the mapping $\psi \colon \mathbb{R} \to \mathbb{R}$ satisfies

$$\psi(x) = \Phi(x) + \hat{\psi}(x) \quad \text{and} \quad \hat{\psi}(x+1) = \hat{\psi}(x), \quad x \in \mathbb{R}.$$

The homeomorphism $h \colon S^1 \to S^1$ we are looking for is represented as

$$h(e^{2\pi i x}) = e^{2\pi i u(x)},$$

with a homeomorphism $u \colon \mathbb{R} \to \mathbb{R}$. In particular, u is a continuous, strictly increasing function satisfying $u(x+1) = u(x) + 1$ and so is of the form

$$u(x) = x + \hat{u}(x), \quad \text{where } \hat{u}(x+1) = \hat{u}(x),\ x \in \mathbb{R}.$$

The inverse mapping $u^{-1}\colon \mathbb{R} \to \mathbb{R}$ is also continuous, strictly increasing and satisfies $u^{-1}(x+1) = u^{-1}(x) + 1$, so that

$$h^{-1}(e^{2\pi i x}) = e^{2\pi i u^{-1}(x)}, \quad x \in \mathbb{R}.$$

Indeed,

$$\begin{aligned} h^{-1}(h(e^{2\pi i x})) &= h^{-1}(e^{2\pi i u(x)}) \\ &= e^{2\pi i u^{-1}(u(x))} \\ &= e^{2\pi i x} \end{aligned}$$

for all $x \in \mathbb{R}$. The *functional equation* to be solved becomes $\varphi \circ h(z) = h \circ \varphi_0(z)$, for all $z \in S^1$, or equivalently,

$$e^{2\pi i \psi(u(x))} = e^{2\pi i u(\Phi(x))}$$

and we shall *solve* the nonlinear equation

$$\psi(u(x)) = u(\Phi(x))$$

for the unknown mapping u. The following theorem is a special case of a general phenomenon encountered in expanding mappings for which we refer to the monograph [113], Chapter 4.11] by W. Szlenk.

Theorem I.13 (Structural stability of the map $\varphi(z) = z^2$). *We consider on $\mathbb{R}$ the mapping $\psi(x) = 2x + \hat{\psi}(x)$ satisfying $\hat{\psi}(x+1) = \hat{\psi}(x)$ and assume $\hat{\psi}$ to be Lipschitz small in the sense that $|\hat{\psi}(x) - \hat{\psi}(y)| \le L|x-y|$ for all $x, y \in \mathbb{R}$ with a Lipschitz constant satisfying $0 < L < 1$. Then, there exists a unique, strictly increasing homeomorphism $u\colon \mathbb{R} \to \mathbb{R}$ satisfying $u(x+1) = u(x) + 1$ for every $x \in \mathbb{R}$ and solving the equation*

$$\psi(u(x)) = u(\Phi(x)) = u(2x)$$

for all $x \in \mathbb{R}$. For the 1-periodic function $\hat{u}$, defined by $u(x) = x + \hat{u}(x)$, the estimate

$$|\hat{u}|_\infty \le |\hat{\psi}|_\infty$$

holds in the supremum norm.

Postponing the proof we observe that the homeomorphism u is unique, so that one can ask whether u is a C^1-diffeomorphism if the map ψ is continuously differentiable. In general, this is *not* the case, as we will convince ourselves next. We assume that $u\colon \mathbb{R} \to \mathbb{R}$ is a diffeomorphism *and* that the function $\hat{\psi}$ is of class C^1 and satisfies $\hat{\psi}(0) = 0$ and $\hat{\psi}'(0) \neq 0$. Using the equation $\psi(u(x)) = u(2x)$ we obtain

$$u(u^{-1}(0)) = 0 = \hat{\psi}(0) = \psi(0) = \psi(u(u^{-1}(0))) = u(2u^{-1}(0))$$

and conclude that $2u^{-1}(0) = u^{-1}(0)$. Therefore, $u(0) = 0$ and differentiating the equation $\psi(u(x)) = u(2x)$ in x at $x = 0$, results in

$$\psi'(0)u'(0) = 2u'(0).$$

Since u is a diffeomorphism, $u'(0) \neq 0$ and therefore $\psi'(0) = 2$ contradicting $\psi'(0) = 2 + \hat{\psi}'(0) \neq 2$. We see that the *eigenvalue* of ψ' at the fixed point $\psi(0) = 0$ is an invariant under a *differentiable* conjugation.

Proof of Theorem I.13 [Expansion $\rightsquigarrow$ Contraction]. (1) In the first step we shall show that the map $\psi\colon \mathbb{R} \to \mathbb{R}$ *is bijective and Lipschitz-continuous*, and that its inverse mapping ψ^{-1} *is a contraction.*

Due to $\psi(x) - \psi(y) = 2(x - y) + \hat{\psi}(x) - \hat{\psi}(y)$ we obtain for all $x \geq y$ the estimate

$$(*) \quad \underbrace{(2 - L)}_{=:r_1}(x - y) \leq \psi(x) - \psi(y) \leq \underbrace{(2 + L)}_{=:r_2}(x - y).$$

From $0 \leq L < 1$ we deduce $1 < r_1 < r_2 < 3$ and hence ψ is Lipschitz-continuous and strictly increasing, and therefore bijective in view of the intermediate value theorem. Consequently, there exists the inverse map $\psi^{-1}\colon \mathbb{R} \to \mathbb{R}$. By inserting $\psi^{-1}(x') = x$ and $\psi^{-1}(y') = y$ into the estimate $(*)$, we find

$$r_1\big(\psi^{-1}(x') - \psi^{-1}(y')\big) \leq x' - y' \leq r_2\big(\psi^{-1}(x') - \psi^{-1}(y')\big)$$

for all $x' \geq y'$, so that

$$\frac{1}{r_2}(x' - y') \leq \psi^{-1}(x') - \psi^{-1}(y') \leq \frac{1}{r_1}(x' - y').$$

It follows that $|\psi^{-1}(x') - \psi^{-1}(y')| \leq K|x' - y'|$ where the constant K is defined as

$$K := \max\{r_1^{-1}, r_2^{-1}\} = r_1^{-1} < 1.$$

We have proved that the mapping ψ^{-1} is a contraction. (The inverse ψ^{-1} is a map in $\mathbb{R}$, it is *not* the covering of a map of S^1.)

(2) In order to *solve the functional equation* $\psi(u(x)) = u(2x)$ *for the mapping* u we shall solve the equivalent equation

$$u(x) = \psi^{-1}(u(2x))$$

by the contraction principle and introduce the metric space

$$X = \{u \in C^0(\mathbb{R}, \mathbb{R}) \mid u(x + 1) = u(x) + 1 \text{ and increasing}\}$$

equipped with the metric

$$d_X(u, v) = \sup_{x \in \mathbb{R}} |u(x) - v(x)| = \max_{0 \leq x \leq 1} |u(x) - v(x)| =: |u - v|_\infty.$$

We have used that the difference $u - v$ is 1-periodic. As one can easily verify, the metric space(X, d_X) is *complete.* We note that the elements of X are chosen to be increasing functions (and not strictly increasing), since otherwise the space X would not be complete. We claim that the mapping T, defined by the formula

$$T\colon X \to C^0(\mathbb{R}, \mathbb{R}), \quad (Tu)(x) = \psi^{-1}(u(2x)), \quad x \in \mathbb{R},$$

maps the space X into itself and satisfies $d_X(Tu, Tv) \leq K d_X(u, v)$, for all u and v in X, so that it is a contraction with the contraction constant $K < 1$ introduced above. Indeed, since u and ψ^{-1} are increasing, the function Tu is also increasing. From $\psi(x+1) = 2(x+1) + \hat{\psi}(x) = \psi(x) + 2$ one concludes that the inverse map satisfies $\psi^{-1}(y+2) = \psi^{-1}(y) + 1$. Using this, we compute,

$$\begin{aligned} (Tu)(x+1) &= \psi^{-1}(u(2x+2)) \\ &= \psi^{-1}(u(2x)+2) \\ &= \psi^{-1}(u(2x)) + 1 \\ &= (Tu)(x) + 1. \end{aligned}$$

Consequently, $Tu \in X$ and T maps our metric space into itself. It remains to show that T is a contraction. This follows immediately from the contraction property of ψ^{-1},

$$\begin{aligned} d_X(Tu, Tv) &= \max_{0 \leq x \leq 1} |\psi^{-1}(u(2x)) - \psi^{-1}(v(2x))| \\ &\leq K \max_{0 \leq x \leq 1} |u(2x) - v(2x)| \\ &= K d_X(u, v). \end{aligned}$$

By the contraction principle of Banach there exists a *unique fixed point* $u \in X$ satisfying $Tu = u$, so that $u(x) = \psi^{-1}(u(2x))$ for all x.

(3) In order to show that u *is strictly increasing* we argue by contradiction and assume that there exist real numbers $\alpha < \beta$ in the interval $[0, 1]$ satisfying $u(\alpha) = u(\beta)$. Since u is increasing,

$$u(x) = \text{const.}, \quad \alpha \leq x \leq \beta.$$

The interval contains a binary interval and so we find integers $n \geq 1$ and $k \in \{0, 1, \dots, 2^n - 1\}$, for which

$$u(x) = \text{const.}, \quad \frac{k}{2^n} \leq x \leq \frac{k+1}{2^n}.$$

Using the equation $\psi(u(x)) = u(2x)$ we deduce from $u(\frac{k}{2^n}) = u(\frac{k+1}{2^n})$, that also the equation

$$u\Big(\frac{k}{2^{n-1}}\Big) = u\Big(\frac{k+1}{2^{n-1}}\Big)$$

holds true. Indeed, $u(\frac{k}{2^{n-1}}) = u(2\frac{k}{2^n}) = \psi(u(\frac{k}{2^n})) = \psi(u(\frac{k+1}{2^n})) = u(2\frac{k+1}{2^n}) = u(\frac{k+1}{2^{n-1}})$. We repeat this procedure, until in the denominator $2^{n-n} = 1$ shows up. At this point we have $u(k) = u(k+1)$ in *contradiction* to $u(k+1) = u(k) + 1$ and hence proving that u must be strictly increasing.

(4) From $u(x+1) = u(x) + 1$ it follows that $u(x+n) = u(x) + n$ for every $n \in \mathbb{Z}$ from which we conclude that $\lim_{x\to\pm\infty} u(x) = \pm\infty$. Since u is continuous by construction, the surjectivity of u follows from the intermediate value theorem. The injectivity of u is a consequence of the strict monotonicity. Hence u is a strictly increasing bijection of $\mathbb{R}$ onto itself and therefore a homeomorphism of $\mathbb{R}$.

(5) In order to verify the announced *estimate* we deduce from the equation $\psi(u(x)) = u(2x)$ and the definition $\psi(x) = 2x + \hat{\psi}(x)$ that $2u(x) + \hat{\psi}(u(x)) = u(2x)$. Recalling $u(x) = x + \hat{u}(x)$ we obtain the equation

$$2x + 2\hat{u}(x) + \hat{\psi}(u(x)) = 2x + \hat{u}(2x)$$

and estimate

$$\begin{aligned} |\hat{u}(x)| &= \left|\frac{1}{2}\hat{u}(2x) - \frac{1}{2}\hat{\psi}(u(x))\right| \\ &\le \frac{1}{2}|\hat{u}|_\infty + \frac{1}{2}|\hat{\psi}|_\infty \end{aligned}$$

for every $x \in \mathbb{R}$. Taking the supremum on the left-hand side, the desired estimate $|\hat{u}|_\infty \le |\hat{\psi}|_\infty$ follows and Theorem I.13 is proved. □

I.5 Measure preserving maps and the ergodic theorem

The previous examples (with the exception of the contractions) are *measure preserving* with respect to the Lebesgue measure. This section deals with the *part played by the measures* in dynamics.

A measure space is a triple $(X, \mathcal{A}, m)$, in which X is a set, $\mathcal{A}$ a σ-algebra of subsets of X (called *measurable* sets), and $m\colon \mathcal{A} \to [0,\infty]$ a measure. In the following we assume the measure space to be *finite*, assuming that $m(X) < \infty$. We denote by $L = L(X, \mathcal{A}, m)$ the vector space of integrable functions $f : X \to \mathbb{R} \cup \{\pm\infty\}$. These are the measurable functions, for which the (Lebesgue-)integral is defined and finite. To facilitate the notation, we sometimes omit the measures in the integrals and suppress the integration domain, if it is the whole space. We also suppress the variable over which it is integrated and write $\int f := \int_X f(x)\,dm(x)$ for the integral. To avoid an accumulation of brackets, we simply write, e.g. Tx instead of $T(x)$ or $T^j x$ instead of $T^j(x)$.

Definition. A mapping $T\colon X \to X$ is called *measurable*, if $T^{-1}(A) \in \mathcal{A}$ for every $A \in \mathcal{A}$, where $T^{-1}(A) = \{x \in X \mid T(x) \in A\}$ is the preimage of A. A measurable mapping $T\colon X \to X$ is called *measure preserving*, if

$$m(T^{-1}(A)) = m(A)$$

for every $A \in \mathcal{A}$. We note that T does not need to be bijective.

It is useful to observe that a mapping T is measure preserving precisely if

$$(*) \qquad \int f \circ T = \int f \quad \text{for } all\ f \in L(X, \mathcal{A}, m).$$

Indeed, if $A \in \mathcal{A}$, then

$$\int \chi_A(Tx)\, dm = \int \chi_{T^{-1}(A)}(x)\, dm = m(T^{-1}(A)),$$
$$\int \chi_A(x)\, dm = m(A).$$

The equation $(*)$ holds true, in particular, for the characteristic function $f = \chi_A$ of the set A and hence $m(T^{-1}(A)) = m(A)$. If, conversely, the map T is measure preserving, then the equation $(*)$ follows for the characteristic functions $f = \chi_A$ of measurable sets $A \in \mathcal{A}$. But then, it holds true for all step functions, and so for every integrable function $f \in L(X, \mathcal{A}, m)$.

Example. We recall the expanding map $\varphi\colon S^1 \to S^1$ of the circle defined by $z \mapsto z^2$ and consider the restriction T of its covering map to the fundamental domain (0,1] which is equipped with the Lebesgue measure. The map $T\colon (0,1] \to (0,1]$ is defined by

$$T(x) = \begin{cases} 2x, & 0 < x \le 1/2, \\ 2x - 1, & 1/2 < x \le 1. \end{cases}$$

For the open interval $(a,b) \subset (0,1]$ we have $T^{-1}((a,b)) = \left(\frac{a}{2}, \frac{b}{2}\right) \cup \left(\frac{1}{2}(a+1), \frac{1}{2}(b+1)\right)$, where the union of the sets is disjoint, so that

$$m(T^{-1}(a,b)) = \frac{1}{2}(b-a) + \frac{1}{2}(b-a) = b - a = m((a,b)),$$

as illustrated in Figure I.7. This holds true for every interval in (0, 1] and hence also for every open subset. It follows that $m(T^{-1}A) = m(A)$ for every Lebesgue measurable set $A \subset (0,1]$, because a measurable set is the countable intersection of open sets up to a null set. Therefore, the map T is *measure preserving*. Alternatively we can also check the criterion $(*)$ above. If $f\colon (0,1] \to \mathbb{R}$ if integrable, then $\int_0^1 f(Tx)\,dx = \int_0^{1/2} f(2x)\,dx + \int_{1/2}^1 f(2x-1)\,dx = \frac{1}{2}\int_0^1 f(y)\,dy + \frac{1}{2}\int_0^1 f(y)\,dy = \int_0^1 f(y)\,dy$, where we have used the substitutions $y = 2x$ respectively $y = 2x - 1$, proving once more that T is measure preserving.

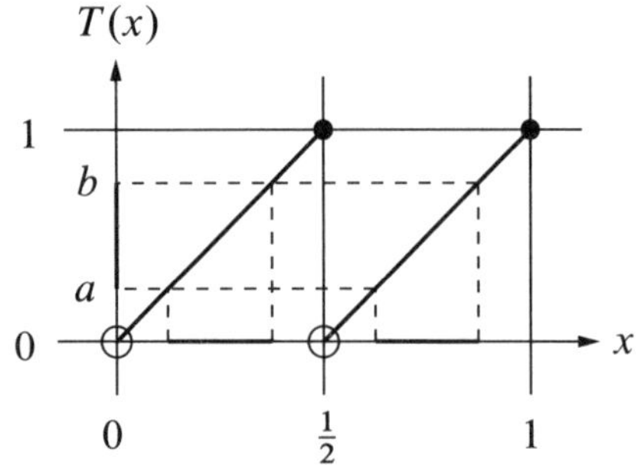

Figure I.7. Preimages of (a,b) under the mapping T in the example.

If the measure is finite the measure preserving maps have strong *recurrence properties* as was already known to H. Poincaré.

Theorem I.14 (Recurrence theorem of Poincaré). *Let $(X,\mathcal{A},m)$ be a **finite** measure space and $T\colon X\to X$ a **measure preserving** map. For a measurable set $A\in\mathcal{A}$ we define the subset $A_0\subset A$ by*

$$A_0 := \{x\in A \mid T^j x\in A \text{ for infinitely many integers } j\ge 0\}.$$

Then the set A_0 is measurable and its measure is equal to $m(A_0)=m(A)$.

The theorem shows that almost every point in A returns to A infinitely often. The theorem is only valid for finite measure spaces, as the translation $x\mapsto x+1$ on $\mathbb{R}$ shows.

Proof. For the integers $n\ge 1$ we introduce the sets $C_n := \{x\in A \mid T^j x\notin A \text{ for all } j\ge n\}$, so that

$$A_0 = A\setminus\bigcup_{n\ge 1} C_n.$$

In order to prove the theorem it suffices to show that $C_n\in\mathcal{A}$ is measurable and $m(C_n)=0$ for every $n\ge 1$. Since the set A belongs to $\mathcal{A}$ and since T is measurable, we conclude that $C_n = A\setminus\bigcup_{j\ge n}T^{-j}(A)\in\mathcal{A}$ is, indeed, a measurable set. Moreover, using the notation $T^{-0}(A) := A$,

$$C_n\subset\bigcup_{j\ge 0}T^{-j}(A)\setminus\bigcup_{j\ge n}T^{-j}(A)$$

from which we conclude, because the measure $m(X)$ is finite, that

$$m(C_n)\le m\Big(\bigcup_{j\ge 0}T^{-j}(A)\Big)-m\Big(\bigcup_{j\ge n}T^{-j}(A)\Big).$$

Since T is measure preserving, both unions have the same measure, in view of

$$\bigcup_{j\geq n} T^{-j}(A) = T^{-n}\Big(\bigcup_{j\geq 0} T^{-j}(A)\Big).$$

Thus $m(C_n) = 0$ and the theorem is proved. □

We shall estimate when the orbit returns for the first time. The time of return is determined by the measure of the set. We assume that $(X, \mathcal{A}, m)$ is a *finite* measure space and $T\colon X \to X$ a *measure preserving* map and let $A \in \mathcal{A}$ be a measurable set satisfying $m(A) > 0$. We claim that if $N := [m(X)/m(A)]$ then there exists an integer j in $1 \leq j \leq N$ for which

$$m(T^{-j}(A) \cap A) > 0.$$

Arguing indirectly, we assume that $m(T^{-j}(A) \cap A) = 0$ for $1 \leq j \leq N$. Then, the sets $T^{-n}(A)$ and $T^{-m}(A)$ are *almost disjoint* for $0 \leq m < n \leq N$, in view of

$$\begin{aligned} m\big(T^{-n}(A) \cap T^{-m}(A)\big) &= m\big(T^{-m}(T^{-(n-m)}(A) \cap A)\big) \\ &= m\big(T^{-(n-m)}(A) \cap A)\big) \\ &= 0. \end{aligned}$$

Since T is measure preserving, it follows that

$$m(X) \geq \sum_{j=0}^{N} m(T^{-j}(A)) = m(A)(1+N) > m(A)(1 + \frac{m(X)}{m(A)} - 1) = m(X),$$

leading to the contradiction $m(X) > m(X)$ and hence proving the claim.

Theorem I.15. *Assume the triple $(X, \mathcal{A}, m)$ to be a finite measure space and the map $T\colon X \to X$ to be measure preserving. Assume in addition that (X, d) is a metric space possessing a countable basis of open sets and assume that all open sets are measurable and of positive measure. Then, there exists for almost every point $x \in X$ a sequence $j_k \to \infty$ of integers so that*

$$T^{j_k}(x) \to x.$$

In this sense, almost every point is recurrent.

Proof. In view of the postulated countable basis of open sets we find a dense sequence $(x_k)_{k\geq 1}$ in X. For every $n \geq 1$ the open balls $B(x_k, 1/n)$ cover the set X. By Theorem I.14 we find a null set $N = N(k, n)$ having the property that every point $x \in B(x_k, 1/n) \setminus N$ returns infinitely often to the ball $B(x_k, 1/n)$. We denote the countable union of these null sets over all k and n by the same letter N. Thus a point $x \in X \setminus N$ returns infinitely often into every ball $B(x_k, 1/n)$ to which it belongs. Since every neighborhood of x contains such a ball, the theorem is proved. □

In the following, we investigate how the points of an orbit $\mathcal{O}^+(x)$ under a measure preserving map T are *statistically* distributed in the space X. Given $A \in \mathcal{A}$ and $x \in X$, we ask, how often does the orbit $(T^j x)_{j \geq 0}$ visit the set A *on the average*? Note that

$$\frac{1}{n}\#\{0 \leq j \leq n-1 \mid T^j x \in A\} = \frac{1}{n}\sum_{j=0}^{n-1} \chi_A(T^j x).$$

We are interested in the convergence of the sum as $n \to \infty$.

The theorem by G. Birkhoff (1932) provides an answer to the question. According to this theorem the pointwise limit exists (in $\mathbb{R}$) for *almost all* $x \in X$, and, in addition, we also have L^1-convergence towards the limit function. We recall that in the equidistribution theorem of H. Weyl (Theorem I.2), we have convergence for every point, assuming the functions to be Riemann integrable instead of Lebesgue integrable.

Theorem I.16 (Ergodic theorem of G. Birkhoff). *We consider the* ***finite*** *measure space* $(X, \mathcal{A}, m)$ *and assume the map* $T\colon X \to X$ *to be* ***measure preserving***. *For every integrable function* $f \in L(X, \mathcal{A}, m)$ *there exist a function* $f^* \in L(X, \mathcal{A}, m)$ *and a null set* $N \subset X$ *(*N ***depending*** *on* f*) satisfying* $T^{-1}(N) = N$ *and*

(i) $\lim_{n\to\infty} \frac{1}{n}\sum_{j=0}^{n-1} f(T^j x) = f^*(x)$ *for all* $x \in X \setminus N$,

(ii) $f^*(Tx) = f^*(x)$ *for all* $x \in X$,

(iii) $\int_X f^* = \int_X f$,

(iv) $\int_X \left|\frac{1}{n}\sum_{j=0}^{n-1} f(T^j x) - f^*(x)\right| dm \to 0$, $n \to \infty$.

Postponing the *proof* to the end of this section, we first introduce the concept of ergodicity and draw some consequences from the ergodic theorem. We recall that the subset $A \subset X$ is called T-invariant, if $T^{-1}(A) = A$.

Definition. Assume $(X, \mathcal{A}, m)$ to be a finite measure space. A measure preserving map $T\colon X \to X$ is called *ergodic*, if, for *every* T-invariant set $A \in \mathcal{A}$, the following holds true:

$$m(A) = m(X) \quad \text{or} \quad m(A) = 0.$$

In an ergodic system it is *not* possible to split X into two *invariant subsets* of *positive* measure.

The following proposition characterizes the ergodicity in a different way; the map T is ergodic, precisely if the T-invariant (measurable) functions are constant almost everywhere.

Proposition I.17 (Criterion for ergodicity). *We consider the finite measure space* $(X, \mathcal{A}, m)$ *and the measure preserving map* $T\colon X \to X$. *The following two statements are equivalent.*

(i) *T is ergodic.*

(ii) *For every* $f \in L(X, \mathcal{A}, m)$ *we conclude from* $f(T(x)) = f(x)$ *for all* $x \in X$ *that* $f(x) =$ *constant almost everywhere.*

Proof. (ii) $\Longrightarrow$ (i): We assume that the set $A \in \mathcal{A}$ satisfies $T^{-1}(A) = A$. Its characteristic function satisfies $\chi_A(Tx) = \chi_{T^{-1}(A)}(x) = \chi_A(x)$. In view of the assumption, χ_A is constant almost everywhere and we conclude that either $m(A) = 0$ or $m(A) = m(X)$.

(i) $\Longrightarrow$ (ii): We assume that the map T is ergodic and that the function $f \in L(X, \mathcal{A}, m)$ satisfies $f(Tx) = f(x)$ for all $x \in X$. If f is *not constant almost everywhere,* there exists a real number $c \in \mathbb{R}$ such that the set $A := \{x \in X \mid f(x) \geq c\}$ has the measure $0 < m(A) < m(X)$. Since f is invariant, also the set A is invariant, in contradiction to the ergodicity of T. □

The proof shows that in the statement (ii) of Proposition I.17 in place of $f \in L(X, \mathcal{A}, m)$ we can also take $f \in L^p(X, \mathcal{A}, m)$ for any $1 \leq p \leq \infty$ (remember that $m(X) < \infty$) or the measurable functions $f : X \to \overline{\mathbb{R}}$. Another characterization of ergodicity follows from the ergodic theorem.

Theorem I.18 (Ergodicity criterion). *Let* $(X, \mathcal{A}, m)$ *be a finite measure space and let the map* $T : X \to X$ *be measure preserving. Then the following two statements are equivalent.*

(i) T *is ergodic.*

(ii) *For every* $f \in L(X, \mathcal{A}, m)$ *there exists a null set* $N = N(f)$ *such that*

$$\lim_{n\to\infty} \frac{1}{n} \sum_{j=0}^{n-1} f(T^j x) - \frac{1}{m(X)} \int_X f$$

for every $x \in X \setminus N$, *i.e., for almost every orbit the mean value of the function* $f \in L$ ***over the orbit*** *equals the mean value of* f ***over the space*** X.

Proof [Theorem I.16, Proposition I.17]. (i) $\Longrightarrow$ (ii): If the map T is ergodic and $f \in L$, then, also $f^* \in L$ and $f^*(T(x)) = f^*(x)$ for every $x \in X$, in view of Theorem I.16. Hence, it follows from Proposition I.17 that $f^*(x) = c \in \mathbb{R}$ almost everywhere and consequently,

$$\int_X f = \int_X f^* = \int_X c = c\, m(X),$$

hence $c = \frac{1}{m(X)} \int f$ and the statement (ii) follows from Theorem I.16.

(ii) $\Longrightarrow$ (i): If $f \in L$ satisfies $f(Tx) = f(x)$ for all $x \in X$, we show that $f(x)$ is constant almost everywhere. Then, the statement (i) follows from Proposition I.17. In view of Theorem I.16, there exists a null set $N \subset X$, such that for $x \in X \setminus N$

we have

$$\begin{aligned} f(x) &\overset{\text{Invar.}}{=} \frac{1}{n}\sum_{j=0}^{n-1} f(T^j(x)) \quad \text{for } \textit{all } n \geq 1 \\ &= \lim_{n\to\infty} \frac{1}{n}\sum_{j=0}^{n-1} f(T^j(x)) \\ &\overset{\text{Vor.}}{=} \frac{1}{m(X)}\int f \end{aligned}$$

showing that $f(x)$ is constant for almost all x. □

Corollary I.19 (Equidistribution). *Let $(X, \mathcal{A}, m)$ be a finite measure space and let the map $T: X \to X$ be* **ergodic**. *If $A \in \mathcal{A}$ satisfies $m(A) > 0$, then there exists a null set $N = N(A) \subset X$ such that for $x \in X \setminus N$ the following holds true:*

$$\frac{1}{n}\#\{0 \leq j \leq n-1 \mid T^j x \in A\} \to \frac{m(A)}{m(X)}.$$

In general, $N \neq \emptyset$ and N depends on the set A.

Proof. We take the characteristic function $f = \chi_A$. Then $f \in L$ and according to Theorem I.18 there exists a null set $N = N(A)$ such that for $x \in X \setminus N$,

$$\frac{1}{n}\sum_{j=0}^{n-1}\chi_A(T^j x) \to \frac{1}{m(X)}\int \chi_A = \frac{m(A)}{m(X)}$$

as $n \to \infty$. □

Concerning the next consequence from the ergodic theorem, we note that for every *finite* measure space $(X, \mathcal{A}, m)$ satisfying $m(X) > 0$ a new standardized measure μ can be defined by $\mu(A) = m(A)/m(X)$ for every set $A \in \mathcal{A}$. We then have $\mu(X) = 1$ so that $(X, \mathcal{A}, \mu)$ is a *probability space*.

Corollary I.20. *Let $(X, \mathcal{A}, m)$ be a measure space satisfying $m(X) = 1$ and let $T: X \to X$ be a measure preserving map. Then, the following two statements are equivalent:*

(i) *T is ergodic.*

(ii) *For* ***all*** *$A, B \in \mathcal{A}$ we have*

$$\frac{1}{n}\sum_{j=0}^{n-1} m(T^{-j}(A) \cap B) \to m(A)m(B).$$

Proof [Ergodic theorem, convergence theorem of Lebesgue]. (i) $\Longrightarrow$ (ii): Taking the characteristic function $f = \chi_A$, we conclude from the ergodic theorem (Theorem I.16) that

$$\frac{1}{n}\sum_{j=0}^{n-1} \chi_A(T^j x) \to m(A) \quad \text{for almost every } x.$$

Therefore,

$$(*) \quad \frac{1}{n}\sum_{j=0}^{n-1} \chi_A(T^j x)\chi_B(x) \to m(A)\chi_B(x) \quad \text{for almost every } x.$$

In view of $\chi_A(T^j x)\chi_B(x) = \chi_{T^{-j}(A)}(x)\chi_B(x) = \chi_{T^{-j}(A)\cap B}(x)$ the left side function in the formula $(*)$ is majorized by the characteristic function $\chi_X \equiv 1$ of the whole space. By integrating $(*)$ and using the convergence theorem of Lebesgue we obtain

$$\frac{1}{n}\sum_{j=0}^{n-1} m(T^{-j}(A) \cap B) \to m(A)m(B).$$

(ii) $\Longrightarrow$ (i): If $E \in \mathcal{A}$ is a T-invariant set, we take $A = B = E$ in the statement (ii) and find

$$m(E) = \frac{1}{n}\sum_{j=0}^{n-1} m(E) \to m(E)^2$$

from which we conclude that $m(E) = 0$ or $m(E) = 1$. Therefore, the map T is ergodic. □

Ergodicity is a property of a measure preserving map on a measure space, while transitivity is a property of a continuous map on a topological space. Next, we equip a measure space with a topology and show that under additional assumptions an ergodic system is also transitive.

Proposition I.21 (Transitivity of ergodic systems). *Let $(X, \mathcal{A}, m)$ be a finite measure space and let $T : X \to X$ be an* **ergodic** *map. Moreover, assume that (X, d) is a metric space possessing a countable basis of open sets. Furthermore, assume that every open set $U \neq \emptyset$ is measurable and of positive measure $m(U) > 0$. Under these assumptions, there exists a null set $N \subset X$ such that*

$$\overline{\mathcal{O}^+(x)} = X \text{ for } x \in X \setminus N,$$

i.e., almost all orbits are dense.

If, in addition, the map $T : X \to X$ is continuous and if the metric space (X, d) is complete, the set

$$R^+ := \{x \in X \mid \overline{\mathcal{O}^+(x)} = X\}$$

is dense in X and of second Baire category.

The set R^+ of initial points of dense orbits is large in the sense of the measure ($m(R^+) = m(X)$) and, under the additional assumptions, it is also large in the functional analytic sense of the Baire category.

Example. The assumptions on the space X are fulfilled for the circle S^1 and the torus T^n equipped with the Lebesgue measure and we shall later on look at these examples again.

Proof of Proposition I.21 [Ergodic theorem, category theorem of Baire]. (1) We take the countable base $(V_k)_{k\geq 1}$ of open sets of X satisfying $V_k \neq \emptyset$. According to Corollary I.19 (*equidistribution*) there exists for V_k a null set $N_k \subset X$ such that for all $x \in X \setminus N_k$ we have $T^j x \in V_k$ for infinitely many integers j, using that $m(V_k) > 0$. In particular, $\mathcal{O}^+(x) \cap V_k \neq \emptyset$ for $x \in X \setminus N_k$. This holds true for every $k \geq 1$. The countable union $N := \bigcup_{k\geq 1} N_k$ is a null set, so that $m(N) = 0$ and, of course, $\mathcal{O}^+(x) \cap V_k \neq \emptyset$ for $x \in X \setminus N$ and for every $k \geq 1$. If $\emptyset \neq V \subset X$ is any open set, then, according to the definition of the base, there exists an index k such that $V_k \subset V$ and therefore,

$$\mathcal{O}^+(x) \cap V \neq \emptyset, \quad x \in X \setminus N.$$

This holds true for every open set $\emptyset \neq V$ so that

$$\overline{\mathcal{O}^+(x)} = X, \quad x \in X \setminus N.$$

Moreover, $m(R^+) = m(X)$, since N is a null set.

(2) We assume now that the map T is in addition *continuous*. Then for every $k \geq 1$ the set

$$\mathcal{O}^-(V_k) := \bigcup_{j\geq 1} T^{-j}(V_k)$$

is open. If $\emptyset \neq U$ is open, then $m(U) > 0$, and due to $m(V_k) > 0$, we can argue as above to find a point $x \in U$ satisfying $T^j x \in V_k$ for infinitely many j, in particular,

$$U \cap \mathcal{O}^-(V_k) \neq \emptyset.$$

Therefore, the open set $\mathcal{O}^-(V_k)$ is dense, and this holds true for every $k \geq 1$. If X is *complete*, then, according to the category theorem of Baire (Lemma I.10) the countable intersection

$$R := \bigcap_{j\geq 1} \mathcal{O}^-(V_j)$$

of open and dense sets is dense in X and, by definition, of second Baire category. As in the proof of Theorem I.9 one shows for $p \in R$ that

$$\overline{\mathcal{O}^+(p)} = X.$$

Therefore, $R \subset R^+$ and the theorem is proved. □

Before we prove the ergodicity of our dynamical systems on the circle and on the torus described above, it is useful to recall in a short interlude some results about Fourier series. The proofs for most of the statements can be found in standard textbooks on analysis as e.g. in the textbook of K. Stromberg [111], or in the classic [31] by H. Dym and H. P. McKean. The theorem by L. Carleson, as well as the example by A. N. Kolmogorov below, can be found in the book [46] by L. Grafakos.

Facts about Fourier series. In the following we consider integrable functions $f \in L^1(S^1)$ which are measurable functions $x \mapsto f(e^{2\pi i x})$, periodic of period 1 and, abbreviating $f(e^{2\pi i x}) \equiv f(x)$, satisfy

$$\int_0^1 |f(x)|\,dx < \infty.$$

Fourier coefficients. With $f \in L^1(S^1)$ one can associate the *sequence* $(\hat{f}(n))_{n\in\mathbb{Z}}$ of *numbers* (called *Fourier coefficients*) defined by

$$\hat{f}(n) := \int_0^1 f(x)e^{-2\pi i n x}\,dx, \quad n \in \mathbb{Z}.$$

Two classical statements about Fourier coefficients are the following.

- *Riemann–Lebesgue lemma:* If $f \in L^1(S^1)$, then

$$\hat{f}(n) \to 0, \quad |n| \to \infty.$$

- If $f, g \in L^1(S^1)$, the following statement holds true:

$$\hat{f}(n) = \hat{g}(n) \text{ for all } n \in \mathbb{Z} \Longrightarrow f = g \text{ almost everywhere.}$$

  In other words an element of L^1 is uniquely determined by its Fourier coefficients so that the map $f \mapsto \hat{f}$ is *injective*.

The Hilbert space $L^2(S^1)$. The *Hilbert space* $L^2(S^1)$ is equipped with the scalar product

$$(f, g) = \int_0^1 f(x)\overline{g(x)}\,dx$$

and possesses the orthonormal system $(e_n)_{n\in\mathbb{Z}}$ defined by

$$e_n := e^{2\pi i n x}, \quad n \in \mathbb{Z},$$

and satisfying $(e_n, e_k) = \delta_{ij}$. Thus, for $f \in L^2(S^1)$ the Fourier coefficients can be written as

$$\hat{f}(n) = (f, e_n), \quad n \in \mathbb{Z}.$$

Fourier series. If $f \in L^1(S^1)$, we abbreviate $S_n(f) := \sum_{|k| \le n} \hat{f}(k) e_k$. The trigonometrical polynomial $S_n(f)$ belongs to $C^\infty(S^1)$ and is equal to

$$S_n(f)(x) = \sum_{|k| \le n} \hat{f}(k) e^{2\pi i k x}.$$

A classical result in Hilbert space is as follows.

- *Theorem of Riesz–Fischer*: If $f \in L^2(S^1)$, then its Fourier series

  $$f = \sum_{k \in \mathbb{Z}} \hat{f}(k) e_k$$

  converges in $L^2(S^1)$,

  $$\|f - S_n(f)\|_{L^2(S^1)} \to 0 \quad \text{for } n \to \infty,$$

  where $\|f\|_{L^2(S^1)} := (f, f)^{1/2}$ is the norm in a Hilbert space.

In this sense the functions $(e_n)_{n \in \mathbb{Z}}$ constitute a *Hilbert basis* of $L^2(S^1)$, i.e., $\overline{\operatorname{span}}((e_n)_{n \in \mathbb{Z}}) = L^2(S^1)$. A Hilbert space is separable precisely if it possesses a countable Hilbert base. Such a space is *isometrically isomorphic* to the sequence space

$$\ell^2 := \{c = (c_n)_{n \in \mathbb{Z}} \mid \sum_{k \in \mathbb{Z}} |c_k|^2 =: \|c\|_2^2 < \infty\}.$$

The isomorphism is the linear map $f \mapsto \hat{f} = (\hat{f}(n))_{n \in \mathbb{Z}}$ which satisfies

$$\|f\|_{L^2(S^1)}^2 = \sum_{k \in \mathbb{Z}} |\hat{f}(k)|^2 = \|(\hat{f}(n))_{n \in \mathbb{Z}}\|_2^2,$$

called *Plancherel identity*.

Pointwise convergence. The question of *pointwise* convergence is subtle, as the following results show.

- Given a null set $N \subset [0, 1]$, there exists a continuous function $f \in C(S^1)$ such that $S_n(f)(x)$ *diverges* for *all* $x \in N$.
- *A. N. Kolmogorov* (1926): There exists an element $f \in L^1(S^1)$ such that $S_n(f)(x)$ *diverges* for *every* $x \in S^1$!
- *L. Carleson* (1966): If $f \in L^2(S^1)$, then

  $$S_n(f)(x) \to f(x) \quad \text{almost everywhere.}$$

- *R. A. Hunt* (1968): If $p > 1$ and $f \in L^p(S^1)$, then

$$S_n(f)(x) \to f(x) \quad \text{almost everywhere.}$$

- Assume that $f \in C^p(S^1)$ and $p \geq 1$. Then

$$\sup_{x \in [0,1]} |f(x) - S_n(f)(x)| \leq c(f) n^{-p+\frac{1}{2}},$$

where $c(f)$ is a constant. The convergence is *uniform* and the faster, the smoother the function is.

After these recollections of Fourier series, we return to our simple examples of dynamical systems and investigate their ergodicity.

Examples. (1) *Rigid rotations.* We return to the mapping

$$\varphi \colon S^1 \to S^1, \quad \varphi(z) = \vartheta z,$$

where $\vartheta = e^{2\pi i \alpha}$ and $z = e^{2\pi i x}$ for a real number $x \in \mathbb{R}$.

Claim. We first claim that the rigid rotation φ is *measure preserving* with respect to the Lebesgue measure.

Proof. We assume $f : S^1 \to \mathbb{C}$ to be integrable and define

$$F(x) := f(e^{2\pi i x}).$$

Then $F(x+1) = F(x)$ for $x \in \mathbb{R}$ and F is locally integrable. Recalling the translation $\Phi(x) = x + \alpha$ we calculate

$$\begin{aligned}
\int_0^1 F(\Phi(x))\, dx &= \int_0^1 F(x+\alpha)\, dx \qquad (y = x + \alpha) \\
&= \int_\alpha^{\alpha+1} F(y)\, dy \\
&= \int_\alpha^1 F(y)\, dy + \int_1^{\alpha+1} F(y-1)\, dy \\
&= \int_\alpha^1 F(y)\, dy + \int_0^\alpha F(y)\, dy \\
&= \int_0^1 F(y)\, dy.
\end{aligned}$$

One concludes that $\int_{S^1} f \circ \varphi = \int_{S^1} f$ for every $f \in L(S^1)$, so that the map is indeed measure preserving as claimed. □

- The case $\alpha = p/q \in \mathbb{Q}$.
  In this case there exist invariant functions $f: S^1 \to \mathbb{C}$ which are not constant, as the example $f(z) = z^q = e^{2\pi i q x}$ shows. Indeed, due to $\vartheta^q = 1$ we find
$$f(\varphi(z)) = (\vartheta z)^q = z^q = f(z), \quad z \in S^1.$$
  Having found an invariant function which is not constant we conclude from Proposition I.17 that the map φ is *not ergodic*.
- The case $\alpha \in \mathbb{R} \setminus \mathbb{Q}$.
  In this case φ is *ergodic*. To prove this, it suffices to show that all invariant functions $f \in L^2(S^1)$ are constant almost everywhere, cf. the remark following Proposition I.17. If $f \in L^2(S^1)$ we look at its *Fourier series*
$$f(z) = \sum_{k \in \mathbb{Z}} f_k z^k, \quad z = e^{2\pi i x}$$
  which converges in L^2. Then,
$$f(\varphi(z)) = f(\vartheta z) = \sum_{k \in \mathbb{Z}} (f_k \vartheta^k) z^k.$$
  If f is invariant, $f(\varphi(z)) = f(z)$, we conclude from the *uniqueness of the Fourier coefficients* that
$$f_k = \vartheta^k f_k, \quad k \in \mathbb{Z}.$$
  Due to $\vartheta^k \neq 1$ if $k \neq 0$ (since α irrational) it follows that $f_k = 0$ for all $k \neq 0$, so that $f(z) = f_0$ is constant almost everywhere. This proves that the irrational rotation of the circle is ergodic.

Exercise. Consider the map $\varphi\colon T^n \to T^n$ on the torus which in the covering space $\mathbb{R}^n$ is the translation
$$\Phi(x) = x + \omega, \quad x \in \mathbb{R}^n,$$
and assume that
$$\langle \omega, j \rangle \notin \mathbb{Z}, \quad j \in \mathbb{Z}^n \setminus \{0\}.$$
Prove that φ is *ergodic* with respect to the Lebesgue measure on T^n.

(2) *Expansion.* Next, we return to our expanding map $\varphi\colon S^1 \to S^1$ defined by $\varphi(z) = z^2$. The map φ is measure preserving with respect to the Lebesgue measure, as we have already seen, and we prove the ergodicity of the map φ by demonstrating that the integrable invariant functions are constant almost everywhere. Let $f \in L(S^1)$ and introduce the function $g(z) := f(\varphi(z)) = f(z^2)$. If $F(x) :=$

$f(e^{2\pi ix})$, then $F(x+1) = F(x)$ for all $x \in \mathbb{R}$, and $G(x) := g(e^{2\pi ix}) = F(2x)$. Computing the *Fourier coefficients* g_{2k} of order $2k$ of the function G, we obtain

$$\begin{aligned} g_{2k} &:= \int_0^1 G(x)e^{-2\pi i2kx}\,dx \\ &= \int_0^1 F(2x)e^{-2\pi i2kx}\,dx \\ &= \int_0^{1/2} F(2x)e^{-2\pi i2kx}\,dx + \int_{1/2}^1 F(2x-1)e^{-2\pi i2kx}\,dx \\ &= \frac{1}{2}\int_0^1 F(y)e^{-2\pi iky}\,dy + \frac{1}{2}\int_0^1 F(y)e^{-2\pi iky}\,dy \\ &= \int_0^1 F(y)e^{-2\pi iky}\,dy \\ &=: f_k. \end{aligned}$$

If f is invariant, then $G(x) = F(2x) = F(x)$ and hence $g_k = f_k$. In view of the above calculation, $f_{2k} = g_{2k} = f_k$ for all $k \in \mathbb{Z}$ and we see that

$$f_k = f_{2k} = f_{2^2k} = \cdots = f_{2^nk} = \cdots.$$

Since $\lim_{|k|\to\infty} f_k = 0$ in view of the Riemann–Lebesgue lemma, we conclude that

$$f_k = 0, \quad \text{for all } k \neq 0.$$

Therefore, $F(x) = f_0$ is constant almost everywhere and according to Proposition I.17, our map is ergodic.

Remark. We have verified that the map $\varphi\colon z \mapsto z^2$ on the circle S^1 satisfies all the assumptions of Proposition I.21, from which *the statement* (iii) *in Proposition* I.12 *now follows.* Namely, the set $R^+ := \{z \in S^1 \mid \overline{\mathcal{O}^+(z)} = S^1\}$ of initial points of dense orbits has full measure,

$$m(R^+) = m(S^1),$$

and, in addition, the set R^+ is *dense* in S^1 and of second Baire category.

Exercise. Every real number $0 < x \leq 1$ can be represented by a unique decimal expansion

$$x = 0.x_1x_2x_3\cdots = \sum_{j\geq 1} \frac{x_j}{10^j}$$

containing infinitely many non zero digits $x_j \in \{0, 1, 2, \ldots, 9\}$. Demonstrate that on the average, the number of zeros in the decimal expansion is equal to $1/10$ for *almost all* $0 < x \le 1$. Hint: Consider the mapping $\varphi\colon S^1 \to S^1$ defined by $\varphi(z) = z^{10}$ in the fundamental domain of the covering space. It is the map $T\colon (0,1] \to (0,1]$ defined by $T(x) = 10x \mod 1$. The map T is measure preserving and ergodic. Consider now the real number $x = 0.x_1x_2x_3\ldots$ and the interval $A := (0, 1/10)$, then

$$T^j x \in A \Longleftrightarrow x_{j+1} = 0.$$

Apply the equidistribution theorem.

Proof of the ergodic theorem. We conclude this chapter with the proof of the ergodic theorem following the arguments of A. M. Garsia (1965) and using the maximal ergodic lemma. We recall the statement of the theorem.

Theorem I.16. *We assume that* $(X, \mathcal{A}, m)$ *is a* ***finite*** *measure space and* $T\colon X \to X$ *a* ***measure preserving*** *map. For every integrable function* $f \in L(X, \mathcal{A}, m)$ *there exist a function* $f^* \in L(X, \mathcal{A}, m)$ *and a null set* $N \subset X$ *(*N ***depending*** *on* f*) satisfying* $T^{-1}(N) = N$ *and*

(i) $\lim_{n\to\infty} \frac{1}{n}\sum_{j=0}^{n-1} f(T^j x) = f^*(x)$ *for all* $x \in X \setminus N$,

(ii) $f^*(Tx) = f^*(x)$ *for all* $x \in X$,

(iii) $\int_X f^* = \int_X f$,

(iv) $\int_X \left|\frac{1}{n}\sum_{j=0}^{n-1} f(T^j x) - f^*(x)\right| dm \to 0,\ n \to \infty$.

Proof [Convergence theorems from integration theory]. If $f \in L(X, \mathcal{A}, m)$ we abbreviate in the following

$$S_n(x) = S_n(f, x) = \sum_{j=0}^{n-1} f(T^j x), \quad n \ge 1.$$

Our first aim is to prove that the limit $\lim_{n\to\infty} \frac{1}{n} S_n(x)$ does exist almost everywhere, by showing that

$$m\Big(\Big\{x \in X \mid \underline{\lim}\, \frac{1}{n} S_n(x) < \overline{\lim}\, \frac{1}{n} S_n(x)\Big\}\Big) = 0.$$

(1) *Maximal ergodic lemma.* We introduce the sequence of integrable functions

$$\begin{aligned} S_0(x) &\equiv 0, \\ S_n^+(x) &= \max_{0\le k\le n} S_k(x) \end{aligned}$$

and observe that $S_n^+(x) \ge 0$ for all $x \in X$.

Lemma I.22. *Let* $f \in L(X, \mathcal{A}, m)$ *and define the subsets*

$$A_n = \{x \in X \mid S_n^+(x) > 0\}.$$

Then

$$\int_{A_n} f\,dm \geq 0, \quad \textit{for all } n \geq 1.$$

Proof. From the definition of S_n^+ it follows that

$$f(x) + S_n^+(Tx) \geq f(x) + S_k(Tx) = S_{k+1}(x), \quad 0 \leq k \leq n.$$

If $x \in A_n$, then there exists an integer $1 \leq k \leq n$ for which $S_k(x) > 0$. Consequently,

$$f(x) + S_n^+(Tx) \geq \max_{1 \leq k \leq n} S_k(x) = \max_{0 \leq k \leq n} S_k(x) = S_n^+(x)$$

and hence $f(x) \geq S_n^+(x) - S_n^+(Tx)$. By using that S_n^+ is equal to zero outside of A_n and $S_n^+(Tx) \geq 0$ everywhere, we conclude from this inequality and from the measure preservation of the map T that

$$\begin{aligned}
\int_{A_n} f &\geq \int_{A_n} S_n^+(x) - \int_{A_n} S_n^+(Tx) \\
&= \int_X S_n^+(x) - \int_{A_n} S_n^+(Tx) \\
&\geq \int_X S_n^+(x) - \int_X S_n^+(Tx) \\
&= \int_X S_n^+(x) - \int_X S_n^+(x) \\
&= 0
\end{aligned}$$

and the lemma is proved. □

(2) For $a < b$ we define the subset Y of X by

$$Y = Y(a,b) = \Big\{x \in X \mid \underline{\lim}\, \frac{1}{n} S_n(x) < a < b < \overline{\lim}\, \frac{1}{n} S_n(x)\Big\}.$$

Lemma I.23. *The set Y is measurable and invariant under the map T, i.e., $T^{-1}(Y) = Y$.*

Proof. The statement $Y \in \mathcal{A}$ follows from elementary measure theory, since the functions S_n are all measurable. To prove the invariance, we have to show that $\underline{\lim}\, \frac{1}{n} S_n(x) = \underline{\lim}\, \frac{1}{n} S_n(Tx)$ and $\overline{\lim}\, \frac{1}{n} S_n(x) = \overline{\lim}\, \frac{1}{n} S_n(Tx)$. However, this follows immediately from the identity

$$\frac{1}{n} S_n(Tx) = \frac{1}{n} S_{n+1}(x) - \frac{1}{n} f(x) = \frac{1}{n+1} S_{n+1}(x) \underbrace{\frac{n+1}{n}}_{\to 1} - \underbrace{\frac{1}{n}}_{\to 0} f(x)$$

by taking the $\underline{\lim}$ respectively $\overline{\lim}$, and the lemma is proved. □

Lemma I.24. *From* $m(X) < \infty$ *it follows that* $m(Y) = 0$.

Proof. We apply Lemma I.22 to the T-invariant set Y (instead of X) and to the function g (instead of f), defined by

$$g(x) := f(x) - b.$$

Since $m(X) < \infty$ the function $g \in L(X, \mathcal{A}, m)$ is integrable. Hence, setting

$$A_n = \{x \in Y \mid S_n^+(g, x) > 0\},$$

we have

$$\int_{A_n} (f - b) = \int_Y \chi_{A_n}(f - b) \geq 0.$$

According to the definition of Y there exists for every point $x \in Y$ an integer j for which

$$\frac{1}{j} S_j(g, x) = \frac{1}{j} S_j(f, x) - b > 0$$

and hence $S_j^+(g, x) > 0$. Consequently every $x \in Y$ is contained in some set A_j, so that $Y = \bigcup_{n \geq 1} A_n$. The monotonicity of the sequence of sets $A_n \subset A_{n+1} \subset \cdots$ implies $\lim \chi_{A_n}(x) = \chi_Y(x)$. Using the convergence theorem of Lebesgue we obtain in the limit as $n \to \infty$,

$$\int_Y (f - b) \geq 0.$$

In exactly the same way one proves $\int_Y (a - f) \geq 0$. Addition of the inequalities results in the inequality

$$(a - b) \int_Y dm = (a - b) m(Y) \geq 0.$$

Since $a < b$, one concludes $m(Y) = 0$ and the lemma is proved. □

(3) *Pointwise convergence.* In view of Lemma I.24 the set

$$N_0 := \left\{x \in X \mid \underline{\lim} \frac{1}{n} S_n(x) < \overline{\lim} \frac{1}{n} S_n(x)\right\},$$

is the countable union of null sets

$$N_0 = \bigcup_{\substack{a<b \\ a,b \in \mathbb{Q}}} Y(a, b),$$

and hence a null set, so that $m(N_0) = 0$. For $x \notin N_0$, hence for *almost all* $x \in X$, we abbreviate the limit

$$\underline{\lim} \frac{1}{n} S_n(x) = \overline{\lim} \frac{1}{n} S_n(x) = \lim_{n \to \infty} \frac{1}{n} S_n(x) =: \varphi(x),$$

where $-\infty \leq \varphi(x) \leq +\infty$. Defining $\varphi(x) = 0$ on the null set N_0 we show next that $\varphi(x) \in \mathbb{R}$ is a real number almost everywhere. This will follow once we have proved that $\varphi \in L(X, \mathcal{A}, m)$. Since T is measure preserving, we have $\int_X |f(T^j x)|\,dm = \int_X |f|\,dm$ for every $j \geq 0$ and therefore,

$$\int_X \left|\frac{1}{n} S_n(x)\right| dm \leq \int_X |f|\,dm.$$

Using the Lemma of Fatou we can estimate

$$\int_X |\varphi| = \int_X \underline{\lim} \left|\frac{1}{n} S_n(x)\right| \leq \underline{\lim} \int_X \left|\frac{1}{n} S_n(x)\right| \leq \int_X |f| < \infty.$$

Hence $\varphi \in L(X, \mathcal{A}, m)$ and therefore, $\varphi(x) \in \mathbb{R}$ almost everywhere. The set

$$N = N_0 \cup \{x \in X \mid |\varphi(x)| = \infty\},$$

on which $\lim_{n\to\infty} \frac{1}{n} S_n(x)$ does not exist in $\mathbb{R}$, is therefore a null set. Since the limits $\underline{\lim}$, $\overline{\lim}$ are invariant under T (due to Lemma I.23), the set N is also invariant under T. We now *define* the function $f^*\colon X \to \mathbb{R}$ showing up in the statement of the theorem as follows:

$$f^*(x) := \begin{cases} \lim_{n\to\infty} \frac{1}{n} S_n(f, x), & x \notin N, \\ 0, & x \in N. \end{cases}$$

Then $f^* \in L(X, \mathcal{A}, m)$ and one sees, as in Lemma I.23, that $f^*(Tx) = f^*(x)$ for all $x \in X$. So far, we have proved the statements (i) and (ii) of the ergodic theorem.

(4) It remains to show that

(a) $\frac{1}{n} S_n(f, x) \to f^*$ in L^1,

(b) $\int_X f^* = \int_X f$.

We begin by proving a *special case*. We assume that f is a bounded function assuming that $|f| \leq K$ for a constant $K > 0$. Then,

$$\left|\frac{1}{n} S_n(x)\right| \leq K,$$

and the statement (a) follows by means of the dominated convergence theorem of Lebesgue. Since the map T is measure preserving,

$$\int_X \frac{1}{n} S_n(x) = \int_X f, \quad n \geq 1,$$

and the statement (b) follows, again using the convergence theorem of Lebesgue,

$$\int_X f = \lim_{n\to\infty} \int_X \frac{1}{n} S_n(x) = \int_X \lim_{n\to\infty} \frac{1}{n} S_n(x) = \int_X f^*.$$

(5) We next prove the *general case by approximation* with the truncated functions of $f \in L(X, \mathcal{A}, m)$, defined by

$$f_N(x) = \begin{cases} f(x), & |f(x)| \le N, \\ 0, & |f(x)| > N. \end{cases}$$

Then, $|f_N| \le N$, and in view of the special case we have the pointwise convergence

$$\frac{1}{n} S_n(f_N, x) \to f_N^*(x), \quad n \to \infty$$

almost everywhere and also the convergence in L^1. Let $\varepsilon > 0$. By the triangle inequality one estimates

$$\int \left| \frac{1}{n} S_n(f) - f^* \right| dm \le \int \left| \frac{1}{n} S_n(f) - \frac{1}{n} S_n(f_N) \right| dm + \int \left| \frac{1}{n} S_n(f_N) - f_N^* \right| dm + \int |f_N^* - f^*| \, dm,$$

and we are going to estimate each term on the right-hand side by $\varepsilon/3$ if N and n are sufficiently large.

Since $f \in L$ we have $|f| < \infty$ almost everywhere, and since $f_N(x) = f(x)\chi_{|f| \le N}(x) \to f(x)$ we conclude by the convergence theorem of Lebesgue that

$$f_N \to f \text{ in } L^1.$$

Therefore, choosing N large enough,

$$\int |f_N - f| \le \varepsilon/3.$$

Using that the map T is measure preserving we can now estimate the first term as follows.

$$\begin{aligned} \int \left| \frac{1}{n} S_n(f_N) - \frac{1}{n} S_n(f) \right| dm &= \int \frac{1}{n} \left| \sum_{j=0}^{n-1} (f_N - f)(T^j x) \right| dm \\ &\le \frac{1}{n} \sum_{j=0}^{n-1} \int |(f_N - f)(T^j x)| \, dm \\ &= \frac{1}{n} \sum_{j=0}^{n-1} \int |(f_N - f)(x)| \, dm \\ &= \int |f_N - f| \, dm \\ &\le \varepsilon/3. \end{aligned}$$

This holds true for *all* n and the *first term* is taken care of. We now apply the *Lemma of Fatou* to the first term to obtain as $n \to \infty$,

$$\int |f_N^* - f^*|\, dm \le \varepsilon/3,$$

which is the desired estimate of the *third term*. For the *second term* we conclude

$$\int \left| \frac{1}{n} S_n(f_N) - f_N^* \right| dm \le \varepsilon/3$$

for n large enough, from the convergence statement in the special case. Altogether,

$$\int \left| \frac{1}{n} S_n(f) - f^* \right| \le \varepsilon$$

for n large enough. This is true for every $\varepsilon > 0$. Therefore,

$$\frac{1}{n} S_n(f) \to f^* \text{ in } L^1$$

and the proof of the statement (a) in the general case is completed. As for the statement (b) we recall that, because of the measure preservation of the map T,

$$\int \frac{1}{n} S_n(f)\, dm = \int f$$

for all $n \ge 1$, and using the L^1-convergence we finally obtain

$$\int f = \lim_{n \to \infty} \int \frac{1}{n} S_n(f) = \int f^*.$$

Herewith, the ergodic theorem of Birkhoff is proved. □

Literature. There are several monographs on dynamical systems which cover most of the topics treated in the first four chapters. Among them [49] by B. Hasselblatt and A. Katok, [113] by W. Szlenk, [63] by R. Mañé and [91] by C. Robinson. The special topic of ergodic theory is treated, for example, in the monographs [23] by I. Cornfield, S. V. Fomin and Ya. G. Sinai, [108] and [109] by Ya. G. Sinai, [88] by M. Pollicott, [60] by U. Krengel, [37] H. Fuerstenberg and [115] by P. Walters. For surveys about specific problems in dynamical systems including historical information and many references we recommend the Handbooks of Dynamical Systems [50] and [35].

Chapter II
Invariant manifolds of hyperbolic fixed points

In this chapter we shall study a diffeomorphism $\varphi\colon \mathbb{R}^n \to \mathbb{R}^n$ of class C^k for $k \geq 1$, near a fixed point which, by translation, can be assumed to be the origin so that $\varphi(0) = 0$. The orbit through the origin is the set $\mathcal{O}(0) = \{0\}$ consisting of the origin. The diffeomorphism φ is of the form

$$\begin{aligned}\varphi(x) &= \varphi(0) + d\varphi(0)x + \hat{\varphi}(x)\\ &= d\varphi(0)x + \hat{\varphi}(x)\\ &= Ax + \hat{\varphi}(x),\end{aligned}$$

where $A \in \mathcal{L}(\mathbb{R}^n)$ is a *linear isomorphism* and where the remainder satisfies $\hat{\varphi}(0) = 0$ and $d\hat{\varphi}(0) = 0$. We shall assume in the following that the fixed point is hyperbolic. From a dynamical point of view such a fixed point is completely unstable and we shall first demonstrate that near such a fixed point, the orbits of the diffeomorphism behave topologically like the orbits of the linearized map at the fixed point (theorem of Hartman–Grobman). There exist two distinguished sets, called the stable, respectively the unstable invariant manifold, issuing from the hyperbolic fixed point. The stable manifold consists of the set of all points in the space that under the iterates of the map converge to the fixed point, while the unstable manifold is the set of points that converge under the iterates of the inverse map to the fixed point. Considering only those points of these sets that stay under the iterates of the map, respectively under the iterates of the inverse map in the neighborhood of the fixed point, we shall show that the stable and unstable invariant manifolds are, locally near the fixed point, embedded submanifolds represented by graphs of functions (theorem of Hadamard–Perron). Globally, however, these stable and unstable invariant manifolds are the images of injective immersions of Euclidean spaces (theorem of S. Smale) which away from the fixed point can intersect each other, giving rise to so-called homoclinic points. Homoclinic points considerably complicate the orbit structure of the diffeomorphism nearby and imply the existence of orbits with quite unpredictable long-time behavior, as the next chapter will show.

In the following we shall denote by $\mathcal{L}(X, Y)$ the vector space of the linear and continuous maps between the two normed spaces X and Y and abbreviate $\mathcal{L}(X) = \mathcal{L}(X, X)$.

II.1 Hyperbolic fixed points

We shall give two equivalent definitions of a hyperbolic fixed point.

Definition. A linear isomorphism $A \in \mathcal{L}(\mathbb{R}^n)$ of $\mathbb{R}^n$ is called hyperbolic, if it leaves a splitting

$$\mathbb{R}^n = E_+ \oplus E_-$$

invariant so that

$$A = \begin{pmatrix} A_+ & 0 \\ 0 & A_- \end{pmatrix} \quad \text{with respect to } E_+ \oplus E_-$$

where $E_\pm \neq \{0\}$ and $A_+ \in \mathcal{L}(E_+)$ and $A_- \in \mathcal{L}(E_-)$. These linear maps satisfy the inequalities

$$\begin{aligned} |A_+^j x| &\leq c\vartheta_+^j |x| \quad \text{for all } j \geq 0,\ x \in E_+, \\ |A_-^{-j} x| &\leq c\vartheta_-^j |x| \quad \text{for all } j \geq 0,\ x \in E_-, \end{aligned}$$

with a constant $c \geq 0$ and with constants

$$\begin{aligned} 0 < \vartheta_+ < 1, \\ 0 < \vartheta_- < 1. \end{aligned}$$

The fixed point 0 of the above diffeomorphism $\varphi\colon \mathbb{R}^n \to \mathbb{R}^n$ is called hyperbolic if the linearization of the diffeomorphism at the fixed point $A := d\varphi(0) \in \mathcal{L}(\mathbb{R}^n)$ is a hyperbolic isomorphism.

It is easily verified that the *dynamical behavior* of the hyperbolic map A characterizes the spaces E_+ and E_- as follows:

$$\begin{aligned} E_+ &= \{x \in \mathbb{R}^n \mid A^j x \to 0,\ j \to +\infty\} \\ &= \{x \in \mathbb{R}^n \mid \sup_{j\geq 0} |A^j x| < \infty\}, \end{aligned}$$

$$\begin{aligned} E_- &= \{x \in \mathbb{R}^n \mid A^j x \to 0,\ j \to -\infty\} \\ &= \{x \in \mathbb{R}^n \mid \sup_{j\geq 0} |A^{-j} x| < \infty\}. \end{aligned}$$

We shall use the following notation with respect to the splitting $\mathbb{R}^n = E_+ \oplus E_-$:

$$\begin{aligned} x &= (x_+, x_-) = x_+ \oplus x_-, \\ Ax &= (A_+x_+, A_-x_-) = A_+x_+ \oplus A_-x_-. \end{aligned}$$

An alternative definition of the hyperbolicity of an isomorphism A in terms of its spectrum $\sigma(A)$ consisting of the set of eigenvalues of A is the following.

Definition. An isomorphism $A \in \mathcal{L}(\mathbb{R}^n)$ is called *hyperbolic* if its spectrum has the properties

$$|\lambda| \neq 1, \quad \lambda \in \sigma(A),$$

and there exist points λ in the spectrum satisfying $|\lambda| < 1$ *and* also points satisfying $|\lambda| > 1$.

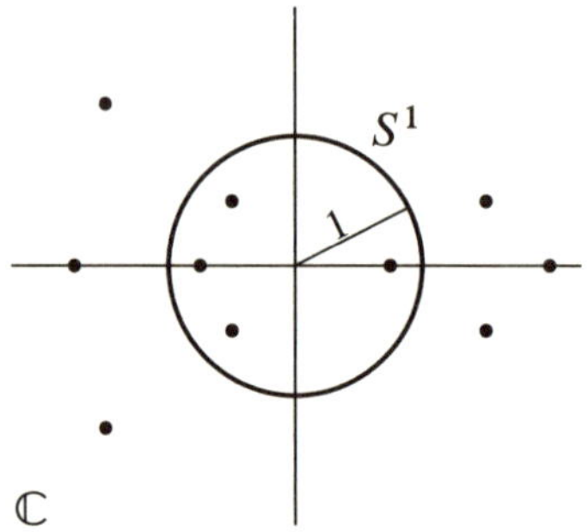

Figure II.1. Spectrum of a hyperbolic isomorphism.

The two definitions of a hyperbolic linear isomorphism are equivalent. Indeed, if A is hyperbolic according to the second definition, we denote by E_+ the *generalized eigenspace* of all eigenvalues λ satisfying $|\lambda| < 1$ and by E_- the generalized eigenspace of all eigenvalues with $|\lambda| > 1$ and obtain the invariant splitting $\mathbb{R}^n = E_+ \oplus E_-$.

The restricted maps $A_+ = A|_{E_+} : E_+ \to E_+$ (and analogously for A_-) have the spectra

$$\sigma(A_+) = \sigma(A) \cap \{|z| < 1\},$$
$$\sigma(A_-) = \sigma(A) \cap \{|z| > 1\}.$$

In order to verify the desired estimates for $|A_+^j x|$ and $|A_-^{-j} x|$ on E_+ and E_- we recall that the *spectral radius* $r_{A_+} := \max\{|\lambda| \mid \lambda \in \sigma(A_+)\}$ of A_+ satisfies

$$\lim_{j\to\infty} \|A_+^j\|^{\frac{1}{j}} = r_{A_+} < 1,$$

where $\|\cdot\|$ denotes the operator norm. Therefore, we find an integer j_0 such that for all $j \geq j_0$ the estimates $\|A_+^j\|^{\frac{1}{j}} \leq \vartheta$ hold true for a constant $\vartheta < 1$ and the inequalities $|A_+^j x| \leq \vartheta^j |x|$ follow from the definition of the operator norm. To ensure the desired estimate also for the finitely many $j < j_0$, one simply puts a sufficiently large constant c in front of the previous estimates on the right-hand side. The estimates for $|A_-^{-j} x|$ follow by the same argument applied to the linear map A_-^{-1}. Therefore, the two definitions of hyperbolicity are *equivalent*.

Example. The linear map $A \in \mathcal{L}(\mathbb{R}^2)$, in the canonical basis $\{e_1, e_2\}$ represented by the matrix

$$A = \begin{pmatrix} \mu & 0 \\ 0 & \nu \end{pmatrix}, \quad 0 < \mu < 1 < \nu,$$

has the spectrum $\sigma(A) = \{\mu, \nu\}$. Moreover, $E_+ = \mathbb{R} \times \{0\}$ and $E_- = \{0\} \times \mathbb{R}$. The orbits are sketched in Figure II.2.

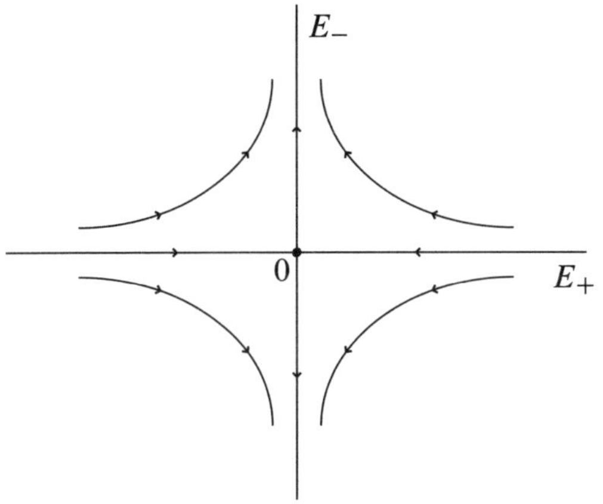

Figure II.2. Schematic representation of the orbits of $A \in \mathcal{L}(\mathbb{R}^2)$ in the example.

It is convenient for the arguments later on to introduce new and equivalent norms with respect to which the constant c in the definition of the hyperbolicity is equal to 1.

Proposition II.1 (Adapted norms). *Assume that the map $A \in \mathcal{L}(\mathbb{R}^n)$ is hyperbolic. Then, there exist equivalent norms $|\cdot|_*$ on E_+ and E_- such that*

$$|A_+x|_* \le \alpha |x|_*, \quad x \in E_+,$$
$$|A_-^{-1}x|_* \le \beta |x|_*, \quad x \in E_-,$$

with constants $\vartheta_+ < \alpha < 1$ and $\vartheta_- < \beta < 1$.

With respect to these norms the linear maps A_+ and A_-^{-1} are ***contractions,*** *so that $\|A_+\|_* < 1$ and $\|A_-^{-1}\|_* < 1$.*

Proof. In order to construct the new norm in the space E_+ we *choose* $\vartheta_+ < \alpha < 1$ and then choose the integer N so large that

$$c(\vartheta_+/\alpha)^N \le 1.$$

For $x \in E_+$ we define the new norm on E_+ by

$$|x|_* := \sum_{j=0}^{N-1} \alpha^{-j} |A_+^j x|.$$

It satisfies

$$|x| \le |x|_* \le C|x|$$

with the constant $C = cN$, so that the two norms are equivalent. It remains to verify that A_+ is a contraction satisfying $\|A_+\|_* \le \alpha < 1$. To see this, we estimate

$$\begin{aligned} |A_+x|_* &= \sum_{j=0}^{N-1} \alpha^{-j} |A_+^{j+1} x| \\ &= \alpha\Big(\sum_{j=1}^{N-1} \alpha^{-j} |A_+^j x| + \alpha^{-N} |A_+^N x|\Big) \\ &\le \alpha\Big(\sum_{j=1}^{N-1} \alpha^{-j} |A_+^j x| + |x|\Big) \\ &= \alpha \sum_{j=0}^{N-1} \alpha^{-j} |A_+^j x| \\ &= \alpha |x|_*, \end{aligned}$$

where we have used, in view of our choice of the integer N, that

$$\alpha^{-N} |A_+^N x| \le c\alpha^{-N} \vartheta_+^N |x| \le |x|.$$

On the subspace E_- one carries out the same construction taking A_-^{-1} instead of A_+. The proposition is proved. □

As already mentioned at the beginning of the chapter, a hyperbolic fixed point gives rise to the so-called stable manifold and the unstable manifold. The stable manifold consists of all the points in the space that converge under the iterates of the map to the fixed point, while the unstable manifold consists of all those points that converge to the fixed point under the iterates of the inverse map.

Definition. The *stable*, respectively *unstable, invariant manifold* of a hyperbolic fixed point x^* of the diffeomorphism φ are the *sets*

$$\begin{aligned} W_+(\varphi, x^*) &:= \{x \in \mathbb{R}^n \mid \varphi^j(x) \to x^*,\ j \to \infty\}, \\ W_-(\varphi, x^*) &:= \{x \in \mathbb{R}^n \mid \varphi^{-j}(x) \to x^*,\ j \to \infty\}. \end{aligned}$$

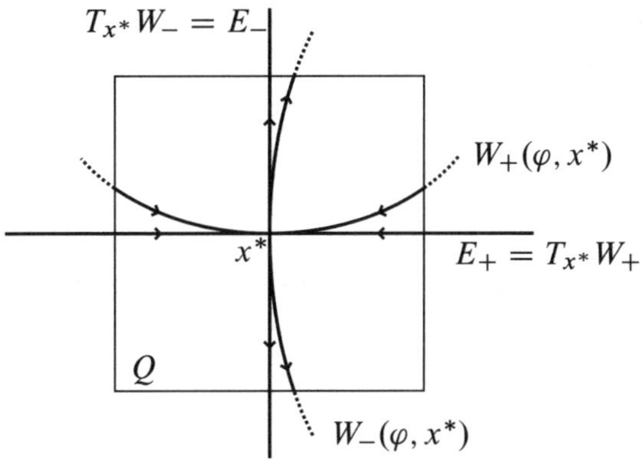

Figure II.3. Hyperbolic fixed point x^* with $W_+(\varphi, x^*)$ and $W_-(\varphi, x^*)$ in a neighborhood Q.

These sets contain the fixed point x^* and are clearly *invariant* under the map,

$$\varphi(W_+) = W_+,$$
$$\varphi(W_-) = W_-.$$

Example. If $x^* = 0$ and $\varphi = A \in \mathcal{L}(\mathbb{R}^n)$ is a hyperbolic linear isomorphism, then

$$W_+(A, 0) = E_+, \quad W_-(A, 0) = E_-.$$

This is the dynamical characterization of the spaces E_+ and E_-.

In order to study the topological behavior of a diffeomorphism locally near the hyperbolic fixed point we start with a technical preparation.

Proposition II.2. *If X is a Banach space and $A \in \mathcal{L}(X, X)$ a linear isomorphism and $\varphi\colon X \to X$ a map of the form*

$$\varphi(x) = Ax + g(x),$$

where $g\colon X \to X$ is a Lipschitz-continuous map whose Lipschitz constant ε satisfies

$$\varepsilon \|A^{-1}\| < 1,$$

then the map φ is a homeomorphism of X and the inverse map φ^{-1} is Lipschitz-continuous.

Proof. (1) We first prove that the map φ *is bijective.* Let $y \in X$. Since A is bijective, there exists a unique $x_0 \in X$ solving $Ax_0 = y$ and we show that there exists a unique solution $x \in X$ of the equation $y = Ax_0 = Ax + g(x)$, and hence solving the equation

$$x = x_0 - A^{-1}g(x).$$

Defining $f\colon X \to X$ by $f(x) := x_0 - A^{-1}g(x)$ it follows from the assumptions of the proposition that

$$|f(x) - f(y)| = |A^{-1}g(x) - A^{-1}g(y)| \le \|A^{-1}\|\,\varepsilon|x - y|,$$

so that f is a contraction. The contraction principle guarantees a unique fixed point $x = f(x) = x_0 - A^{-1}g(x)$. This fixed point x is the unique solution of the equation $y = \varphi(x)$ and we see that φ is bijective.

(2) Next we shall verify the *Lipschitz continuity of the inverse.* From $|x - y| = |A^{-1}Ax - A^{-1}Ay| \le \|A^{-1}\|\|Ax - Ay|$ we conclude

$$|Ax - Ay| \ge \|A^{-1}\|^{-1}|x - y|,$$

and so, using the triangle inequality, we can estimate

$$\begin{aligned}
|\varphi(x) - \varphi(y)| &= |Ax - Ay + g(x) - g(y)| \\
&\ge |Ax - Ay| - |g(x) - g(y)| \\
&\ge (\|A^{-1}\|^{-1} - \varepsilon)|x - y|.
\end{aligned}$$

Using our assumption $\varepsilon < \|A^{-1}\|^{-1}$, one obtains

$$|x - y| \le \frac{|\varphi(x) - \varphi(y)|}{\|A^{-1}\|^{-1} - \varepsilon}$$

and therefore,

$$|\varphi^{-1}(u) - \varphi^{-1}(v)| \le \frac{|u - v|}{\|A^{-1}\|^{-1} - \varepsilon}$$

for all $u, v \in X$. We see that the inverse map φ^{-1} is Lipschitz-continuous with Lipschitz constant equal to

$$L_{\varphi^{-1}} = \frac{1}{\|A^{-1}\|^{-1} - \varepsilon} = \frac{\|A^{-1}\|}{1 - \varepsilon\|A^{-1}\|},$$

and the proposition is proved. □

Our next result describes the *local behavior* of a diffeomorphism φ near a hyperbolic fixed point. The diffeomorphism behaves in an open neighborhood of the fixed point topologically like the *linearized* map at the fixed point.

Theorem II.3 (Hartman–Grobman). *Let 0 be a hyperbolic fixed point of the diffeomorphism $\varphi\colon \mathbb{R}^n \to \mathbb{R}^n$ so that*

$$\varphi(x) = Ax + \hat{\varphi}(x), \quad \frac{\hat{\varphi}(x)}{|x|} \to 0,\ x \to 0,$$

with a linear hyperbolic isomorphism A. Then, there exist open neighborhoods U and V of 0 and a ***homeomorphism*** *$h\colon U \to V$ satisfying $h(0) = 0$ and*

$$h \circ \varphi(x) = A \circ h(x)$$

if $x \in U$ and $\varphi(x) \in U$. Consequently, if defined,

$$h \circ \varphi^j \circ h^{-1}(y) = A^j y, \quad j \in \mathbb{Z}.$$

The *proof* of this local theorem is based on the following *global perturbation theorem* of a hyperbolic map $A \in \mathcal{L}(\mathbb{R}^n)$.

Theorem II.4. *Let $A \in \mathcal{L}(\mathbb{R}^n)$ be a* ***hyperbolic*** *isomorphism and let $\varphi, \psi\colon \mathbb{R}^n \to \mathbb{R}^n$ be two mappings of the form*

$$\varphi(x) = Ax + \hat{\varphi}(x), \quad \psi(x) = Ax + \hat{\psi}(x)$$

where the mappings $\hat{\varphi}$ and $\hat{\psi}$ are bounded and Lipschitz-continuous,

$$|\hat{\varphi}|_\infty := \sup_{x \in \mathbb{R}^n} |\hat{\varphi}(x)| < \infty,$$
$$|\hat{\varphi}(x) - \hat{\varphi}(y)| \le \varepsilon |x - y| \quad \textit{for all } x, y \in \mathbb{R}^n,$$

and the same for the map $\hat{\psi}$ with the same Lipschitz constant ε. Then, if ε is sufficiently small, the maps φ and ψ are homeomorphisms of $\mathbb{R}^n$ and there exists a ***unique*** *homeomorphism $h\colon \mathbb{R}^n \to \mathbb{R}^n$ of the form*

$$h(x) = x + \hat{h}(x), \quad |\hat{h}|_\infty < \infty$$

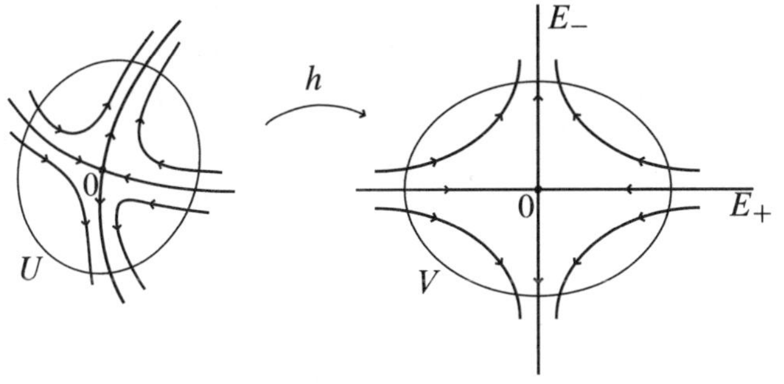

Figure II.4. The theorem by Hartman and Grobman.

satisfying

$$\varphi \circ h(x) = h \circ \psi(x), \quad x \in \mathbb{R}^n.$$

Moreover, $h(0) = 0$ *if we assume, in addition, that* $\varphi(0) = 0 = \psi(0)$.

Proof [Contraction principle]. According to Proposition II.2, the maps φ and ψ are homeomorphisms for small ε. Moreover, they are Lipschitz-continuous and their inverse maps are also Lipschitz-continuous.

If $\varphi = A + \hat{\varphi}$ and $\psi = A + \hat{\psi}$ are the two given maps, we look for a mapping $u = \mathrm{Id} + \hat{u}$, which solves the functional equation $\varphi \circ u = u \circ \psi$ or, more explicitly, the equivalent equation

$$A + A\hat{u} + \hat{\varphi} \circ u = A + \hat{\psi} + \hat{u} \circ \psi.$$

Hence, we first look for a continuous and bounded map $\hat{u} \colon \mathbb{R}^n \to \mathbb{R}^n$ that solves the functional equation

$$A\hat{u}(x) - \hat{u}(\psi(x)) = \hat{\psi}(x) - \hat{\varphi}(u(x)), \quad x \in \mathbb{R}^n.$$

We denote by $X := C_b(\mathbb{R}^n, \mathbb{R}^n)$ the *Banach space* of continuous and bounded mappings $v \colon \mathbb{R}^n \to \mathbb{R}^n$, equipped with the sup-norm

$$|v|_\infty = \sup_{x \in \mathbb{R}^n} |v(x)|,$$

where in $\mathbb{R}^n$ we choose for $x = (x_+, x_-) \in E_+ \oplus E_-$ the maximum norm $|x| = \max\{|x_+|, |x_-|\}$. The norms $|x_+|$, $|x_-|$ are the adapted norms on E_+ and E_- introduced in Proposition II.1.

We now introduce the linear and continuous map $L \in \mathcal{L}(X)$ as follows:

$$L \colon X \to X, \quad v \mapsto Lv,$$
$$(Lv)(x) := Av(x) - v(\psi(x)), \quad x \in \mathbb{R}^n,$$

or in short,

$$Lv = Av - v \circ \psi.$$

Lemma II.5. *If* $A \in \mathcal{L}(\mathbb{R}^n)$ *is a hyperbolic isomorphism, then the map* $L \colon X \to X$ *is also a continuous linear isomorphism of the Banach space and so the inverse map has a finite operator norm* $\|L^{-1}\| < \infty$. *If* $\psi(0) = 0$, *it follows for* $v \in X$ *satisfying* $v(0) = 0$ *that also* $(L^{-1}v)(0) = 0$.

Proof [Hyperbolicity of A]. For a given $g \in X$ we look for a solution $v \in X$ of the equation

$$Av - v \circ \psi = g.$$

According to the splitting $\mathbb{R}^n = E_+ \oplus E_-$, we decompose the hyperbolic isomorphism $A = A_+ \oplus A_-$ satisfying, with respect to the adapted norms, the estimates

$$\|A_+\| \le \vartheta, \quad \|A_-^{-1}\| \le \vartheta, \quad 0 < \vartheta < 1.$$

Decomposing the images, we set

$$v(x) = v_+(x) \oplus v_-(x),$$
$$g(x) = g_+(x) \oplus g_-(x),$$

for functions $v_+ \colon \mathbb{R}^n \to E_+$ and $v_- \colon \mathbb{R}^n \to E_-$ which are elements of the Banach spaces $X_\pm = C_b(\mathbb{R}^n, E_\pm)$. We wish to solve the system of equations

$$A_+ v_+ - v_+ \circ \psi = g_+,$$
$$A_- v_- - v_- \circ \psi = g_-.$$

Rewriting these equations by composing the first equation from the right with $\circ\psi^{-1}$ and applying the linear map $A_-^{-1}\circ$ from the left to the second equation we obtain the equivalent system

$$v_+ - A_+ v_+ \circ \psi^{-1} = -g_+ \circ \psi^{-1},$$
$$v_- - A_-^{-1} v_- \circ \psi = A_-^{-1} g_-.$$

Introducing the linear operators $G_\pm \in \mathcal{L}(X_\pm)$ by

$$(G_+ v_+)(x) := A_+ v_+(\psi^{-1}(x)),$$
$$(G_- v_-)(x) := A_-^{-1} v_-(\psi(x)),$$

we can rewrite the previous equations in $X_\pm$ as

$$(\mathbb{1} - G_+) v_+ = -g_+ \circ \psi^{-1},$$
$$(\mathbb{1} - G_-) v_- = A_-^{-1} g_-.$$

From the hyperbolicity of the isomorphism A we deduce the following estimates in the sup-norm:

$$|G_+ v_+|_\infty \le \|A_+\| \, |v_+|_\infty \le \vartheta |v_+|_\infty,$$
$$|G_- v_-|_\infty \le \|A_-^{-1}\| \, |v_-|_\infty \le \vartheta |v_-|_\infty.$$

For the operator norms we therefore obtain the estimates $\|G_-\|, \|G_+\| \le \vartheta < 1$. Thus the inverse maps are given by the well-known *Neumann series*, namely

$$(\mathbb{1} - G_\pm)^{-1} = \sum_{j \ge 0} G_\pm^j \in \mathcal{L}(X_\pm).$$

Therefore the desired unique solution v, given by

$$v_+ = (\mathbb{1} - G_+)^{-1}(-g_+(\psi^{-1})),$$
$$v_- = (\mathbb{1} - G_-)^{-1}(A_-^{-1} g_-),$$

satisfies the estimates in the sup-norm:

$$|v_+|_\infty \le \frac{1}{1-\vartheta}|g_+|_\infty,$$
$$|v_-|_\infty \le \frac{\vartheta}{1-\vartheta}|g_-|_\infty \le \frac{1}{1-\vartheta}|g_-|_\infty.$$

We have solved the equation $Lv = g$ in the Banach space X for a unique v. From $|L^{-1}g|_\infty/|g|_\infty = |v|_\infty/|g|_\infty \le 1/(1-\vartheta)$ for every $0 \ne g \in X$ we conclude that the operator norm of the inverse map satisfies $\|L^{-1}\| \le 1/(1-\vartheta) < \infty$. The proof of the lemma is complete. □

It is instructive to look at the *explicit formulas* for the inverse maps $v_\pm = L^{-1}(g_\pm)$. They are as follows:

$$v_+(x) = -\sum_{j\ge 0} A_+^j g_+(\psi^{-(j+1)}(x)),$$
$$v_-(x) = \sum_{j\ge 0} A_-^{-(j+1)} g_-(\psi^j(x)).$$

Coming *back to the proof of Theorem* II.4, we want to solve the functional equation $A\hat{u}(x) - \hat{u}(\psi(x)) = \hat{\psi}(x) - \hat{\varphi}(u(x))$. The left-hand side of the equation is equal to $L\hat{u}\,(x)$, and we have to solve the equation

$$\hat{u} = L^{-1}[\hat{\psi} - \hat{\varphi}\circ(\mathrm{Id}+\hat{u})] =: F(\hat{u}),$$

where, on the right-hand side, we also have defined the map $F\colon X \to X$, which is a *contraction* if ε is small enough. Indeed,

$$F(\hat{u}) - F(\hat{v}) = -L^{-1}[\hat{\varphi}\circ(\mathrm{Id}+\hat{u}) - \hat{\varphi}\circ(\mathrm{Id}+\hat{v})],$$

and according to the Lipschitz assumption in Theorem II.4,

$$|\hat{\varphi}(x+\hat{u}(x)) - \hat{\varphi}(x+\hat{v}(x))| \le \varepsilon|\hat{u}(x)-\hat{v}(x)| \le \varepsilon|\hat{u}-\hat{v}|_\infty,$$

so that

$$|\hat{\varphi}\circ(\mathrm{Id}+\hat{u}) - \hat{\varphi}\circ(\mathrm{Id}+\hat{v})|_\infty \le \varepsilon|\hat{u}-\hat{v}|_\infty.$$

Hence, in view of Lemma II.5,

$$|F(\hat{u}) - F(\hat{v})|_\infty \le \|L^{-1}\|\,\varepsilon|\hat{u}-\hat{v}|_\infty < \infty$$

for all $\hat{u}, \hat{v} \in X$. Choosing ε small enough, the map $F\colon X \to X$ is a contraction.

Consequently, there exists a *unique fixed point* $\hat{h}$ of F, and the map $h := \mathrm{Id}+\hat{h}$ is the unique solution of the functional equation from Theorem II.4 in the class of the maps $h\colon \mathbb{R}^n \to \mathbb{R}^n$ satisfying $h - \mathrm{Id} \in X$.

It remains to show that $h = \mathrm{Id} + \hat{h}$ is a *homeomorphism*. We shall conclude this from the uniqueness of the solution of the functional equation $\varphi \circ h = h \circ \psi$. If we exchange φ and ψ, there exists again a unique map $g = \mathrm{Id} + \hat{g}$ satisfying $|\hat{g}|_\infty < \infty$ and solving the functional equation $\psi \circ g = g \circ \varphi$. We then get

$$\varphi \circ (h \circ g) = h \circ \psi \circ g,$$
$$h \circ \psi \circ g = (h \circ g) \circ \varphi,$$

and therefore, $\varphi \circ (h \circ g) = (h \circ g) \circ \varphi$. On the other hand, $\varphi \circ \mathrm{Id} = \mathrm{Id} \circ \varphi$. Since $h \circ g = \mathrm{Id} + w$ satisfies $|w|_\infty < \infty$, it follows from the uniqueness of the solution of the functional equations in this class that $h \circ g = \mathrm{Id}$. In exactly the same way, one sees that also $g \circ h = \mathrm{Id}$. Since the inverse map $h^{-1} = g$ is continuous, the map $h \colon \mathbb{R}^n \to \mathbb{R}^n$ is a homeomorphism. The proof of Theorem II.4 is complete. □

Lemma II.6. *Let X be a Banach space and B_r the closed ball in X centered at* 0 *and of radius r. Then the map $f \colon X \to B_r$ defined by*

$$f(x) = \begin{cases} x, & |x| \le r, \\ r\frac{x}{|x|}, & |x| \ge r \end{cases}$$

is Lipschitz-continuous with a Lipschitz constant equal to 2.

Proof. Let $x, y \in X$. For $x, y \in B_r$ the statement is obvious. Assuming that $|x|, |y| > r$ we estimate, using the triangle inequality,

$$\begin{aligned} |f(x) - f(y)| &= r\left|\frac{x}{|x|} - \frac{y}{|y|}\right| \\ &= \frac{r}{|x|}\left|x - \frac{|x|}{|y|}y\right| \\ &\le \frac{r}{|x|}\left(|x - y| + \left|y - \frac{|x|}{|y|}y\right|\right) \\ &= \frac{r}{|x|}\left(|x - y| + \Big||y| - |x|\Big|\right) \\ &\le \frac{r}{|x|}(|x - y| + |x - y|) \\ &\le 2|x - y|. \end{aligned}$$

The remaining cases are verified similarly. □

After these preparations, we shall prove the Hartman–Grobman theorem.

Proof of Theorem II.3. We consider the diffeomorphism

$$\varphi(x) = Ax + \hat{\varphi}(x)$$

of Theorem II.3 possessing the origin as a hyperbolic fixed point. Fix $\delta > 0$ and let B_r be the closed ball centered at 0 with radius r. According to the assumption, $\hat{\varphi}(0) = 0$ and $\hat{\varphi}'(0) = 0$, where $\hat{\varphi}'$ denotes the derivative $d\hat{\varphi}$. Since $\hat{\varphi}'$ is continuous, we can choose the radius $r = r(\delta) > 0$ so small that $|\hat{\varphi}'(x)| < \delta$ on B_r. Then, the map $\hat{\varphi}$ is Lipschitz-continuous on B_r and has a Lipschitz constant $< \delta$. We now modify the given diffeomorphism *outside* of B_r in such a way that the modified map meets the assumptions of the technical Theorem II.4. For this purpose, we take the map f introduced in Lemma II.6 and define the map Φ by

$$\Phi(x) := Ax + \widehat{\Phi}(x) := Ax + \hat{\varphi}(f(x)), \quad x \in \mathbb{R}^n.$$

Thus, $\Phi = \varphi$ on B_r and $\widehat{\Phi}$ is, according to the above lemma, Lipschitz-continuous with a Lipschitz constant $< 2\delta$. Since $\hat{\varphi}$ is bounded on B_r, we also have $|\widehat{\Phi}|_\infty < \infty$.

Finally we can apply Theorem II.4 (in this theorem we take A instead of φ and Φ instead of ψ). If $\varepsilon > 0$ is the smallness constant occurring in Theorem II.4, we choose $\delta > 0$ so small that $2\delta < \varepsilon$. Then Theorem II.4 guarantees a homeomorphism h satisfying $h(0) = 0$ and solving the functional equation $h \circ \Phi = A \circ h$. By choosing $U \subset B_r$ and introducing $V := h(U)$ and restricting h on U, we conclude that $h \circ \varphi = A \circ h$ on $U \cap \varphi^{-1}(U)$, as claimed in the theorem. This completes the proof of the theorem of Hartman–Grobman. □

II.2 Local invariant manifolds

Having looked at the behavior of orbits near the unstable fixed point from a topological point of view, we shall look at it next from a differentiable point of view and come back to the diffeomorphism $\varphi\colon \mathbb{R}^n \to \mathbb{R}^n$ of class C^k, for $k \geq 1$ possessing the *hyperbolic fixed point* $0 = \varphi(0)$, so that it is of the form

$$\varphi(x) = Ax + g(x),$$

where g is a function satisfying $g(0) = 0$, $g'(0) = 0$ and where $A \in \mathcal{L}(\mathbb{R}^n)$ is the linearized map having the representation

$$A = \begin{pmatrix} A_+ & 0 \\ 0 & A_- \end{pmatrix} \quad \text{with respect to } E_+ \oplus E_- = \mathbb{R}^n$$

and satisfying $\|A_+\| \leq \alpha$ and $\|A_-^{-1}\| \leq \alpha$ for a constant $0 < \alpha < 1$.

Definition. For the *neighborhood* Q of the fixed point 0, defined by

$$Q := \{x = (x_+, x_-) \in E_+ \oplus E_- \mid |x_+| \leq r,\ |x_-| \leq r\},$$

$r > 0$, the *local invariant manifolds* of the fixed point 0 are defined as the *sets*

$$W^+_{\text{loc}}(Q) := \{x \in Q \mid \varphi^j(x) \in Q \text{ for all } j \geq 0\},$$
$$W^-_{\text{loc}}(Q) := \{x \in Q \mid \varphi^{-j}(x) \in Q \text{ for all } j \geq 0\}.$$

These sets are clearly *invariant* under φ resp. φ^{-1},

$$\varphi(W^+_{\text{loc}}) \subset W^+_{\text{loc}}, \quad \varphi^{-1}(W^-_{\text{loc}}) \subset W^-_{\text{loc}}.$$

It turns out that these distinguished sets consist of those points in Q which do converge under the iterates of the map, resp. under the iterates of the inverse map to the fixed point. Moreover, the sets are embedded manifolds represented by graphs of differentiable functions.

Theorem II.7 (Hadamard–Perron). *We assume that the diffeomorphism φ belongs to the differentiability class C^k for some $k \geq 1$. Then if $r > 0$ is sufficiently small,*

$$\begin{aligned} W^+_{\text{loc}}(Q) &= \{x \in Q \mid \varphi^j(x) \in Q,\ j \geq 0 \text{ and } \varphi^j(x) \to 0,\ j \to \infty\} \\ &= \{x \in Q \mid x = (x_+, h(x_+)) \in E_+ \oplus E_-\}, \end{aligned}$$

where $h\colon E_+ \to E_-$ is a function of class C^k satisfying $h(0) = 0$ and $h'(0) = 0$, so that the tangent space at the point 0 is given by

$$T_0 W^+_{\text{loc}}(Q) = E_+.$$

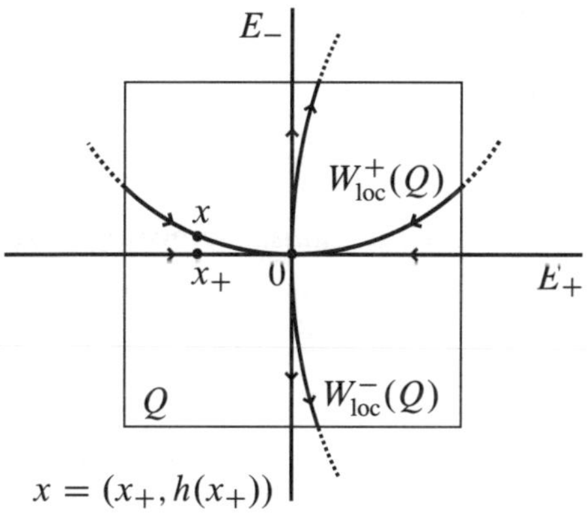

Figure II.5. The theorem of Hadamard and Perron.

Similarly,

$$\begin{aligned} W^-_{\text{loc}}(Q) &= \{x \in Q \mid \varphi^{-j}(x) \in Q,\ j \geq 0 \text{ and } \varphi^{-j}(x) \to 0,\ j \to \infty\} \\ &= \{x \in Q \mid x = (k(x_-), x_-) \in E_+ \oplus E_-\} \end{aligned}$$

with a function $k\colon E_- \to E_+$ from C^k satisfying $k(0) = 0$ and $k'(0) = 0$, so that

$$T_0 W^-_{\text{loc}}(Q) = E_-.$$

The theorem of Hadamard–Perron is of local nature but will be deduced from a global perturbation theorem of the linear map A. To do so we assume, *in addition*, that the map $\varphi(x) = Ax + g(x)$ satisfies

$$|g|_\infty < \infty, \quad |g'|_\infty \leq \delta,$$

so that g can be considered as a *small perturbation* of the linear map $A \in \mathcal{L}(\mathbb{R}^n)$. The locally stable and unstable manifolds of this map are now defined as the sets

$$W^+_{\text{loc}}(\mathbb{R}^n) = \{x \in \mathbb{R}^n \mid \sup_{j \geq 0} |\varphi^j(x)| < \infty\},$$
$$W^-_{\text{loc}}(\mathbb{R}^n) = \{x \in \mathbb{R}^n \mid \sup_{j \geq 0} |\varphi^{-j}(x)| < \infty\}.$$

The sets consist of bounded half orbits of the map φ and we are going to prove that these half orbits converge under the iterates of the map, resp. under the iterates of the inverse map to the hyperbolic fixed point. We shall also see that these sets are embedded submanifolds.

Theorem II.8. *Let $\varphi\colon \mathbb{R}^n \to \mathbb{R}^n$ be a C^k-diffeomorphism ($k \geq 1$) possessing the hyperbolic fixed point $x = 0$ and let φ be of the form $\varphi(x) = Ax + g(x)$ with a map satisfying $g(0) = 0$, $|g|_\infty < \infty$ and $|g'|_\infty \leq \delta$. Then, if δ is small enough,*

$$\begin{aligned} W^+_{\text{loc}}(\mathbb{R}^n) &= \{x \in \mathbb{R}^n \mid \varphi^j(x) \to 0,\ j \to \infty\} \\ &= \{x = (x_+, x_-) \mid x_- = h(x_+)\} \end{aligned}$$

for a function $h\colon E_+ \to E_-$ from C^k satisfying $h(0) = 0$ and $h'(0) = 0$, so that

$$T_0 W^+_{\text{loc}} = E_+$$

is the tangent space at the fixed point. Similarly,

$$\begin{aligned} W^-_{\text{loc}}(\mathbb{R}^n) &= \{x \in \mathbb{R}^n \mid \varphi^{-j}(x) \to 0,\ j \to \infty\} \\ &= \{x = (x_+, x_-) \mid x_+ = k(x_-)\} \end{aligned}$$

for a function $k\colon E_- \to E_+$ from C^k satisfying $k(0) = 0$ and $k'(0) = 0$ and

$$T_0 W^-_{\text{loc}} = E_-.$$

Proof [Contraction principle, implicit function theorem]. (1) We first show that for every $a = x_+ \in E_+$ there exists a unique point $x_0 \in \mathbb{R}^n$ of the form

$$x_0 = a \oplus h(a) \in E_+ \oplus E_-,$$

whose orbit converges to the fixed point, i.e., $\varphi^j(x_0) \to 0$ as $j \to \infty$.

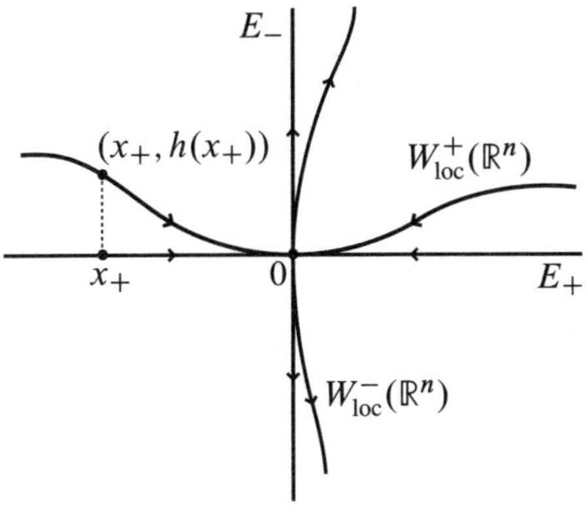

Figure II.6. Theorem II.8: $W^+_{\text{loc}}(\mathbb{R}^n)$ as graph of h.

We denote in the following by $P_\pm \in \mathcal{L}(\mathbb{R}^n)$ the projections onto $E_\pm$, defined by

$$P_\pm(x_+, x_-) = x_\pm \quad \text{for } (x_+, x_-) \in E_+ \oplus E_-.$$

According to the postulated A-invariance of the splitting,

$$AP_\pm = P_\pm A = P_\pm A P_\pm = A_\pm P_\pm.$$

Moreover, the estimates

$$\|AP_+\| \le \alpha, \quad \|A^{-1}P_-\| \le \alpha$$

hold true for a constant $\alpha < 1$.

The *sequence* $(x_j)_{j\ge 0}$ is a *half orbit* of the map φ, if $x_{j+1} = \varphi(x_j)$ for every $j \ge 0$, or explicitly, if

$$x_{j+1} = Ax_j + g(x_j), \quad j \ge 0.$$

We shall next derive successively equivalent but more convenient conditions for a sequence to be a half orbit of φ. In the first step (a) we use the splitting $x_{j+1} = P_+x_{j+1} + P_-x_{j+1}$, in order to get two equations. Then we use in step (b) the commutativity relation $P_\pm A = AP_\pm$. Then in step (c) we apply A^{-1} to the second equation and, finally, in step (d) we use $x_j = P_+x_j + P_-x_j$ and an index-shift to come back to a single equation.

$$x_{j+1} = Ax_j + g(x_j) \overset{\text{(a)}}{\Longleftrightarrow} \begin{cases} P_+x_{j+1} = P_+Ax_j + P_+g(x_j), \\ P_-x_{j+1} = P_-Ax_j + P_-g(x_j), \end{cases}$$

$$\overset{\text{(b)}}{\Longleftrightarrow} \begin{cases} P_+x_{j+1} = AP_+x_j + P_+g(x_j), \\ P_-x_{j+1} = AP_-x_j + P_-g(x_j), \end{cases}$$

$$\overset{\text{(c)}}{\Longleftrightarrow} \begin{cases} P_+x_{j+1} = AP_+x_j + P_+g(x_j), \\ P_-x_j = A^{-1}P_-x_{j+1} - A^{-1}P_-g(x_j), \end{cases}$$

$$\overset{\text{(d)}}{\Longleftrightarrow} \begin{cases} x_j = AP_+x_{j-1} + P_+g(x_{j-1}), \\ \qquad\qquad +A^{-1}P_-x_{j+1} - A^{-1}P_-g(x_j),\ j \geq 1, \\ x_0 = P_+x_0 + A^{-1}P_-x_1 - A^{-1}P_-g(x_0),\ j = 0. \end{cases}$$

In order to reformulate the last representation of a half orbit as a fixed point of a functional in a sequence space, we introduce the *Banach space X of the bounded half sequences* in $\mathbb{R}^n$,

$$X := \{x = (x_j)_{j\geq 0},\ x_j \in \mathbb{R}^n \mid \sup_{j\geq 0} |x_j| < \infty\}$$

equipped with the norm

$$\|x\| := \sup_{j\geq 0} |x_j|.$$

The norm of the point $x_j = x_j^+ \oplus x_j^- \in E_+ \oplus E_- = \mathbb{R}^n$ is the maximum norm

$$|x_j| = \max\{|x_j^+|, |x_j^-|\},$$

where $|x_j^+|$, $|x_j^-|$ are the adapted norms on E_+ and E_- guaranteed by Proposition II.1. We shall characterize the bounded half orbits as *fixed points* of maps. For this purpose, we introduce the following family of maps. For every $a \in E_+$ we define the map F_a by

$$F_a \colon X \to X, \quad x = (x_j)_{j\geq 0} \mapsto F_a(x) = (F_a(x)_j)_{j\geq 0},$$

where the bounded sequence $(F_a(x)_j)_{j\geq 0}$ is defined by the formula

$$F_a(x)_j = \begin{cases} AP_+x_{j-1} + P_+g(x_{j-1}) + A^{-1}P_-x_{j+1} - A^{-1}P_-g(x_j), & j \geq 1, \\ a + A^{-1}P_-x_1 - A^{-1}P_-g(x_0), & j = 0. \end{cases}$$

In view of the above characterization (d) of a half orbit we conclude for the sequence $x = (x_j)_{j\geq 0} \in X$ that

$$F_a(x) = x \Longleftrightarrow \begin{cases} x_{j+1} = Ax_j + g(x_j), \quad j \geq 0, \\ P_+x_0 = a. \end{cases}$$

We see that the fixed points of F_a are the bounded half orbits satisfying $P_+x_0 = a$.

In order to achieve convergence of the half sequence, we introduce *weighted norms*. For $\lambda \geq 1$ we define the subspace $X_\lambda \subset X$ by

$$X_\lambda := \{x \in X \mid \|x\|_\lambda := \sup_{j\geq 0} \lambda^j |x_j| < \infty\}.$$

If $x \in X_\lambda$, then $\lambda^j|x_j| \leq \|x\|_\lambda$ in view of the definition of the norm. Therefore, $|x_j| \leq \lambda^{-j}\|x\|_\lambda$. Hence *if* we choose $\lambda > 1$, it follows that $x_j \to 0$ as $j \to \infty$. In the following *we shall choose*, recalling that $\|A_+\|, \|A_-^{-1}\| \leq \alpha < 1$, the parameter λ in the interval

$$1 \leq \lambda < 1/\alpha.$$

Lemma II.9. *Every map F_a maps the space X_λ into itself,*

$$F_a\colon X_\lambda \to X_\lambda,$$

and satisfies the estimates

$$\|F_a(x) - F_a(y)\|_\lambda \leq \lambda(\alpha + \delta)\|x - y\|_\lambda$$

for all $x, y \in X_\lambda$.

Proof. Let $x, y \in X_\lambda$. For every $j \geq 1$ we have

$$\begin{aligned}
\lambda^j[F_a(x)_j - F_a(y)_j] &= \lambda^j AP_+(x_{j-1} - y_{j-1}) + \lambda^j P_+[g(x_{j-1}) - g(y_{j-1})] \\
&\quad + \lambda^j A^{-1}P_-(x_{j-1} - y_{j-1}) \\
&\quad - \lambda^j A^{-1}P_-[g(x_j) - g(y_j)] \\
&\in E_+ \oplus E_-,
\end{aligned}$$

while if $j = 0$, then

$$F_a(x)_0 - F_a(y)_0 = A^{-1}P_-(x_1 - y_1) - A^{-1}P_-[g(x_0) - g(y_0)].$$

We estimate in the max-norm $|x| = \max\{|x_+|, |x_-|\}$ for $x = (x_+, x_-)$. Then, the operator norms of the projections are equal to 1, $\|P_+\| = 1$ and $\|P_-\| = 1$. According to the assumption of the theorem, $|g'|_\infty \leq \delta$, and it follows by the mean value theorem that

$$|g(\xi) - g(\eta)| \leq \delta|\xi - \eta|$$

for all $\xi, \eta \in \mathbb{R}^n$. Therefore, we obtain for the E_+ part of $\lambda^j[F_a(x)_j - F_a(y)_j]$ if $j \geq 1$, using the (a-priori estimates) of the linear map A,

$$\begin{aligned}
\lambda^j|P_+F_a(x)_j - P_+F_a(y)_j| &\leq \lambda\|AP_+\|\, \lambda^{j-1}|x_{j-1} - y_{j-1}| \\
&\quad + \lambda\delta\lambda^{j-1}|x_{j-1} - y_{j-1}| \\
&\leq (\lambda\alpha + \lambda\delta)\sup_{j\geq 0}\lambda^j|x_j - y_j| \\
&= \lambda(\alpha + \delta)\|x - y\|_\lambda.
\end{aligned}$$

Similarly, we find for the E_- part the estimates

$$\lambda^j |P_- F_a(x)_j - P_- F_a(y)_j| \le \lambda(\alpha + \delta)\|x - y\|_\lambda.$$

For $j = 0$ the same estimate holds true, in view of the following computation:

$$\begin{aligned}|F_a(x)_0 - F_a(y)_0| &= |A^{-1}P_-(x_1 - y_1) - A^{-1}P_-[g(x_0) - g(y_0)]| \\ &\le \alpha|x_1 - y_1| + \alpha\delta|x_0 - y_0| \\ &\le \alpha\lambda|x_1 - y_1| + \alpha\delta|x_0 - y_0| \\ &\le (\alpha + \alpha\delta)\|x - y\|_\lambda \\ &\le (\alpha + \delta)\|x - y\|_\lambda \\ &\le \lambda(\alpha + \delta)\|x - y\|_\lambda.\end{aligned}$$

Altogether,

$$\begin{aligned}\lambda^j |F_a(x)_j - F_a(y)_j| &= \max\big\{\lambda^j |P_+ F_a(x)_j - P_+ F_a(y)_j|, \\ &\qquad\qquad \lambda^j |P_- F_a(x)_j - P_- F_a(y)_j|\big\} \\ &\le \lambda(\alpha + \delta)\|x - y\|_\lambda\end{aligned}$$

for every $j \ge 0$. Consequently, taking on the left-hand side the supremum over $j \ge 0$, we have proved the desired inequality

$$\|F_a(x) - F_a(y)\|_\lambda \le \lambda(\alpha + \delta)\|x - y\|_\lambda.$$

The statement $F_a(X_\lambda) \subset X_\lambda$ follows from this inequality. Indeed, in view of $F_a(0) = (a, 0, 0, \dots)$ we can estimate

$$\begin{aligned}\|F_a(x)\|_\lambda &= \|F_a(x) - F_a(0) + (a, 0, 0, \dots)\|_\lambda \\ &\le \|F_a(x) - F_a(0)\|_\lambda + \|(a, 0, 0, \dots)\|_\lambda \\ &\le \lambda(\alpha + \delta)\|x\|_\lambda + |a| \\ &< \infty.\end{aligned}$$

This finishes the proof of Lemma II.9. □

Continuing with the proof of Theorem II.8, we recall that $0 < \alpha < 1$ and choose the smallness parameter $\delta > 0$ occurring in this theorem, so small that

$$\lambda\alpha + \lambda\delta =: \mu < 1$$

for all parameter values $\lambda \ge 1$ which are contained in the closed interval whose upper bound is strictly larger than 1 and whose lower bound is equal to 1. For these parameter values the maps $F_a \colon X_\lambda \to X_\lambda$ are contractions in view of Lemma II.9

so that we can apply the contraction principle. Because of $X_\lambda \subset X$, there exists a unique fixed point $x(a) = (x(a)_j)_{j\geq 0} \in X$ satisfying

$$F_a(x(a)) = x(a)$$

and belonging to all these spaces X_λ.

We have proved so far that there exists *a unique bounded half orbit satisfying* $P_+x_0 = a$. Since $x(a)$ lies in a space X_λ with $\lambda > 1$, this half orbit converges automatically to the fixed point 0 of the map φ.

The zero component of our fixed point sequence satisfies

$$\begin{aligned} x(a)_0 &= P_+x(a)_0 \oplus P_-x(a)_0 \\ &= a \oplus h(a) \in E_+ \oplus E_- = \mathbb{R}^n, \end{aligned}$$

where the map $h\colon E_+ \to E_-$ is *defined* by

$$h(a) := P_-x(a)_0.$$

In view of our construction, the *half orbit* is the sequence

$$x(a)_j = \varphi^j(x(a)_0), \quad j \geq 0.$$

In the case that $a = 0$ we already have a fixed point of F_0. Indeed, in view of $g(0) = 0$ the zero sequence $0 = (0)_{j\geq 0}$ is a fixed point. Therefore, by the uniqueness, $x(0)_0 = (0,0) \in E_+ \oplus E_-$ and hence $h(0) = 0$.

(2) *Lipschitz continuity of h.* If $x(a)$ and $x(b)$ are the two fixed points of the maps F_a and F_b we obtain, using Lemma II.9, the estimate

$$\begin{aligned} \|x(a) - x(b)\|_\lambda &= \|F_a(x(a)) - F_b(x(b))\|_\lambda \\ &\leq \|F_a(x(a)) - F_a(x(b))\|_\lambda + \|F_a(x(b)) - F_b(x(b))\|_\lambda \\ &\leq \mu\|x(a) - x(b)\|_\lambda + |a - b|. \end{aligned}$$

We therefore conclude $\|x(a) - x(b)\|_\lambda \leq \frac{1}{1-\mu}|a - b|$, and hence, recalling the definition of the norm in the sequence space, $|x(a)_j - x(b)_j| \leq \frac{\lambda^j}{1-\mu}|a - b|$ for every $j \geq 0$. In particular, the following estimate holds true for $j = 0$:

$$|x(a)_0 - x(b)_0| \leq \frac{1}{1-\mu}|a - b|.$$

Consequently, recalling $h(a) := P_-x(a)_0$,

$$|h(a) - h(b)| \leq \frac{1}{1-\mu}|a - b|$$

for all $a, b \in E_+$, so that the function $h\colon E_+ \to E_-$ is *Lipschitz-continuous.*

So far we have proved that the stable invariant manifold is the graph of a Lipschitz-continuous function and that, moreover, the half orbits of all its points converge to the fixed point. Using the implicit function theorem we are going to verify that the function h is differentiable.

(3) *Differentiability of* h. We define the functions F and G by

$$F\colon X \times E_+ \to X, \quad F(x,a) := F_a(x)$$

and

$$G\colon X \times E_+ \to X, \quad G(x,a) := x - F(x,a),$$

and observe that if $g \in C^k$ for an integer $k \geq 1$, then also the maps F and G are of class C^k.The solution set, consisting of the points $(x,a) \in X \times E_+$ which solve the equation $G(x,a) = 0$, is known to us, namely,

$$G(x,a) = 0 \iff (x,a) = (x(a),a),$$

where $x(a) \in X$ is the fixed point of the map F_a. Therefore,

$$G(x(a),a) = 0, \quad a \in E_+.$$

We now use the implicit function theorem as a *regularity theorem*. We denote by $D_1 = \frac{\partial}{\partial x}$ the derivative in the first variable. If $D_1 G(x,a) \in \mathcal{L}(X,X)$ is a continuous isomorphism of the Banach space X, then the implicit function theorem guarantees that the function $a \mapsto x(a)\colon E_+ \to X$ belongs to C^k. We have

$$D_1 G(x,a) = \mathbb{1} - D_1 F(x,a) \in \mathcal{L}(X)$$

and we show that the operator norm satisfies $\|D_1 F(x,a)\| \leq \mu < 1$. From the theorem of *Neumann* it then follows that the bounded linear map $D_1 G(x,a)$ is a continuous isomorphism of the Banach space.

In order to estimate this operator norm we take an element $y = (y_j)_{j\geq 0} \in X$ and express $D_1 F(x,a) y \in X$ in *components*. This means for $j \geq 1$ that

$$[D_1 F(x,a)y]_j = AP_+ y_{j-1} + P_+ g'(x_{j-1}) y_{j-1} + A^{-1} P_- y_{j+1} - A^{-1} g'(x_j) y_j$$

and for $j = 0$,

$$[D_1 F(x,a)y]_0 = A^{-1} P_- y_1 - A^{-1} P_- g'(x_0) y_0.$$

As in the proof of Lemma II.9 one proves (now $\lambda = 1$) the estimates

$$\big|[D_1 F(x,a)y]_j\big| \leq (\alpha + \delta)\|y\| \leq \mu\|y\|$$

for $j \geq 0$. Taking the supremum over $j \geq 0$, we obtain $\|D_1 F(x,a)y\| \leq \mu\|y\|$. This holds true for every $y \in X$. Therefore, taking the supremum over all y satisfying $\|y\| \leq 1$, we obtain the desired estimate of the operator norm:

$$\|D_1 F(x,a)\| \leq \mu < 1.$$

(4) We finally prove that $h'(0) = 0 \in \mathcal{L}(E_+, E_-)$. Recalling the definition $h(a) := P_-[x(a)_0]$, the differentiation of the function h at the point a gives, for every $b \in E_+$,

$$\begin{aligned} h'(a)b &= P_- D_a[x(a)_0]b \\ &= P_-[D_a x(a)b]_0. \end{aligned}$$

In view of the fixed point property of $x(a)$ we have the identity

$$x(a) = F(x(a), a), \quad a \in E_+.$$

Differentiating this identity in the variable a we obtain by the *chain rule*,

$$D_a x(a)b = D_1 F(x(a), a) D_a x(a)b + D_2 F(x(a), a)b, \quad b \in E_+,$$

where D_2 denotes the derivative in the second argument. Writing for simplicity D instead of D_a for the derivative in a, we have $Dx(a) \in \mathcal{L}(E_+, X)$. If $a = 0$ in E_+, then $x(0) = 0 \in X$, so that $x(0)_j = 0$ for all $j \geq 0$. Using the assumption $g'(0) = 0 \in \mathcal{L}(\mathbb{R}^n)$, one concludes

$$[Dx(0)b]_j = AP_+[Dx(0)b]_{j-1} + A^{-1}P_-[Dx(0)b]_{j+1}$$

for $j \geq 1$, and for $j = 0$,

$$[Dx(0)b]_0 = A^{-1}P_-[Dx(0)b]_1 + b.$$

Applying the projection operator P_- to both sides and observing $P_-P_+ = 0$ we obtain the equations

$$P_-[Dx(0)b]_j = A^{-1}P_-[Dx(0)b]_{j+1}, \quad j \geq 0,$$

from which we deduce the estimates

$$|P_-[Dx(0)b]_j| \leq \alpha |P_-[Dx(0)b]_{j+1}|, \quad j \geq 0,$$

so that

$$\|P_- Dx(0)b\| \leq \alpha \|P_- Dx(0)b\|.$$

Since $\alpha < 1$, this means that $P_- Dx(0)b = 0$ and so

$$h'(0)b = P_-[Dx(0)b]_0 = 0$$

for all $b \in E_+$, hence $h'(0) = 0$, as claimed.

We have proved the statement of the theorem for the local stable invariant manifold. The analogous statement for the local unstable invariant manifold $W^-_{\text{loc}}(\mathbb{R}^n)$ is proved the same way. Hence, the proof of Theorem II.8 is complete. □

After these preparations, we shall prove the theorem of Hadamard and Perron.

Proof of Theorem II.7. The proof of the local Theorem II.7 will be a consequence of Theorem II.8. We modify the diffeomorphism φ of Theorem II.7 *outside* of an open neighborhood of the origin. For this, we take the cut-off function $\xi \in C^\infty(\mathbb{R}, [0, 1])$ satisfying

$$\begin{aligned} \xi(t) &= 1, \quad t \le 1/2, \\ \xi(t) &= 0, \quad t \ge 1 \end{aligned}$$

and define the modified maps

$$\tilde{\varphi}(x) := Ax + \tilde{g}(x) := Ax + \xi\Big(\frac{|x|}{\varepsilon}\Big)g(x).$$

Then,

$$\tilde{\varphi}(x) = \begin{cases} \varphi(x), & |x| \le \varepsilon/2, \\ Ax, & |x| \ge \varepsilon. \end{cases}$$

For a given $\delta > 0$ we can, due to our assumptions $g(0) = 0$ and $g'(0) = 0$, find an $\varepsilon > 0$ sufficiently small such that

$$|\tilde{g}'(x)| < \delta, \quad x \in \mathbb{R}^n.$$

Therefore, for a suitable $\delta > 0$, the modified diffeomorphism $\tilde{\varphi}$ meets the assumptions of Theorem II.8. Next, we choose the neighborhood Q of the origin as in Theorem II.7 so small that

$$\tilde{\varphi}(x) = \varphi(x), \quad x \in Q.$$

Lemma II.10. *For the neighborhood* $Q := \{x = (x_+, x_-) \in E_+ \oplus E_- \mid |x_+| \le r,\ |x_-| \le r\}$ *the following holds true:*

$$W^+_{\text{loc}}(Q) = W^+_{\text{loc}}(\mathbb{R}^n, \tilde{\varphi}) \cap Q,$$

if $r > 0$ *is sufficiently small.*

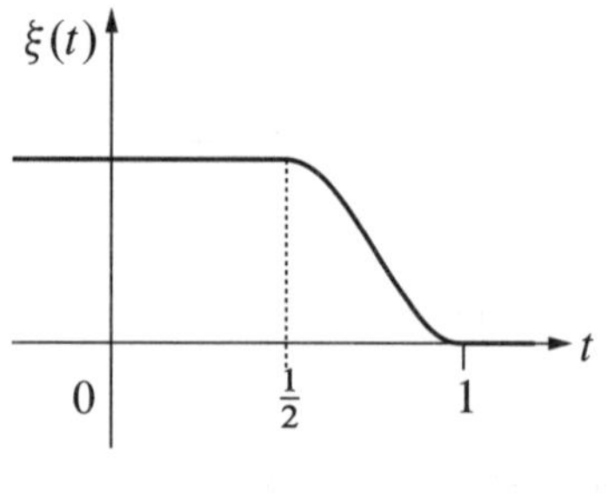

Figure II.7. The cut-off function ξ.

Proof. Due to $\tilde{\varphi} = \varphi$ in Q we clearly have the inclusion

$$W^+_{\text{loc}}(Q) \subset W^+_{\text{loc}}(\mathbb{R}^n, \tilde{\varphi}) \cap Q.$$

In order to prove the opposite inclusion we take $x \in W^+_{\text{loc}}(\tilde{\varphi}) \cap Q$ and show that its image $y = (y_+, y_-) := \varphi(x)$ lies in Q. By repeating the argument it then follows that the positive orbit of x lies in Q and the lemma follows. Recall now that the map $\varphi(x) = Ax + g(x)$ is of the form

$$\begin{aligned} y_+ &= A_+ x_+ + g_+(x), \\ y_- &= A_- x_- + g_-(x). \end{aligned}$$

Since $x \in W^+_{\text{loc}}(\tilde{\varphi})$, we know from Theorem II.8 that $x_- = h(x_+)$. Because the maps h, h', g, g' all vanish in 0, there exists for every $\varepsilon > 0$ an $r > 0$ such that

$$|g(x)| \le \varepsilon r \quad \text{and} \quad |h(x_+)| \le \varepsilon r$$

if $|x| \le r$ (note that from $|x| \le r$, it follows that $|x_+| \le r$). Since $\alpha < 1$, we can choose ε so small that

$$|y_+| \le \|A_+\||x_+| + |g_+(x)| \le \alpha r + \varepsilon r < r$$

and

$$|y_-| \le \|A_-\||h(x_+)| + |g_-(x)| \le \|A_-\|\varepsilon r + \varepsilon r < r.$$

Hence, $y \in Q$, and the lemma follows. □

In view of this lemma we finally obtain from Theorem II.8,

$$\begin{aligned} W^+_{\text{loc}}(Q) &= W^+_{\text{loc}}(\mathbb{R}^n, \tilde{\varphi}) \cap Q \\ &= \operatorname{graph}(h) \cap Q \\ &= \{x \in Q \mid x = (x_+, h(x_+)) \in E_+ \oplus E_-\}. \end{aligned}$$

This is the statement of Theorem II.7 for the stable local invariant manifold. The proof for the unstable local invariant manifold W^-_{loc} is analogous. This completes the proof of Theorem II.7. □

II.3 Stable and unstable invariant manifolds

Theorem II.7 demonstrates that locally the stable and unstable manifolds of the diffeomorphism $\varphi\colon \mathbb{R}^n \to \mathbb{R}^n$ issuing from the hyperbolic fixed point 0 are embedded submanifolds in an open neighborhood of the fixed point. Globally, this is not necessarily true anymore. Indeed, in contrast to the statement of Theorem II.8 the stable and unstable invariant manifolds can globally pile up on themselves, as we shall see

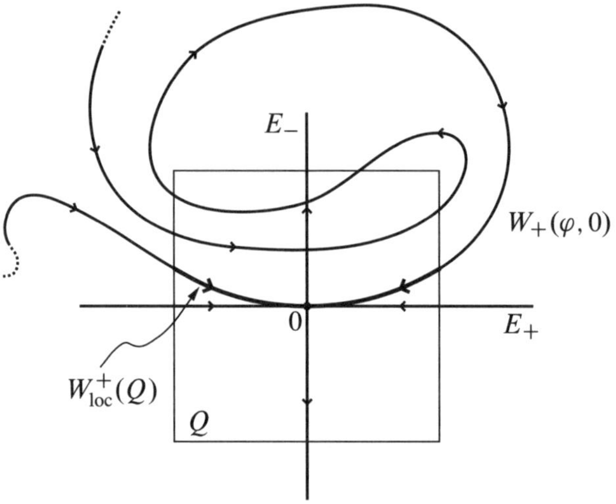

Figure II.8. The stable invariant manifold $W_+(\varphi, 0)$.

later on. Let us first recall the definitions of the stable invariant manifold $W_+(0,\varphi)$ and the unstable invariant manifold $W_-(0,\varphi)$ making use of Theorem II.7,

$$\begin{aligned} W_+(0,\varphi) &:= \{x \in \mathbb{R}^n \mid \varphi^j(x) \to 0,\ j \to +\infty\} \\ &= \bigcup_{n \geq 0} \varphi^{-n}(W^+_{\mathrm{loc}}(Q)), \end{aligned}$$

$$\begin{aligned} W_-(0,\varphi) &:= \{x \in \mathbb{R}^n \mid \varphi^{-j}(x) \to 0,\ j \to +\infty\} \\ &= \bigcup_{n \geq 0} \varphi^{n}(W^-_{\mathrm{loc}}(Q)). \end{aligned}$$

It turns out that these two sets can be parametrized globally by their tangent spaces at the hyperbolic fixed point, hence, in the notation of Theorem II.7, by the Euclidean spaces E_+ and E_-. This is the statement of the following theorem of S. Smale in which we make use of the notation of Theorem II.7.

Theorem II.11 (S. Smale). *Assume* 0 *to be a hyperbolic fixed point of the* C^k*-diffeomorphism* $\varphi\colon \mathbb{R}^n \to \mathbb{R}^n$, *for* $k \geq 1$. *Then, the stable invariant manifold* $W_+(0,\varphi) \subset \mathbb{R}^n$ *is the image of an* ***injective*** C^k***-immersion*** j *of the tangent space* $E_+ := T_0 W^+_{\mathrm{loc}}(Q)$ *into* $\mathbb{R}^n$, *i.e., there exists a map*

$$j\colon E_+ \to \mathbb{R}^n$$

of class C^k *having the following properties:*

(i) $W_+(0,\varphi) = j(E_+)$,
(ii) *j is injective, that is,* $x_+ \neq x'_+ \Longrightarrow j(x_+) \neq j(x'_+)$,
(iii) $\dim dj(x_+)E_+ = \dim E_+$ *for all* $x_+ \in E_+$.

In precisely the same way, the unstable invariant manifold $W_-(0,\varphi)$ is parametrized by E_-.

The subsets W_+ and W_- are immersed manifolds without boundary and without self intersections, and in general, they are *not submanifolds of* $\mathbb{R}^n$. The *tangent space* of $W_+(0,\varphi)$ in the point $p = j(x_+)$ is equal to

$$T_p W_+(0,\varphi) = dj(x_+)E_+,$$

so that $\dim T_p W_+(0,\varphi) = \dim E_+$.

Proof of Theorem II.11 [Constructive]. Theorem II.11 is an easy consequence of Theorem II.8 and its proof, from which we also adopt the notation. We consider the diffeomorphism φ and the modified diffeomorphism $\tilde{\varphi}$ from the proof of Theorem II.7 and recall the following facts:

$$\tilde{\varphi}(x) = \varphi(x), \quad x \in Q, \tag{1}$$

$$\begin{aligned} W_+(0,\varphi) &= \{x \mid \varphi^j(x) \to 0,\ j \to +\infty\} \\ &= \bigcup_{n \geq 0} \varphi^{-n}(W^+_{\text{loc}}(Q)), \end{aligned} \tag{2}$$

$$\begin{aligned} \widetilde{W}_+ &:= W^+_{\text{loc}}(\mathbb{R}^n, \tilde{\varphi}) \\ &= \{x \mid \tilde{\varphi}^j(x) \to 0,\ j \to +\infty\} \\ &= \{x \mid x = (x_+, h(x_+)),\ x \in E_+\}, \end{aligned} \tag{3}$$

$$W^+_{\text{loc}}(Q) = \widetilde{W}_+ \cap Q, \tag{4}$$

$$\tilde{\varphi}(\widetilde{W}_+) = \widetilde{W}_+, \tag{5}$$

$$\tilde{\varphi}(\widetilde{W}_+ \cap Q) \subset Q. \tag{6}$$

Remark. For a given point $x = (x_+, h(x_+)) \in \widetilde{W}_+$, there exists, due to (3) and (4) an integer $s \geq 0$ satisfying

$$\tilde{\varphi}^s(x) \in \widetilde{W}_+ \cap Q = W^+_{\text{loc}}(Q).$$

We define the point $y^{(s)} \in \mathbb{R}^n$ by

$$y^{(s)} := \varphi^{-s} \circ \tilde{\varphi}^s(x) \in W_+(0,\varphi),$$

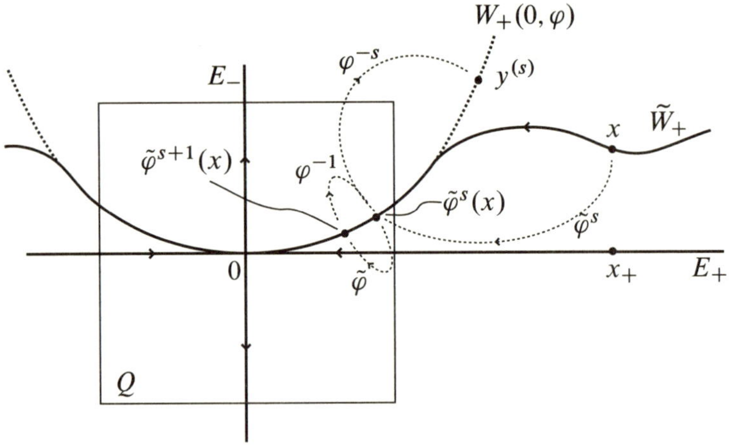

Figure II.9. Illustrating the proof of the theorem by S. Smale.

where we have used the definition (2) of the stable manifold. As long as $\tilde{\varphi}^s(x)$ lies in Q the point $y^{(s)}$ does not depend on the choice of the integer s. Indeed, using (1), (2) and (6) we conclude for $s+1$ that

$$\begin{aligned} y^{(s+1)} &= \varphi^{-(s+1)} \circ \tilde{\varphi}^{s+1}(x) \\ &= \varphi^{-s} \circ \varphi^{-1} \circ \tilde{\varphi} \circ \tilde{\varphi}^s(x) \\ &= \varphi^{-s} \circ \varphi^{-1} \circ \varphi \circ \tilde{\varphi}^s(x) \\ &= \varphi^{-s} \circ \tilde{\varphi}^s(x) \\ &= y^{(s)}, \end{aligned}$$

where, in view of (1), we have used that $\tilde{\varphi} \circ \tilde{\varphi}^s(x) = \varphi \circ \tilde{\varphi}^s(x)$ if $\tilde{\varphi}^s(x) \in Q$.

We now *define* the map $j: E_+ \to \mathbb{R}^n$. Given $x_+ \in E_+$, we choose by means of (3) an integer $s \geq 0$ so large that $\tilde{\varphi}^s(x_+, h(x_+)) \in Q$ and set

$$j(x_+) := \varphi^{-s} \circ \tilde{\varphi}^s(x_+, h(x_+)) \in W_+(0,\varphi).$$

According to the above remark, $j(x_+)$ *does not depend on the choice of the integer* s, as long as $\tilde{\varphi}^s(x) \in Q$. We check the properties of this map j.

(i) The equation $W_+(0,\varphi) = j(E_+)$ follows from the statements (2)–(5).

(ii) In order to prove the injectivity of the map j, namely that $x_+ \neq x'_+$ implies $j(x_+) \neq j(x'_+)$, we take two points $x_+, x'_+ \in E_+$. Then, there exists, due to (3) and (6) a large integer $s \geq 0$ such that *simultaneously* $\tilde{\varphi}^s(x_+, h(x_+)) \in Q$ and

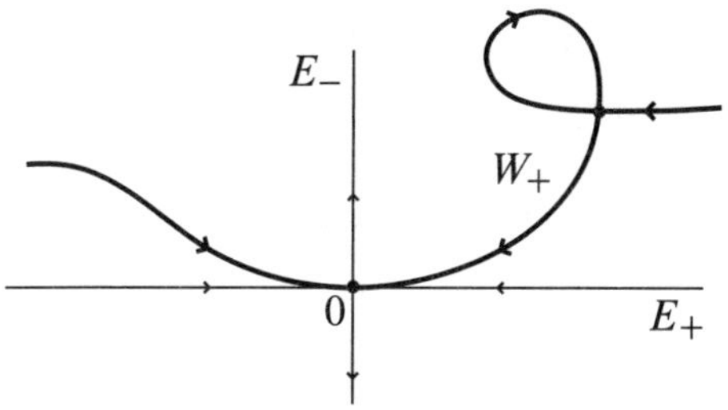

Figure II.10. Does *not* happen according to the theorem by Smale: W_+ having a self intersection.

$\tilde{\varphi}^s(x'_+, h(x'_+)) \in Q$ and hence by definition,

$$j(x_+) = \varphi^{-s} \circ \tilde{\varphi}^s(x_+, h(x_+)),$$
$$j(x'_+) = \varphi^{-s} \circ \tilde{\varphi}^s(x'_+, h(x'_+)).$$

We assume now that $j(x_+) = j(x'_+)$. Since φ^{-s} and $\tilde{\varphi}^s$ are diffeomorphisms, in particular bijective, it follows that $(x_+, h(x_+)) = (x'_+, h(x'_+))$, and therefore $x_+ = x'_+$.

(iii) It remains to show that $\dim dj(x_+)E_+ = \dim E_+$ for all $x_+ \in E_+$. With the injection map $j_0 \colon E_+ \to \mathbb{R}^n$ defined by $j_0(x_+) := (x_+, h(x_+))$ we can write

$$j(x_+) = \varphi^{-s} \circ \tilde{\varphi}^s \circ j_0(x_+),$$

and conclude by the chain rule for the derivative $dj(x_+) \in \mathcal{L}(E_+, \mathbb{R}^n)$ that

$$dj(x_+)\xi = d\varphi^{-s}(\tilde{\varphi}^s \circ j_0(x_+)) \circ d\tilde{\varphi}^s(j_0(x_+)) \circ dj_0(x_+)\xi, \quad \xi \subset E_+.$$

Due to $dj_0(x_+)\xi = (\xi, dh(x_+)\xi)$,

$$\dim dj_0(x_+)E_+ = \dim E_+.$$

Since $d\varphi^{-s}(p), d\tilde{\varphi}^s(p) \in \mathcal{L}(\mathbb{R}^n)$ are linear isomorphisms in every point $p \in \mathbb{R}^n$, it follows that $\dim dj(x_+)E_+ = \dim E_+$ and the theorem of S. Smale is proved. □

In the following we shall abbreviate the stable and unstable invariant manifolds issuing from the hyperbolic fixed point $0 = \varphi(0)$ by simply writing

$$W_+(0) = \{x \mid \varphi^j(x) \to 0,\ j \to \infty\},$$
$$W_-(0) = \{x \mid \varphi^{-j}(x) \to 0,\ j \to \infty\}.$$

We recall that these two immersed manifolds are *invariant* under the diffeomorphism,

$$\varphi(W_+) = W_+, \quad \varphi(W_-) = W_-,$$

contain the fixed point 0 and intersect *transversally* in 0, so that with the tangent spaces $T_0 W_+ = E_+$ and $T_0 W_- = E_-$ we have

$$T_0 W_+ \oplus T_0 W_- = T_0 \mathbb{R}^n.$$

The stable and unstable invariant manifolds cannot intersect themselves but they can intersect each other in points other than the fixed point, namely in so-called homoclinic points, discovered by H. Poincaré in the restricted 3-body problem of celestial mechanics.

Definition. A point of intersection

$$\nu \in W_-(0) \cap W_+(0) \setminus \{0\}$$

is called a *homoclinic point* (to the fixed point 0).

In view of the invariance of the stable and unstable invariant manifolds the orbit $\mathcal{O}(\nu)$ of such a homoclinic point consists of homoclinic points, and every homoclinic point converges to the hyperbolic fixed point under the iterates of the diffeomorphism as well as under the iterates of the inverse diffeomorphism. Hence the homoclinic orbit converges in forward and backward direction to the fixed point.

$$\begin{aligned} \varphi^j(\nu) &\to 0, \quad j \to +\infty, \\ \varphi^j(\nu) &\to 0, \quad j \to -\infty, \\ \varphi^j(\nu) &\in W_+(0) \cap W_-(0), \quad j \in \mathbb{Z}. \end{aligned}$$

The *closure* of the homoclinic orbit $\mathcal{O}(\nu)$ is the set

$$\Lambda := \bigcup_{j \in \mathbb{Z}} \varphi^j(\nu) \cup \{0\}.$$

It consists of two orbits, namely $\mathcal{O}(\nu)$ and $\mathcal{O}(0) = \{0\}$. The set Λ is *compact* and *invariant*, $\varphi(\Lambda) = \Lambda$.

Definition. The homoclinic point ν is called *transversal* (or nondegenerate), if the manifolds $W_+(0)$ and $W_-(0)$ *intersect transversally* in ν, i.e.,

$$T_\nu W_+(0) \oplus T_\nu W_-(0) = T_\nu \mathbb{R}^n.$$

Since the immersed manifolds are invariant and cannot intersect themselves, the transversal homoclinic orbit forces them to oscillate faster and faster as they approach the hyperbolic fixed point, doubling back and piling up on themselves as illustrated in Figure II.11 and one expects quite a complicated orbit structure nearby.

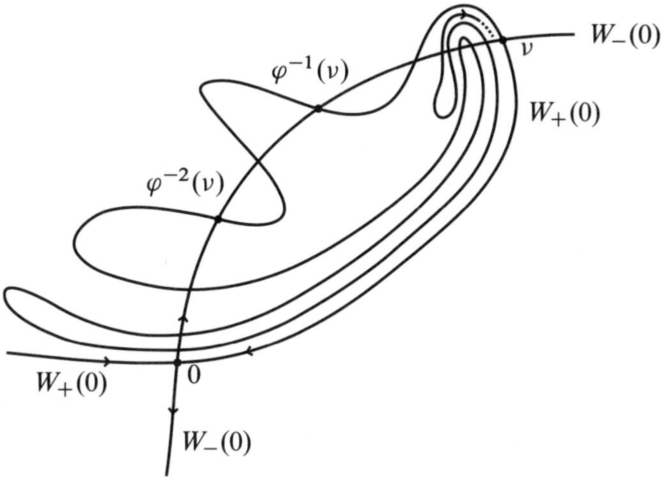

Figure II.11. Stable and unstable invariant manifolds possessing a transversal homoclinic point.

Example. As an illustration we shall describe a simple example of a diffeomorphism in the plane exhibiting homoclinic points. An effective way to produce diffeomorphisms is to take flow maps of vector fields and so we start with the differential equation of second order on the real line,

$$\ddot{x} = x - x^3 = -\frac{d}{dx}V(x),$$

where the potential function is given by

$$V(x) = -\frac{1}{2}x^2 + \frac{1}{4}x^4.$$

This differential equation is equivalent to the system of first-order differential equations

$$\begin{pmatrix} \dot{x} \\ \dot{y} \end{pmatrix} = \begin{pmatrix} y \\ -\frac{d}{dx}V(x) \end{pmatrix} =: X(x, y),$$

defined by the smooth *time independent* vector field X in the plane. The vector field possesses the *integral* $H\colon \mathbb{R}^2 \to \mathbb{R}$, defined as the function

$$H(x, y) = \frac{1}{2}y^2 + V(x).$$

As a consequence, the solution curves of the vector field lie on the compact *level sets* $\{(x, y) \in \mathbb{R}^n \mid H(x, y) = \text{const.}\}$ and hence do exist for all times. The level sets are illustrated in Figure II.12.

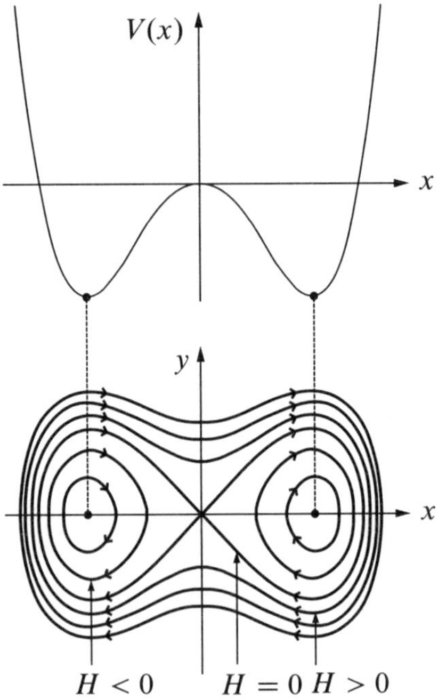

Figure II.12. The potential V and the level lines of the integral H.

The flow φ^t of the vector field X is defined by the unique solutions of the Cauchy initial value problem,

$$\begin{cases} \frac{d}{dt}\varphi^t(z) = X(\varphi^t(z)), & t \in \mathbb{R}, \\ \varphi^0(z) = z, \end{cases}$$

where we have abbreviated the points of the plane by $z = (x, y)$. For a given point z, the continuously differentiable curve $t \mapsto \varphi^t(z)$ in the plane is the unique solution of the vector field X passing through the point $\varphi^0(z) = z$ (called the initial condition) at the time $t = 0$. For fixed time $t = T > 0$ the *time T map*

$$z \mapsto \varphi^T(z)$$

is a *diffeomorphism* of the plane $\mathbb{R}^2$.

The equilibrium points (or rest points) of the vector field are the points z^* at which the vector field vanishes, $X(z^*) = 0$. In view of the uniqueness of the Cauchy initial value problem the solution through the rest point is constant, that is, $\varphi^t(z^*) = z^*$ for all times t so that the equilibrium point is a fixed point of all the time T-maps, $\varphi^T(z^*) = z^*$. How do we know whether such a fixed point is a hyperbolic point?

Lemma II.12. *Let φ^t be the flow of the time independent vector field*

$$\dot{z} = Y(z)$$

where $Y\colon \mathbb{R}^n \to \mathbb{R}^n$ is of class C^1. Then the equilibrium point z^* of Y (i.e., $Y(z^*) = 0$) is a hyperbolic fixed point of the time T-map φ^T of the flow for $T > 0$, if the linearized vector field $dY(z^*)$ at the equilibrium point possesses eigenvalues whose real part is > 0 as well as eigenvalues with real part < 0, but no eigenvalue having real part equal to 0.

Proof. Differentiating the Cauchy initial value problem in the space variable at the equilibrium point z^* while keeping the time t fixed, we see that the linearized flow $d\varphi^t(z^*) =: \Phi(t)$ is the unique solution of the Cauchy initial value problem

$$\dot{\Phi}(t) = dY(\varphi^t(x^*))\Phi(t), \quad \Phi(0) = \mathbb{1}.$$

Since z^* is an equilibrium point, $\varphi^t(z^*) = z^*$ and hence $dY(\varphi^t(z^*)) = dY(z^*)$ is time independent, so that the solution is given by the exponential function

$$\Phi(t) = d\varphi^t(z^*) = e^{t.dY(z^*)}.$$

For the spectrum we conclude

$$\sigma(d\varphi^t(z^*)) = \exp\big(\sigma(t \cdot (dY(z^*)))\big),$$

as is readily verified by looking at the Jordan normal forms of the linear maps. Hence the result follows if we choose $t = T \neq 0$, and the proof is complete. □

Using this lemma one verifies immediately that the origin $0 = (0,0)$ in our example above is a hyperbolic fixed point of the flow map φ^T belonging to the vector field X on $\mathbb{R}^2$. Looking at the level set containing this fixed point, one sees that the stable invariant manifold coincides with the unstable invariant manifold, so that

$$W_+(0) = W_-(0) = \{H(x,y) = 0\}.$$

In this example the stable manifold $W_+(0)$ is a compact set in $\mathbb{R}^2$ and, in view of Smale's theorem, an injective immersion of the real line (see Figure II.13). Every point $\nu \in W_+(0) \setminus \{0\} = W_-(0) \setminus \{0\}$ is a homoclinic point, but clearly not a transversal one, since

$$T_\nu W_+(0) = T_\nu W_-(0).$$

These tangent spaces are spanned by the vector $X(\nu) \neq 0$. We observe that, in general, a diffeomorphism $\psi = \varphi^T$ which is the time T-map of a flow generated

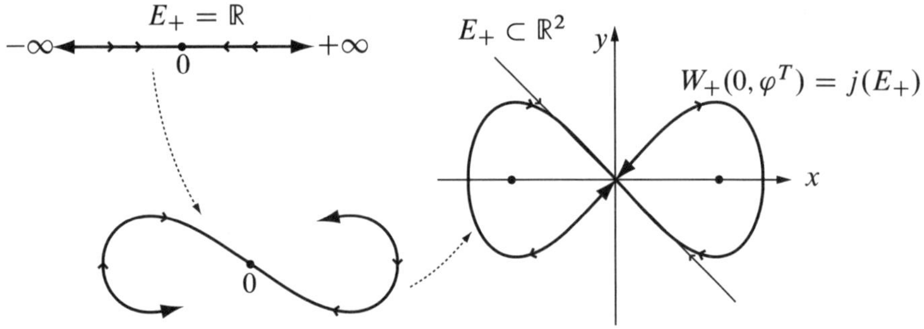

Figure II.13. The injective immersion j in the example.

by a time independent vector field, can only admit degenerate homoclinic points. Indeed, if ψ possesses the hyperbolic fixed point 0 giving rise to the homoclinic point $\nu \in W_+(0) \cap W_-(0)$ we take the unique solution $\varphi^t(\nu)$ passing through the point ν and defined on an open interval containing $t = 0$. It consists of homoclinic points, because $\psi^n(\varphi^t(\nu)) = \varphi^t(\psi^n(\nu)) \to \varphi^t(0) = 0$ converges to the fixed point as $n \to \pm\infty$. Consequently,

$$\frac{d}{dt}\varphi^t(\nu)\Big|_{t=0} = X(\nu)$$

is a tangent vector in $T_\nu W_+$ and also in $T_\nu W_-$ and since $X(\nu) \neq 0$ we obtain

$$T_\nu W_+ \cap T_\nu W_- \neq \{0\},$$

so that the homoclinic point ν is not transversal, as claimed.

In order to create transversal homoclinic points we shall take in the next chapter a *time dependent perturbation* of our time independent vector field X in the plane and start off from the equation

$$\ddot{x} = x - x^3 + \varepsilon f(t, x, \dot{x}),$$

where f is a smooth function that is periodic in time of period $T > 0$, such that $f(t + T, x, \dot{x}) = f(t, x, \dot{x})$ for all $t, x, \dot{x}$. The associated vector field $X(t, z, \varepsilon)$ is now time periodic and depends on the parameter ε. The hyperbolic fixed point $z^* = 0$ for $\varepsilon = 0$ has a continuation to a hyperbolic fixed point $z^*(\varepsilon)$ of the time T-flow map $\varphi^T(z, \varepsilon)$ of the perturbed system, if $\varepsilon \neq 0$ is small. This follows immediately from the implicit function theorem. Indeed, we consider the smooth function $F\colon \mathbb{R}^2 \times \mathbb{R} \to \mathbb{R}$ defined as

$$F(z, \varepsilon) := \varphi^T(z, \varepsilon) - z,$$

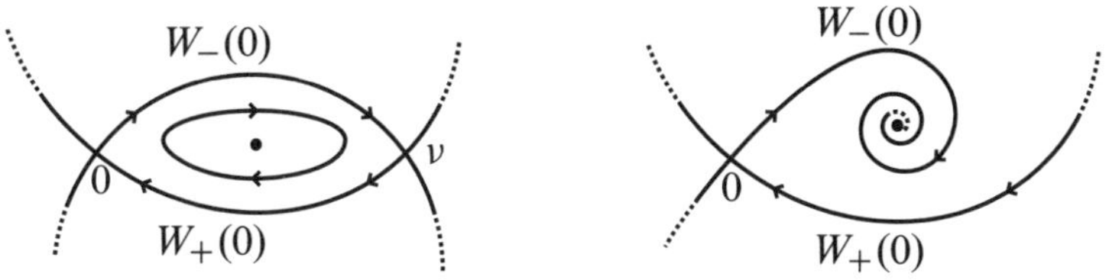

Figure II.14. Two possible scenarios after perturbation: transversal homoclinic point ν (left) and $W_+(0) \cap W_-(0) = \{0\}$ (right).

so that $F(z,\varepsilon) = 0$, if and only if $\varphi^T(z,\varepsilon) = z$ is a fixed point. In order to apply the implicit function theorem we observe that for $\varepsilon = 0$ we have the fixed point $z = 0$, so that $F(0,0) = 0$. Denoting by D_1 the partial derivative in the variable z, we obtain

$$D_1 F(0,0) = D_1\varphi^T(0,0) - \mathbb{1}.$$

If $\varepsilon = 0$, the linear map $D_1\varphi^T(0,0)$ is hyperbolic, as we know, and hence does not admit an eigenvalue equal to 1. Therefore, $D_1 F(0,0)$ is a linear isomorphism and the implicit function theorem guarantees a unique differentiable function $\varepsilon \to z^*(\varepsilon)$ defined for small ε and satisfying $z^*(0) = 0$ and $F(z^*(\varepsilon),\varepsilon) = \varphi^T(z^*(\varepsilon)) - z^*(\varepsilon) = 0$. The fixed point $z^*(\varepsilon)$ remains hyperbolic if ε is small, since the eigenvalues of the Jacobi matrix $D_1\varphi^T(z^*(\varepsilon),\varepsilon)$ depend continuously on ε. In the next chapter, we will see that under additional assumptions on the perturbation these fixed points $z^*(\varepsilon)$ can give rise to transversal homoclinic points for small $\varepsilon \neq 0$.

Definition. We assume that x^* and y^* are two different hyperbolic fixed points of the diffeomorphism φ. A point ν is called *heteroclinic* to x^* and y^* if

$$\nu \in W_+(x^*) \cap W_-(y^*).$$

A heteroclinic point ν is called *transversal* if

$$T_\nu W_+(x^*) \oplus T_\nu W_-(y^*) = T_\nu \mathbb{R}^n.$$

The orbit of a heteroclinic point ν satisfies

$$\varphi^j(\nu) \to x^*, \quad j \to +\infty$$

and

$$\varphi^j(\nu) \to y^*, \quad j \to -\infty$$

in view of the definitions of $W_+(x^*)$ and $W_-(y^*)$.

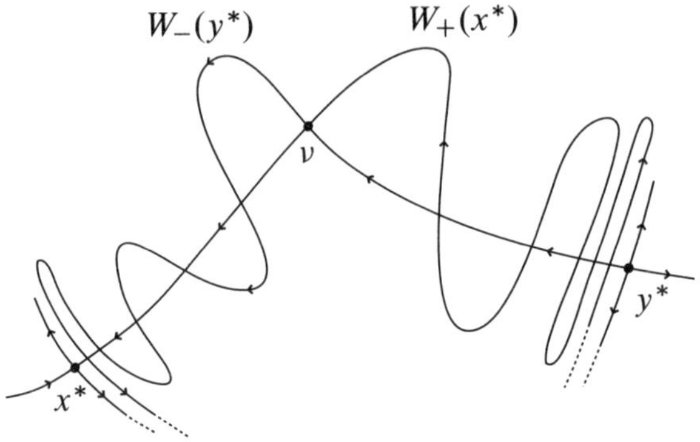

Figure II.15. Transversal heteroclinic point ν associated with the hyperbolic fixed points x^*, y^* and the invariant manifolds.

Literature. The above theorem about local invariant manifolds is named after J. Hadamard, in view of[47] (1910) and after O. Perron in view of [82] (1928). The purpose of the paper [29] (1976) by J. J. Duistermaat is to show that "the results of Perron on perturbed linear systems imply without much effort most of the geometric results on stable manifolds and hyperbolic sets ...". Proofs of invariant manifolds can be found in almost all books on dynamical systems and ordinary differential equations, for example (apart from the references in Chapter I) in the books [53] by M. Irwin, and [5] by H. Amann. For more recent proofs we refer to M. Chaperon in [15] and in [16] and the references therein. Our proofs of the invariant manifold theorem and Smale's theorem make use of the lectures [119] by J. C. Yoccoz. Homoclinic points were discovered by H. Poincaré in [87] in the restricted 3-body problem of celestial mechanics. This interesting discovery is described in [6] by K. Anderson. "Homoclinic trajectories: from Poincaré to the present" is the title of the article [102] by L. P. Shil'nikov. As for recent existence results of homoclinic points and further references we refer to [118] by J. Xia. He demonstrates that for a generic area preserving diffeomorphism on a compact surface, homotopic to the identity, every hyperbolic fixed point possesses a transversal homoclinic point. On the other hand, an old result by E. Zehnder [120] shows that generically even an elliptic fixed point of an area preserving diffeomorphism in two dimensions is a cluster point of transversal homoclinic points.

Chapter III
Hyperbolic sets

The existence of a transversal homoclinic point considerably complicates the orbit structure of a diffeomorphism. In order to describe this complexity, which is sometimes called deterministic chaos, we introduce the concept of a hyperbolic set and prove the shadowing lemma by means of the contraction principle. An application of the shadowing lemma shows that a transversal homoclinic point is a cluster point of homoclinic points (H. Poincaré) and a cluster point of periodic points (G. Birkhoff). In addition, the shadowing lemma allows the construction of embedded Bernoulli systems as subsystems in a neighborhood of a homoclinic orbit (S. Smale). In this way we establish orbits that are characterized by random sequences. The interpretation of such stochastic orbits will be illustrated in the simple system of the periodically perturbed mathematical pendulum.

III.1 Definition of a hyperbolic set

Hyperbolic sets are related to the dynamically unstable behavior of dynamical systems. The tangent spaces split into two invariant subspaces along which there is a contraction, respectively an expansion. The concept goes back to S. Smale and D. V. Anosov in 1967. It turns out that it is hard to get rid of hyperbolic sets by a perturbation because they are structurally stable.

Definition. The subset $\Lambda \subset \mathbb{R}^n$ is called a *hyperbolic set* of the diffeomorphism $\varphi\colon \mathbb{R}^n \to \mathbb{R}^n$ if it has the following properties.

(i) Λ is compact and invariant under φ, i.e., $\Lambda = \varphi(\Lambda)$.

(ii) There exists a splitting of the tangent space in every $x \in \Lambda$,

$$\mathbb{R}^n = T_x\mathbb{R}^n = E_x^+ \oplus E_x^-, \quad x \in \Lambda,$$

which is invariant under the linearization of φ,

$$d\varphi(x)E_x^+ = E_{\varphi(x)}^+,$$
$$d\varphi(x)E_x^- = E_{\varphi(x)}^-,$$

and there exist constants $c > 0$ and $0 < \vartheta < 1$ that are independent of x such that the following estimates hold true:

$$|d\varphi^j(x)\xi| \le c\vartheta^j|\xi|, \quad \xi \in E_x^+,\ j \ge 0,$$
$$|d\varphi^{-j}(x)\xi| \le c\vartheta^j|\xi|, \quad \xi \in E_x^-,\ j \ge 0.$$

(iii) The splitting $E_p^+ \oplus E_p^-$ depends continuously on $p \in \Lambda$. In other words, defining the projections P_p^+ and $P_p^- = \mathbb{1} - P_p^+ \in \mathcal{L}(\mathbb{R}^n)$ onto the subspaces E_p^+ and E_p^- by

$$P_p^{\pm}(v_+ \oplus v_-) = v_{\pm},$$

the mapping

$$\Lambda \to \mathcal{L}(\mathbb{R}^n), \quad p \mapsto P_p^{\pm}$$

is continuous.

Examples. If $\varphi(0) = 0$ is a hyperbolic fixed point of the diffeomorphism φ, then $\Lambda := \{0\}$ is a hyperbolic set. Every closed and invariant subset of a hyperbolic set Λ is again a hyperbolic set. In particular, the periodic orbits in Λ are hyperbolic sets.

We shall demonstrate next that the continuity of the splitting already follows from the postulated estimates in the definition of a hyperbolic set, and begin by drawing some conclusions from the estimates characterizing the subspaces $E_p^{\pm}$ of the tangent spaces.

Lemma III.1. $E_p^{\pm} = \operatorname{Im}(P_p^{\pm}) = \{v \in T_p\mathbb{R}^n \mid \sup_{j \geq 0} |d\varphi^{\pm j}(p)v| < \infty\}$.

Proof. The inclusion $\subset$ immediately follows from the definition. On the other hand, if $v = P_p^+ v + P_p^- v$ belongs to $T_p\mathbb{R}^n$ then, using

$$\begin{aligned}|P_p^- v| &= |d\varphi^{-j}(\varphi^j(p)) \circ d\varphi^j(p) \circ P_p^- v| \\ &\leq c\vartheta^j |d\varphi^j(p) \circ P_p^- v|,\end{aligned}$$

we obtain

$$\begin{aligned}|d\varphi^j(p)v| &= |d\varphi^j(p)P_p^- v + d\varphi^j(p)P_p^+ v| \\ &\geq |d\varphi^j(p)P_p^- v| - |d\varphi^j(p)P_p^+ v| \\ &\geq c^{-1}\vartheta^{-j}|P_p^- v| - c\vartheta^j|P_p^+ v|.\end{aligned}$$

Therefore, $\lim_{j\to\infty} |d\varphi^j(p)v| = \infty$, if $P_p^- v \neq 0$. If $\sup_{j\geq 0} |d\varphi^j(p)v| < \infty$, it also follows that $P_p^- v = 0$ and hence $v \in E_p^+$. The same argument proves that $P_p^+ v = 0$, if $\sup_{j\geq 0} |d\varphi^{-j}(p)v| < \infty$, and the lemma is proved. □

Corollary III.2. *If these estimates hold true, the sums $E_p^+ + E_p^-$ are automatically direct.*

Lemma III.3. *The projections onto E_p^+ and E_p^- are uniformly bounded, i.e., there exists a constant K such that the operator norms satisfy $\|P_p^{\pm}\| \leq K < \infty$ for all $p \in \Lambda$.*

Proof. We choose the integer N so large that $\lambda = c\vartheta^N < 1$ and look at the iterated map $\psi = \varphi^N$. Since Λ is a compact set there is a constant $M > 0$, such that the operator norms satisfy $\|d\psi(p)\| \leq M$ for all p in Λ. Fix p in Λ. Then, in view of the estimates in the definition of a hyperbolic set, for every $v = v_+ + v_-$ in $E_p^+ \oplus E_p^- = T_p\mathbb{R}^n$,

$$\begin{aligned} M|v| &\geq |d\psi(p)v| \\ &= |d\psi(p)v_+ + d\psi(p)v_-| \\ &\geq |d\psi(p)v_-| - |d\psi(p)v_+| \\ &\geq \lambda^{-1}|v_-| - \lambda|v_+|. \end{aligned}$$

By the triangle inequality $|v_+| = |v_+ + v_- - v_-| = |v - v_-| \leq |v| + |v_-|$ and so

$$M|v| \geq \lambda^{-1}|v_-| - \lambda(|v| + |v_-|),$$

and consequently,

$$(M + \lambda)|v| \geq (\lambda^{-1} - \lambda)|v_-|.$$

Therefore, $|P_p^- v| = |v_-| \leq K|v|$ with the constant $K = (M + \lambda)/(\lambda^{-1} - \lambda)$, and we obtain for the operator norm the estimate $\|P_p^-\| \leq K$. This holds true for every p in Λ. A similar argument shows that also $\|P_p^+\|$ is uniformly bounded and the lemma is proved. □

Lemma III.4. *The map* $\Lambda \to \mathcal{L}(\mathbb{R}^n)$ *defined by* $p \to P_p^{\pm}$ *is* continuous.

Proof. (1) We first show that the graph

$$\Gamma := \{(p, P_p^+) \mid p \in \Lambda\} \subset \Lambda \times \mathcal{L}(\mathbb{R}^n)$$

is a *closed* set. If p_n is a sequence in Λ satisfying $p_n \to p \in \Lambda$ and $P_{p_n}^+ \to Q^+$ in $\mathcal{L}(\mathbb{R}^n)$, we have to prove that $Q^+ = P_p^+$. Since every linear map $P_{p_n}^+$ is a projection, i.e., satisfies $(P_{p_n}^+)^2 = P_{p_n}^+$, the linear map Q^+ is also a projection. In particular,

$$\operatorname{Im} Q^+ \oplus \ker Q^+ = \mathbb{R}^n.$$

From Lemma III.3 we know that $\|P_p^+\| \leq K$ for all p, so that $\|d\varphi^j(p_n)P_{p_n}^+\| \leq cK\vartheta^j$. Hence, since $d\varphi^j(p)$ depends continuously on p, we obtain the estimate

$$\|d\varphi^j(p)Q^+\| \leq cK\vartheta^j, \quad j \geq 0.$$

Using Lemma III.1 we, therefore, conclude for the image sets of the operators that $\operatorname{Im}(Q^+) \subset \operatorname{Im}(P_p^+)$. In an analogous way, one finds $\operatorname{Im}(\mathbb{1} - Q^+) \subset \operatorname{Im}(\mathbb{1} - P_p^+)$, so that $\operatorname{Im} Q^+ = \operatorname{Im} P_p^+$ and $\ker Q^+ = \ker P_p^+$. Since Q^+ is a projection, it is thus uniquely determined and we have proved that $Q^+ = P_p^+$. Hence, Γ is a closed set.

(2) According to (1) and Lemma III.3 the set Γ is closed and bounded, hence *compact*. The projection $\pi_1 \colon \Gamma \to \Lambda$ onto Λ defined by $(p, P_p^+) \mapsto p$ is bijective and continuous. Because Γ is a compact set, the inverse map π_1^{-1} is also continuous. Since the projection $\pi_2 \colon \Gamma \to \mathcal{L}(\mathbb{R}^n)$ defined by $(p, P_p^+) \mapsto P_p^+$ is continuous, the composition $p \mapsto P_p^+ = \pi_2 \circ \pi_1^{-1}(p)$ is continuous.

The same arguments show that also the map $p \mapsto P_p^-$ is continuous and the lemma is proved. □

The study of the orbit structure near a homoclinic orbit will be based on the following crucial observation.

Proposition III.5. *Let* 0 *be a hyperbolic fixed point of the diffeomorphism* φ *and let* $\nu \in \mathbb{R}^n$ *be a* ***transversal*** *homoclinic point to* 0. *Then, the closure of its orbit,*

$$\Lambda := \bigcup_{j \in \mathbb{Z}} \varphi^j(\nu) \cup \{0\},$$

is a hyperbolic set.

Proof. The set Λ is invariant and compact. We define the splitting by means of the *tangent spaces* of the stable invariant manifold $W_+(0)$ and the unstable invariant manifold $W_-(0)$ which do intersect along the homoclinic orbit,

$$E_x^+ := T_x W_+(0), \quad E_x^- := T_x W_-(0).$$

Due to the transversality of the homoclinic point ν the tangent spaces split,

$$\mathbb{R}^n = T_x \mathbb{R}^n = E_x^+ \oplus E_x^-, \quad x \in \Lambda.$$

If $v \in T_p W_\pm(0)$, there exists a curve $x \colon I \to W_\pm(0)$ defined on an interval and satisfying $x(0) = p$ and $\dot{x}(0) = v$. Since the manifolds are invariant under the diffeomorphism φ the image curve $t \mapsto \varphi(x(t)) \in \mathbb{R}^n$ satisfies

$$\varphi(x(t)) \in W_\pm \quad \text{and} \quad \varphi(x(0)) = \varphi(p).$$

From

$$\frac{d}{dt}\varphi(x(t)) = d\varphi(x(t))\dot{x}(t)$$

it follows for $t = 0$ that

$$d\varphi(p)v \in T_{\varphi(p)} W_\pm(0).$$

In short, from the invariance $W_\pm = \varphi(W_\pm)$ one concludes

$$d\varphi(x) E_x^\pm = E_{\varphi(x)}^\pm.$$

The required *estimates* will be deduced from the estimates of the linearized map $d\varphi(0)$ at the hyperbolic fixed point, using that $\varphi^j(\nu) \to 0$ as $|j| \to \infty$. For this

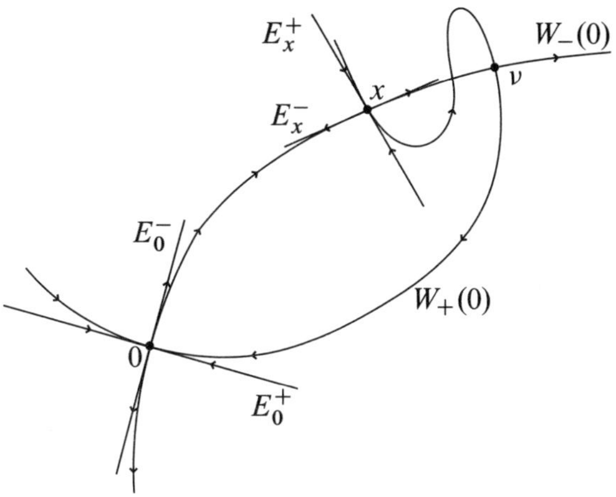

Figure III.1. Definition of the splittings $E_x^+ \oplus E_x^-$ by the tangent spaces.

purpose we introduce new coordinates near 0, for which $W_{\text{loc}}^+ = E_+$ and $W_{\text{loc}}^- = E_-$ holds true. Using the *local* representation in Theorem II.7 of the invariant manifolds (in the neighborhood Q of the fixed point) as graphs of functions,

$$y = h_+(x), \quad (x, y) \in W_{\text{loc}}^+(Q),$$
$$x = h_-(y), \quad (x, y) \in W_{\text{loc}}^-(Q),$$

we define the *coordinate transformation* $\psi(x, y) = (\xi, \eta)$ by

$$\xi = x - h_-(y),$$
$$\eta = y - h_+(x).$$

Then,

$$\psi(0) = 0, \quad d\psi(0) = \mathbb{1}$$

in view of $h_-(0) = 0$, $h_+(0) = 0$, $h'_-(0) = 0$ and $h'_+(0) = 0$. Hence, ψ is a *local diffeomorphism* near 0 by the inverse function theorem. In the new coordinates, the mapping φ is represented by

$$\hat{\varphi} = \psi \circ \varphi \circ \psi^{-1} = (\xi_1, \eta_1),$$

$$\xi_1 = f(\xi, \eta),$$
$$\eta_1 = g(\xi, \eta),$$

and the local invariant manifolds are the sets

$$W_{\text{loc}}^+(Q') = \{(\xi, 0)\} = E_+,$$
$$W_{\text{loc}}^-(Q') = \{(0, \eta)\} = E_-.$$

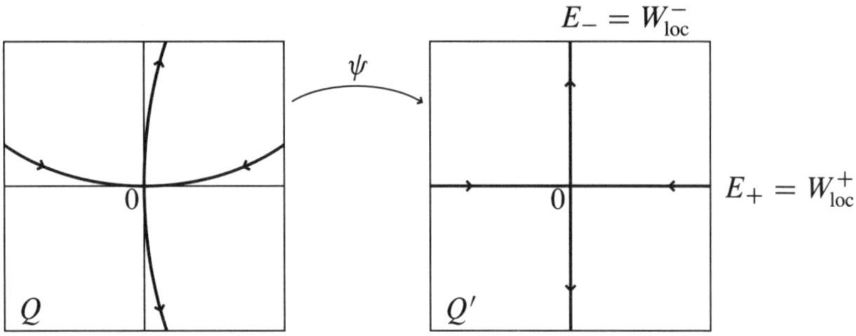

Figure III.2. The coordinate transformation in the proof of Proposition III.5.

From the ±-invariance of these manifolds we deduce

$$f(0,\eta) = 0, \quad g(\xi, 0) = 0,$$

for all ξ and η small enough. Therefore, near 0, using that

$$d\varphi(0) = \begin{pmatrix} A_+ & 0 \\ 0 & A_- \end{pmatrix},$$

the diffeomorphism $\hat{\varphi}$ is of the form

$$\begin{aligned} \xi_1 &= (A_+ + O(\xi,\eta))\xi, \\ \eta_1 &= (A_- + O(\xi,\eta))\eta. \end{aligned}$$

In particular,

$$d\hat{\varphi}(0) = \begin{pmatrix} A_+ & 0 \\ 0 & A_- \end{pmatrix},$$

and along the local stable manifold $W^+_{\text{loc}}(Q')$,

$$d\hat{\varphi}(\xi,0) = \begin{pmatrix} A_+ + O(\xi) & O(\xi) \\ 0 & A_- + O(\xi) \end{pmatrix}.$$

From $\|A_+\|, \|A_-^{-1}\| \le \alpha < 1$ it follows for $v_+ = (\xi_+, 0) \in E_+$ and $|\xi|$ small that

$$|d\hat{\varphi}(\xi,0)v_+| \le \vartheta |v_+|$$

with a constant $\alpha < \vartheta < 1$.

In the original coordinates we have $\varphi = \psi^{-1} \circ \hat{\varphi} \circ \psi$ and therefore $\varphi^j(p) = \psi^{-1} \circ \hat{\varphi}^j \circ \psi(p)$. Let K be a constant, satisfying $\|d\psi\| \le K$ and $\|d\psi^{-1}\| \le K$.

For a point p in the (perhaps smaller) neighborhood Q of 0 we find in view of

$$d\varphi^j(p)v_+ = d\psi^{-1} \circ d\hat{\varphi}^j \circ d\psi(p)v_+, \quad v_+ \in T_p W^+_{\text{loc}}(Q) = E^+_p,$$

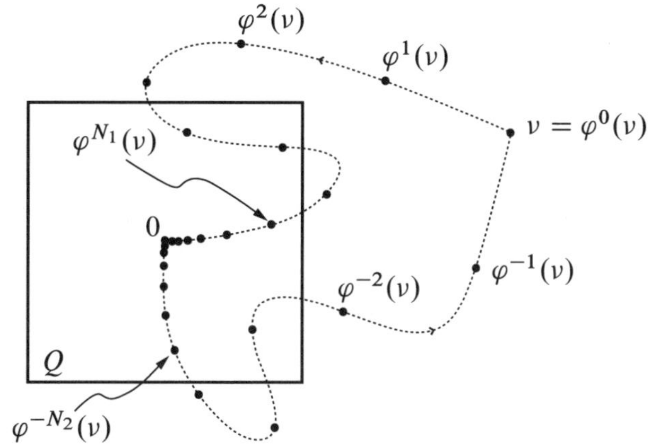

Figure III.3. The estimate in the proof of Proposition III.5.

the estimates

$$|d\varphi^j(p)v_+| \le \vartheta^j K^2 |v_+|, \quad j \ge 0.$$

We have verified the desired estimates for the points $p \in \Lambda \cap Q$. Recalling once more the definition of a homoclinic point, we have $\varphi^j(\nu) \to 0$ as $|j| \to \infty$. Consequently, we find two large integers $N_1, N_2 \ge 0$ such that $\varphi^j(\nu) \in Q$ for all $j \ge N_1$ and $\varphi^{-j}(\nu) \in Q$ for all $j \ge N_2$.

Hence, introducing the integer $N = N_1 + N_2$ we conclude

$$\varphi^N(p) \in Q, \quad p \in \Lambda \setminus Q.$$

From the estimates in Q, we now obtain for all points $p \in \Lambda \setminus Q$ the estimates

$$\begin{aligned}|d\varphi^{N+n}(p)v_+| &= |d\varphi^n(\varphi^N(p)) \circ d\varphi^N(p)v_+| \\ &\le \vartheta^n K^2 |d\varphi^N(p)v_+| \\ &\le \max_{p \in \Lambda \setminus Q} \|d\varphi^N(p)\|\, \vartheta^n K^2 |v_+|.\end{aligned}$$

Setting $c := \max_{p \in \Lambda \setminus Q} \|d\varphi^N(p)\|\, K^2 \vartheta^{-N}$ we therefore find

$$|d\varphi^j(p)v_+| \le c\vartheta^j |v_+|, \quad j \ge N,\ p \in \Lambda \setminus Q.$$

Since $\Lambda \setminus Q$ is a finite set of points we can choose a sufficiently large constant $C \ge c$, for which these estimates hold true also for the integers $0 \le j < N$, and we have verified the desired estimates

$$|d\varphi^j(p)v_+| \le C\vartheta^j |v_+|, \quad v_+ \in E_p^+,\ j \ge 0$$

for all $p \in \Lambda$ and with a constant C independent of p. The estimates for E_p^- are proved analogously and the proof of Proposition III.5 is complete. □

As in the special case of a hyperbolic fixed point, it is very convenient also for hyperbolic sets to introduce new norms in the tangent spaces with respect to which the constant c showing up in the definition of a hyperbolic set is equal to 1.

Proposition III.6 (Adapted norms). *Assume Λ to be a hyperbolic set of the diffeomorphism φ and choose a constant $\vartheta < \mu < 1$. Then, there exist equivalent norms $|\cdot|_x^*$ in $T_x\mathbb{R}^n$ for every $x \in \Lambda$, which depend* ***continuously*** *on x and satisfy*

$$\begin{aligned}|d\varphi(x)v_+|^*_{\varphi(x)} &\le \mu|v_+|^*_x, \quad v_+ \in E_x^+,\\ |d\varphi^{-1}(x)v_-|^*_{\varphi^{-1}(x)} &\le \mu|v_-|^*_x, \quad v_- \in E_x^-.\end{aligned}$$

In these norms, the linear map $d\varphi(x)$ is a contraction *in E_x^+ and an* ***expansion*** *in E_x^-.*

Proof. We choose N so large that $c(\vartheta/\mu)^N < 1$. If $v = v_+ + v_- \in E_x^+ \oplus E_x^- = T_x\mathbb{R}^n$ we define the new norm by

$$|v_+|_x^* := \sum_{j=0}^{N-1} \mu^{-j}|d\varphi^j(x)v_+|.$$

Using $\mu^{-N}|d\varphi^N(x)v_+| \le \mu^{-N}c\vartheta^N|v_+| \le |v_+|$, one obtains the desired estimate

$$\begin{aligned}|d\varphi(x)v_+|^*_{\varphi(x)} &= \sum_{j=0}^{N-1}\mu^{-j}|d\varphi^j(\varphi(x))d\varphi(x)v_+|\\ &= \mu\Big(\sum_{j=1}^{N-1}\mu^{-j}|d\varphi^j(x)v_+| + \mu^{-N}|d\varphi^N(x)v_+|\Big)\\ &\le \mu\Big(\sum_{j=0}^{N-1}\mu^{-j}|d\varphi^j(x)v_+|\Big)\\ &= \mu|v_+|_x^*.\end{aligned}$$

Since φ belongs to the class C^1, the new norms depend continuously on the point x. Analogously, we define the new norm for $v_- \in E_x^-$ using the inverse map φ^{-1} instead of φ. We introduce on the tangent space $T_x\mathbb{R}^n$ the max-norm

$$|v|_x^* := \max\{|v_+|_x^*, |v_-|_x^*\},$$

and the theorem is proved. □

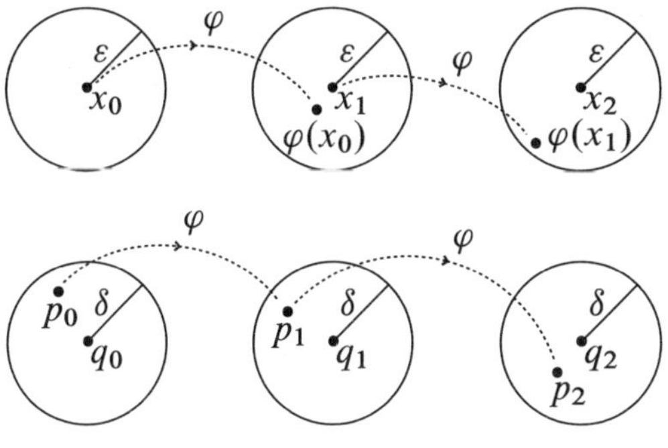

Figure III.4. An ε-pseudo orbit x (above) and a δ shadow orbit p for a sequence q (below).

III.2 The shadowing lemma

In the following, the shadowing lemma will be our main tool. If only an approximate orbit on a hyperbolic set is known, the shadowing lemma guarantees a real orbit nearby which shadows the approximate orbit. This way we shall construct orbits which are determined by their prescribed long-time behavior and not by their initial conditions. To formulate the shadowing lemma, we need some definitions.

Definition. Let $\varphi\colon \mathbb{R}^n \to \mathbb{R}^n$ be a diffeomorphism.

(i) The *sequence* $(x_j)_{j\in\mathbb{Z}}$ in $\mathbb{R}^n$ is an *orbit* of φ, if $x_{j+1} = \varphi(x_j)$ for $j \in \mathbb{Z}$.

(ii) For a given real number $\varepsilon > 0$, the sequence $(x_j)_{j\in\mathbb{Z}}$ is called an *ε-pseudo orbit* of φ, if $|x_{j+1} - \varphi(x_j)| \le \varepsilon$ for all $j \in \mathbb{Z}$.

(iii) If $\delta > 0$ and $q = (q_j)_{j\in\mathbb{Z}}$ is a sequence in $\mathbb{R}^n$, then a *δ-shadowing orbit of q* is an orbit $p = (p_j)_{j\in\mathbb{Z}}$, satisfying $|p_j - q_j| \le \delta$ for all $j \in \mathbb{Z}$.

The following theorem goes back to D. Anosov.

Theorem III.7 (Shadowing lemma). *Let Λ be a hyperbolic set of the diffeomorphism φ. Then, there exists a constant $\delta_0 > 0$ such that for every $0 < \delta \le \delta_0$ there exists an $\varepsilon = \varepsilon(\delta) > 0$ having the following property.*

*For **every** ε-pseudo orbit $q = (q_j)_{j\in\mathbb{Z}}$ of φ **on** the set Λ,*

$$q_j \in \Lambda, \quad |q_{j+1} - \varphi(q_i)| \le \varepsilon, \quad j \in \mathbb{Z},$$

*there exists **a unique** δ-shadowing orbit $p = (p_j)_{j\in\mathbb{Z}}$ of the pseudo, orbit q (for φ) in a neighborhood of Λ.*

Remark. (i) The bracket (for φ) can be replaced by the bracket (for ψ), if ψ is a diffeomorphism satisfying $|\varphi - \psi|_{C^1(U)} \le \mu$ on an open neighborhood U of Λ and if μ is sufficiently small.

(ii) The ε-pseudo orbit q does not have to lie on Λ, it is enough to require that the pseudo orbit $q = (q_j)_{j\in\mathbb{Z}}$ belongs to a sufficiently small neighborhood $V(\Lambda)$ of the hyperbolic set Λ.

Proof of Theorem III.7 [Contraction principle]. We make use of the adapted norms guaranteed by Proposition III.6.

(1) *Formulation of the problem.* If the ε-pseudo orbit $q = (q_j)_{j\in\mathbb{Z}} \subset \Lambda$ is given, we look for an orbit $p = (p_j)_{j\in\mathbb{Z}}$ satisfying $|p_j - q_j| \leq \delta$ for all $j \in \mathbb{Z}$. For this purpose, we look for *corrections* $x = (x_j)_{j\in\mathbb{Z}}$, so that the sequence

$$p = q + x$$

is an *orbit*, hence satisfies

$$q_{j+1} + x_{j+1} = \varphi(q_j + x_j), \quad j \in \mathbb{Z}.$$

Rewriting this equation we look for a sequence $x = (x_j)_{j\in\mathbb{Z}}$ solving the equation

$$x_{j+1} - d\varphi(q_j)x_j = \varphi(q_j + x_j) - q_{j+1} - d\varphi(q_j)x_j =: f_j(x_j).$$

The right-hand side is *small*, if ε is small, and if $\|x\| = \sup_{j\in\mathbb{Z}} |x_j|$ is small. Indeed, due to $f_j(0) = \varphi(q_j) - q_{j+1}$ we have, by assumption, $|f_j(0)| \leq \varepsilon$. In addition, the derivative satisfies $df_j(0) = d\varphi(q_j) - d\varphi(q_j) = 0$ and $df_j(x_j) = d\varphi(q_j + x_j) - d\varphi(q_j)$.

We shall solve the equation $x_{j+1} - d\varphi(q_j)x_j = f_j(x_j)$ by means of the contraction principle.

(2) *The linear problem.* We abbreviate $A_j := d\varphi(q_j) \in \mathcal{L}(\mathbb{R}^n)$. Given a sequence $(g_j)_{j\in\mathbb{Z}}$ in $\mathbb{R}^n$ we look for the sequence $x = (x_j)_{j\in\mathbb{Z}}$ solving

$$x_{j+1} - A_j x_j = g_{j+1}, \quad j \in \mathbb{Z}.$$

For this purpose, we introduce a *sequence space*. Setting $E_j = T_{q_j}\mathbb{R}^n = \mathbb{R}^n$, we define the *Banach space* of bounded sequences by

$$E = \{x = (x_j)_{j\in\mathbb{Z}} \mid x_j \in E_j,\ \|x\| < \infty\}$$

equipped with the norm $\|x\| = \sup_{j\in\mathbb{Z}} |x_j|$. We define the linear map $A \in \mathcal{L}(E)$ by its restrictions $A|_{E_j} := A_j \colon E_j \to E_{j+1}$, as

$$(A(x))_{j+1} := A_j x_j.$$

We want to solve the operator equation $(\mathbb{1} - A)x = g$ in the Banach space E.

Lemma III.8. *If $(q_j)_{j\in\mathbb{Z}}$ is an ε-pseudo orbit on Λ, and if ε is sufficiently small, then the linear map $\mathbb{1} - A \in \mathcal{L}(E)$ is a continuous isomorphism whose inverse map $L := (\mathbb{1} - A)^{-1} \in \mathcal{L}(E)$ is also continuous and has the finite norm $\|L\| < \infty$.*

Proof. We introduce the notation $E_j = E_j^+ \oplus E_j^- = P_{q_j}^+ E_j \oplus P_{q_j}^- E_j$. Given the sequence $g = (g_j) \in E$ we look for a sequence $x = (x_j) \in E$ solving the equation $x_{j+1} - d\varphi(q_j)x_j = g_{j+1}$ for $j \in \mathbb{Z}$, or

$$x_{j+1} = d\varphi(q_j)x_j + g_{j+1}, \quad j \in \mathbb{Z}.$$

With respect to the above splitting we obtain the equivalent equations

$$(*) \quad \begin{cases} P_{q_{j+1}}^+ x_{j+1} = P_{q_{j+1}}^+ d\varphi(q_j)x_j + P_{q_{j+1}}^+ g_{j+1}, \\ P_{q_{j+1}}^- x_{j+1} = P_{q_{j+1}}^- d\varphi(q_j)x_j + P_{q_{j+1}}^- g_{j+1}. \end{cases}$$

The splitting $E_j^+ \oplus E_j^-$ is not invariant under the linearized map $d\varphi(q_j)$, since q is not an orbit. However, along the orbit we know from the definition of the hyperbolicity of the set Λ that

$$P_{\varphi(q_j)}^{\pm} d\varphi(q_j)x_j = d\varphi(q_j)P_{q_j}^{\pm} x_j, \quad j \in \mathbb{Z}.$$

Thus, the equation $(*)$ is equivalent to the following two equations of $(**)$ and $(***)$,

$$(**) \quad P_{q_{j+1}}^+ x_{j+1} = d\varphi(q_j)P_{q_j}^+ x_j + P_{q_{j+1}}^+ g_{j+1} + [P_{q_{j+1}}^+ - P_{\varphi(q_j)}^+]d\varphi(q_j)x_j,$$

$$(***) \quad \begin{aligned} P_{q_j}^- x_j = {} & d\varphi(q_j)^{-1} P_{\varphi(q_j)}^- x_{j+1} - d\varphi(q_j)^{-1} P_{q_{j+1}}^- g_{j+1} \\ & + d\varphi(q_j)^{-1}[P_{q_{j+1}}^- - P_{\varphi(q_j)}^-](x_{j+1} - d\varphi(q_j)x_j). \end{aligned}$$

We introduce the map

$$\Phi\colon E \to E, \quad x = (x_j) \mapsto (\Phi(x)_j), \quad \Phi(x)_j := P_{q_j}^+ \Phi(x)_j + P_{q_j}^- \Phi(x)_j,$$

where $P_{q_j}^+ \Phi(x)_j$ is defined by the right-hand side of the equation $(**)$ and $P_{q_j}^- \Phi(x)_j$ by the right-hand side of the equation $(***)$. By construction the desired sequence is a fixed point of this map,

$$\Phi(x) = x \Longleftrightarrow x_{j+1} - d\varphi(q_j)x_j = g_{j+1}.$$

Since Λ is compact,

$$\sup_{q\in\Lambda} \|d\varphi(q)\| \le K, \quad \sup_{q\in\Lambda} \|d\varphi(q)^{-1}\| \le K$$

for a constant K and the mappings $q \mapsto P_q^{\pm}\colon \Lambda \to \mathcal{L}(\mathbb{R}^n)$ are not only continuous, but *uniformly continuous*. Hence, for every given $\varepsilon' > 0$ there exists an $\varepsilon > 0$ such that

$$\|P_{q_{j+1}}^{\pm} - P_{\varphi(q_j)}^{\pm}\| \le \varepsilon' \text{ for all } j \in \mathbb{Z}, \quad \text{if } |q_{j+1} - \varphi(q_j)| \le \varepsilon \text{ for all } j \in \mathbb{Z},$$

i.e., if the sequence q is an ε-pseudo orbit. Since Λ is hyperbolic, we have (in the adapted norms) the estimates

$$|d\varphi(q_j)P^+_{q_j}x_j| \le \vartheta|x_j|,$$
$$|d\varphi(q_j)^{-1}P^-_{q_j}x_{j+1}| \le \vartheta|x_{j+1}|,$$

with a constant $0 \le \vartheta < 1$. Using this, we shall estimate the Lipschitz constant of the map Φ. Recalling the definition of the norms and using the notation $a \vee b := \max\{a, b\}$, we have

$$\begin{aligned}\|\Phi(x) - \Phi(y)\| &= \sup_{j\in\mathbb{Z}} |\Phi(x)_j - \Phi(y)_j| \\ &= \sup_{j\in\mathbb{Z}} \big[\,|P^+_{q_j}\Phi(x)_j - P^+_{q_j}\Phi(y)_j| \vee |P^-_{q_j}\Phi(x)_j - P^-_{q_j}\Phi(y)_j|\,\big].\end{aligned}$$

The stable part is estimated as

$$\begin{aligned}|P^+_{q_j}\Phi(x)_j - P^+_{q_j}\Phi(y)_j| &= \big|d\varphi(q_j)P^+_{q_j}(x_j - y_j) \\ &\qquad + [P^+_{q_{j+1}} - P^+_{\varphi(q_j)}]d\varphi(q_j)(x_j - y_j)\big| \\ &\le \vartheta|x_j - y_j| + \varepsilon' K|x_j - y_j|.\end{aligned}$$

For the unstable part we get

$$\begin{aligned}|P^-_{q_j}\Phi(x)_j - P^-_{q_j}\Phi(y)_j| &= \Big|d\varphi(q_j)^{-1}P^-_{\varphi(q_j)}(x_{j+1} - y_{j+1}) \\ &\qquad + d\varphi(q_j)^{-1}[P^-_{q_{j+1}} - P^-_{\varphi(q_j)}] \\ &\qquad \cdot \big[x_{j+1} - y_{j+1} - d\varphi(q_j)(x_j - y_j)\big]\Big| \\ &\le \vartheta|x_{j+1} - y_{j+1}| + \varepsilon' K|x_j - y_j| + \varepsilon' K^2|x_j - y_j|.\end{aligned}$$

Taking the supremum over $j \in \mathbb{Z}$,

$$\|\Phi(x) - \Phi(y)\| \le (\vartheta + \varepsilon' K + \varepsilon' K^2)\|x - y\|$$

for all $x, y \in E$. If we choose $\varepsilon' > 0$ so small that $(\vartheta + \varepsilon' K + \varepsilon' K^2) =: \alpha^* < 1$, the map $\Phi\colon E \to E$ is a *contraction*. The unique fixed point $x = (x_j)_{j\in\mathbb{Z}} \in E$ of the map satisfies, in view of the equations (∗∗), (∗∗∗) and of Lemma III.3, the estimate

$$\|x\| = \|\Phi(x)\| \le \|\Phi(x) - \Phi(0)\| + \|\Phi(0)\| \le \alpha^*\|x\| + K'\|g\|,$$

with a constant $K' > 0$ and therefore,

$$\|x\| \le \frac{K'}{1-\alpha^*}\|g\|.$$

In view of $x = (\mathbb{1} - A)^{-1}g$, we have verified the estimate

$$\|(\mathbb{1} - A)^{-1}\| \leq \frac{K'}{1 - \alpha^*},$$

and Lemma III.8 is proved. □

(3) *The nonlinear problem.* Let $r > 0$. We denote the closed balls of radius r in E_j and in E by $B_j(r) := \{x_j \in E_j \mid |x_j| \leq r\}$ and by $B(r) := \{x \in E \mid \|x\| \leq r\}$. We want to solve the equations

$$x_{j+1} - A_j x_j = f_j(x_j),$$

for a sequence $x = (x_j)_{j \in \mathbb{Z}}$ satisfying $x_j \in E_j$, while the sequence of maps $f_j \colon B_j(r) \subset E_j \to E_{j+1}$ is given. Introducing the mapping

$$F \colon B(r) \subset E \to E \quad \text{by} \quad F(x)_{j+1} = f_j(x_j),$$

our equation can be written as $(\mathbb{1} - A)x = F(x)$ or as

$$x = LF(x), \quad x \in B(r),$$

with the continuous linear map $L = (\mathbb{1} - A)^{-1}$. In the following, we write $|\cdot|$ instead of $\|\cdot\|$ for the norm on E and reserve the notation $\|\cdot\|$ for the operator norm.

Lemma III.9. *Let $F \colon B(r) \subset E \to E$ be a map. Assume that the real number $\alpha > 0$ is so small that $\alpha\|L\| \leq 1/2$. If $|F(0)| \leq \alpha r$ and $|F(x) - F(y)| \leq \alpha|x - y|$ for all $x, y \in B(r)$, then the equation $x = LF(x)$ has a* ***unique*** *solution $x \in B(r)$. This solution satisfies the estimate*

$$|x| \leq 2\|L\|\,|F(0)|.$$

Proof. Set $G(x) := LF(x)$. We claim that

(i) $G \colon B(r) \to B(r)$, and
(ii) $|G(x) - G(y)| \leq \frac{1}{2}|x - y|$ for all $x, y \in B(r)$.

In order to prove the claim we take $x, y \in B(r)$ and estimate, using the assumptions,

$$|G(x) - G(y)| \leq \|L\|\,|F(x) - F(y)| \leq \alpha\|L\|\,|x - y| \leq \frac{1}{2}|x - y|.$$

Observing that $|G(0)| = |LF(0)| \leq \|L\|\,|F(0)| \leq \|L\|\alpha r \leq r/2$, we obtain

$$|G(x)| \leq |G(x) - G(0)| + |G(0)| \leq \frac{1}{2}|x| + \frac{r}{2} \leq r,$$

and the claim is proved. Since the metric space $B(r)$ is *complete*, there exists a unique fixed point $x = G(x)$ satisfying $|x| \leq r$ and due to $|x| = |G(x)| \leq |G(x) - G(0)| + |G(0)| \leq \frac{1}{2}|x| + |G(0)|$, we arrive at the desired estimate

$$|x| \leq 2|G(0)| \leq 2\|L\|\,|F(0)|.$$

This concludes the proof of Lemma III.9. □

Finally, we apply the lemma to our situation and complete the proof of the shadowing lemma. We recall that

$$|F(0)| = \sup_j |f_j(0)| = \sup_j |\varphi(q_j) - q_{j+1}| \leq \varepsilon.$$

We choose α so small that $\alpha\|L\| \leq \frac{1}{2}$. Since Λ is compact and $df_j(0) = 0$, we find a radius $r_0 = \delta_0$ such that $\|df_j(x_j)\| \leq \alpha$ for every $x_j \in B_j(r_0)$ and all $j \in \mathbb{Z}$. By the mean value theorem we conclude $|F(x) - F(y)| \leq \alpha|x - y|$ for $x, y \in B(r_0)$. If now $r \leq \delta_0$ and if $\varepsilon \leq \alpha r$, we conclude from Lemma III.9 that the statement of the shadowing lemma holds true with the constant $\delta = r$. This completes the proof of Theorem III.7. □

Proof of the remark following the shadowing lemma. Let $|\varphi - \psi|_{C^1(U)} \leq \mu$ where U is a neighborhood of Λ. Replacing the maps $f_j(x_j)$ in the above proof by the maps

$$f_j'(x_j) = \psi(q_j - x_j) - q_{j+1} - d\psi(q_j)x_j,$$

we can argue as above, if μ is sufficiently small. As for the second part of the remark, we choose $\eta > 0$ such that the $\hat{\varepsilon}$-pseudo orbit $q = (q_j)_{j\in\mathbb{Z}}$ lies in the η-neighborhood of Λ. Choosing a sequence q' *on* Λ satisfying $|q_j - q_j'| \leq \eta$ for all $j \in \mathbb{Z}$, it follows that

$$\begin{aligned}
|q_{j+1}' - \varphi(q_j')| &\leq |q_{j+1}' - q_{j+1}| + |q_{j+1} - \varphi(q_j)| + |\varphi(q_j) - \varphi(q_j')| \\
&\leq \eta + \hat{\varepsilon} + \eta \sup_{x\in\Lambda} \|d\varphi(x)\| \\
&=: \varepsilon,
\end{aligned}$$

so that q' is an ε-pseudo orbit *on* Λ. If $\eta, \hat{\varepsilon}$ are sufficiently small, we can apply the first part of the theorem to the pseudo orbit $q' \subset \Lambda$ to obtain a $(\delta + \eta)$-shadowing orbit for the pseudo orbit q. □

As a first application of the shadowing lemma, we shall prove the closing lemma of Anosov.

Theorem III.10 (Closing lemma of Anosov). *We consider the hyperbolic set Λ of the diffeomorphism φ and let ε, δ be as in the shadowing lemma. If there exist a point $x \in \Lambda$ and an integer $N \geq 1$ satisfying*

$$|\varphi^N(x) - x| \leq \varepsilon,$$

then there exists a point y *in a* δ *neighborhood* $U_\delta(\Lambda)$ *of* Λ *satisfying*

$$\varphi^N(y) = y.$$

Moreover, the periodic orbit $y, \varphi(y), \dots, \varphi^N(y) = y$ *lies in a* δ*-neighborhood of the set* $\{x, \varphi(x), \dots, \varphi^N(x)\}$.

Proof [Uniqueness of the δ-shadowing orbit]. We define the ε-pseudo orbit $q = (q_j)_{j\in\mathbb{Z}}$ by the N-periodic continuation of the finite piece of the orbit

$$\begin{array}{ccccc} x & \varphi(x) & \varphi^2(x) & \dots & \varphi^{N-1}(x) \\ \| & \| & \| & & \| \\ q_0 & q_1 & q_2 & \dots & q_{N-1}, \end{array}$$

so that $q_{j+N} = q_j$ for all $j \in \mathbb{Z}$. By the shadowing lemma there exists a unique δ-shadowing orbit $p = (p_j)_{j\in\mathbb{Z}}$ of the pseudo orbit q and we claim that

$$p_{j+N} = p_j, \quad j \in \mathbb{Z}.$$

To prove the claim, we introduce the shifted orbit sequence $\hat{p} = (\hat{p}_j)_{j\in\mathbb{Z}}$ by $\hat{p}_j = p_{j+N}$. Then, also $\hat{p}$ is a δ-shadowing orbit of the pseudo orbit q, since

$$|\hat{p}_j - q_j| = |p_{j+N} - q_j| = |p_{j+N} - q_{j+N}| \le \delta$$

holds true for all $j \in \mathbb{Z}$. From the uniqueness of the δ-shadowing orbit which shadows the pseudo orbit q, we conclude that $\hat{p} = p$, so that the orbit p is indeed the desired periodic orbit, as claimed in the theorem. □

III.3 Orbit structure near a homoclinic orbit, chaos

In the following we consider a transversal homoclinic point ν at which, by definition, the stable and unstable invariant manifolds issuing form a hyperbolic fixed point of the diffeomorphism φ intersect transversally. Assuming as before the fixed point to be the origin 0 we denote by

$$\Lambda = \overline{\mathcal{O}(\nu)} = \bigcup_{j\in\mathbb{Z}} \varphi^j(\nu) \cup \{0\} = \mathcal{O}(\nu) \cup \mathcal{O}(0)$$

the closure of the homoclinic orbit which consists of two orbits. The compact set Λ is a hyperbolic set of the diffeomorphism φ in view of Proposition III.5 and so we can use the shadowing lemma in order to prove first that the homoclinic point ν is a *cluster point* of other *homoclinic points* belonging to 0 and at the same time also a cluster point of *periodic points*.

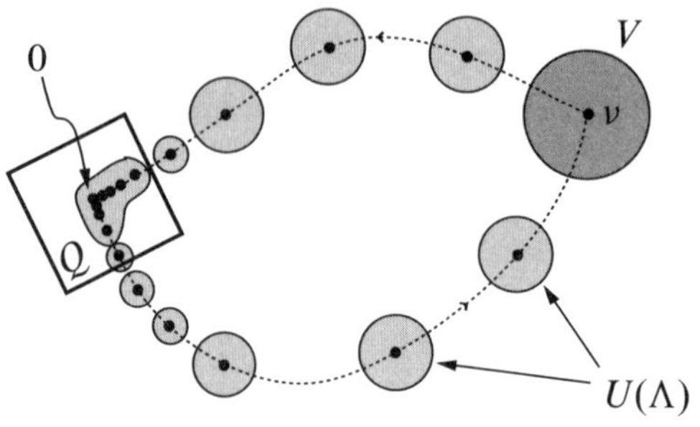

Figure III.5. Homoclinic orbit with neighborhood

Theorem III.11 (G. Birkhoff). *We assume that ν is a transversal homoclinic point belonging to the hyperbolic fixed point 0 of the diffeomorphism φ. Let V be an open neighborhood of ν and let $U = U(\Lambda)$ be an open neighborhood of $\Lambda = \overline{\mathcal{O}(\nu)}$.*

Then, there exist infinitely many periodic points in V whose orbits are contained in the open set U. More precisely, there exists an integer $N_0 = N_0(U, V)$ such that for every integer $N \geq N_0$ there exists a periodic point $p \in V$ having the minimal period N.

Proof [Shadowing lemma]. The hyperbolic set $\Lambda = \mathcal{O}(\nu) \cup \mathcal{O}(0)$ consists of two orbits of φ. If ε and δ_0 are as in the shadowing lemma we choose $0 < \delta \leq \delta_0$ so small that the δ neighborhood of ν lies in V and the δ neighborhood of Λ in the prescribed open set $U(\Lambda)$. We denote by Q the ε neighborhood of the fixed point 0. According to the definition of a homoclinic point ν there exists an integer j_0 such that

$$\varphi^j(\nu) \in Q, \quad |j| \geq j_0.$$

The crucial observation is now the following. Inside of Q it is possible to *jump* from the homoclinic orbit to the fixed point orbit of 0 and back to the homoclinic orbit by committing only an error smaller than $\leq \varepsilon$ as illustrated in Figure III.7.

We use this to construct the ε-pseudo orbits $q = (q_j)_{j \in \mathbb{Z}} \subset \Lambda$ *on* the hyperbolic set Λ having a prescribed *minimal* period, as follows.

$$\begin{array}{ccccccccc} \nu & \varphi(\nu) & \dots & \varphi^{j_0}(\nu) & 0 & \dots & 0 & \varphi^{-j_0}(\nu) & \dots & \varphi^{-1}(\nu) \\ \| & \| & & \| & \| & & \| & \| & & \| \\ q_0 & q_1 & \dots & q_{j_0} & q_{j_0+1} & \dots & q_{j_0+k} & q_{j_0+k+1} & \dots & q_{2j_0+k}\,, \end{array}$$

where the hyperbolic fixed point 0 is visited k-times in succession ($k \geq 1$) and where the scheme is repeated periodically. The integer k can be chosen arbitrarily and determines the minimal period of the pseudo orbit q. The sequence q is a

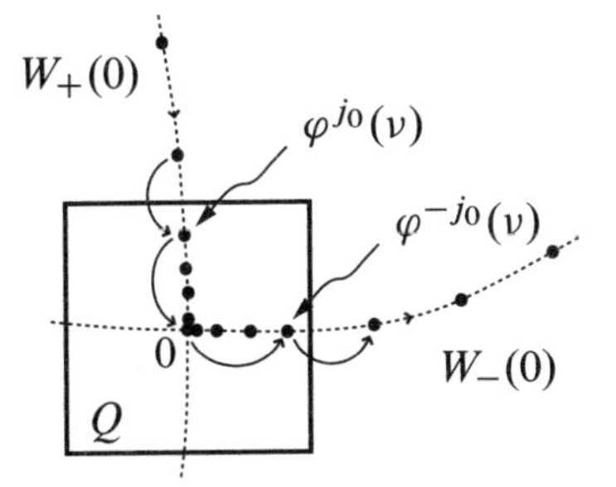

Figure III.6. Jump of the ε-pseudo orbit from $W_+(0)$ onto 0 and then into $W_-(0)$.

periodic ε-pseudo orbit on Λ, namely,

$$q_{j+N} = q_j, \quad j \in \mathbb{Z},$$

where $N = 2j_0 + k + 1$.

In view of the shadowing lemma the pseudo orbit q is shadowed by the unique δ-shadowing orbit $p = (p_j)_{j\in\mathbb{Z}}$ given by $p_j = \varphi^j(p_0)$. By construction, $p_0 \in V$ and $\mathcal{O}(p_0) \subset U(\Lambda)$. Since also the shifted sequence $\hat{p} = (\hat{p}_j)_{j\in\mathbb{Z}}$, defined by $\hat{p}_j = p_{j+N}$, is a δ-shadowing orbit of the pseudo orbit q, it follows from the *uniqueness* that $\hat{p} = p$, hence

$$p_{j+N} = p_j, \quad j \in \mathbb{Z}.$$

Therefore, the shadowing orbit p is a *periodic* orbit of φ having the minimal period N. The theorem follows if we set $N_0 = 2j_0 + 2$.

□

The next result goes back to H. Poincaré. It explains why the transversal homoclinic point forces the invariant manifolds $W_+(0)$ and $W_-(0)$ issuing from the hyperbolic fixed point to double back and pile up on themselves as illustrated in Figure III.7.

Theorem III.12 (H. Poincaré). *We assume that ν is a transversal homoclinic point associated with the hyperbolic fixed point 0 of the diffeomorphism φ. Let V be an open neighborhood of ν and let $U = U(\Lambda)$ be an open neighborhood of $\Lambda = \overline{\mathcal{O}(\nu)}$. Then there exist infinitely many homoclinic points associated with 0 in V, whose orbits run in U, and which are distinguished by two* **rotation numbers** *$r^\pm$.*

Proof [Shadowing lemma]. We again construct a suitable ε-pseudo orbit on Λ using the same notation as in the previous proof and let $r^\pm \in \mathbb{N}_0$ be two integers. Set

$$\begin{matrix} \nu & \varphi(\nu) & \dots & \varphi^{j_0}(\nu) & 0 & \varphi^{-j_0}(\nu) & \dots & \varphi^{-1}(\nu) \\ \| & \| & & \| & \| & \| & & \| \\ q_0 & q_1 & \dots & q_{j_0} & q_{j_0+1} & q_{j_0+2} & \dots & q_{2j_0+1}\,. \end{matrix}$$

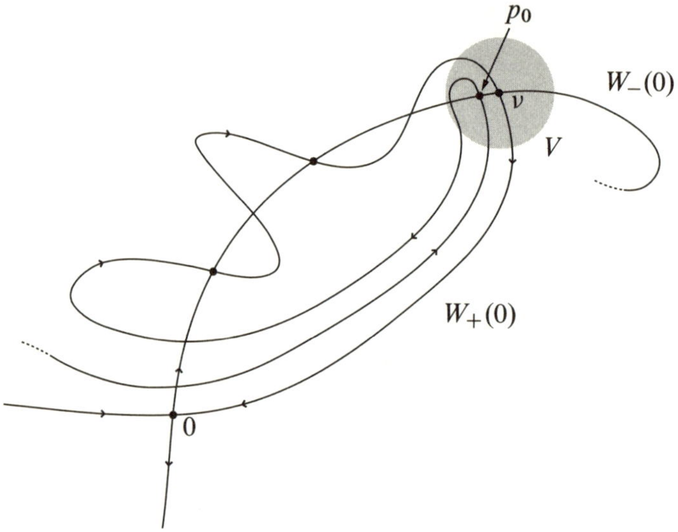

Figure III.7. One of the infinitely many homoclinic points p_0 near ν.

We repeat this finite sequence on the right r^+-times and on the left r^--times, then we add on the left $\varphi^{-j_0}(\nu), \dots, \varphi^{-1}(\nu)$ and on the right $\nu, \varphi(\nu), \dots, \varphi^{j_0}(\nu)$. Finally, we add on the left respectively on the right the infinite sequences $\dots, 0, 0, 0$ resp. $0, 0, 0, \dots$, which belong to the orbit of the fixed point 0. Hence, after choosing the integer j_0 and the open neighborhood Q as in the proof of the previous theorem, we have constructed an ε-pseudo orbit q on the hyperbolic set Λ, for which it holds true that

$$q_j = 0, \quad |j| \geq M,$$

for a suitable constant M.

This pseudo orbit is shadowed by the unique δ-shadowing orbit $p = (p_j)_{j \in \mathbb{Z}}$ which satisfies, by construction, $\mathcal{O}(p_0) \subset U$ and $p_0 \in V$ and, *in addition*,

$$|p_j| \leq \delta, \quad |j| \geq M.$$

Therefore, for $|j|$ large, all the orbit points lie in the δ neighborhood Q_δ of the hyperbolic fixed point 0. Hence, they lie on the *local manifolds* $W^+_{\text{loc}}(Q_\delta)$ respectively $W^-_{\text{loc}}(Q_\delta)$, introduced in II.2. Due to Theorem II.7, $\varphi^j(p_0) \to 0$ as $|j| \to \infty$, if we choose δ sufficiently small. Therefore, p_0 is a *homoclinic* point in V, and, by construction, $p_0 \neq \nu$. Shadowing orbits belonging to different rotation numbers are different from each other. The proof of the theorem is complete. □

More generally, we can consider two hyperbolic fixed points $x^* \neq y^*$ of the diffeomorphism φ. If the invariant manifolds intersect *transversally* in the heteroclinic

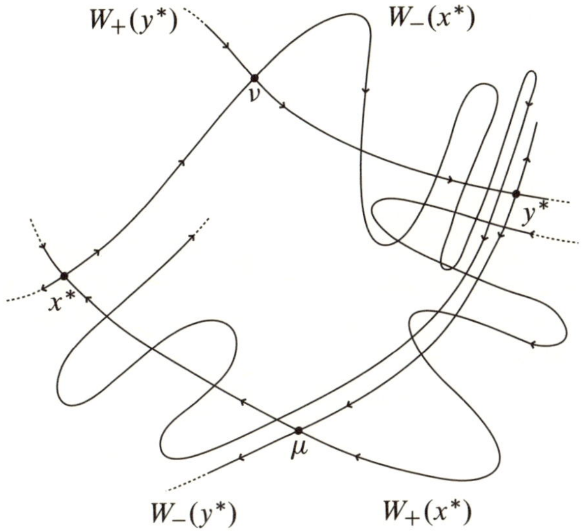

Figure III.8. Transversal heteroclinic points ν, μ belonging to the hyperbolic fixed points x^*, y^*.

points

$$\nu \in W_-(x^*) \cap W_+(y^*),$$
$$\mu \in W_+(x^*) \cap W_-(y^*),$$

as illustrated in Figure III.9, then $\varphi^j(\nu) \to y^*$ as $j \to \infty$ and $\varphi^j(\nu) \to x^*$ as $j \to -\infty$; and analogously for the intersection point μ. The set

$$\Lambda := \mathcal{O}(\nu) \cup \mathcal{O}(\mu) \cup \{x^*\} \cup \{y^*\},$$

consisting of four orbits is a *hyperbolic set*. Let $U(\Lambda)$ an open neighborhood of Λ. Then ν is a *cluster point* of

- heteroclinic points belonging to x^* and y^*,
- homoclinic points belonging to x^* and homoclinic points belonging to y^*,
- periodic points,

whose orbits *all* run in $U(\Lambda)$. The proof of this statement is left to the reader (one constructs suitable ε-pseudo orbits and then applies the shadowing lemma).

We are going to demonstrate that the *complexity* of the orbit structure near a homoclinic orbit can be described *statistically* by the embedding of *Bernoulli systems*. Thus, we shall obtain orbits that are determined by *random sequences*. Introducing the *finite alphabet*

$$A = \{1, 2, \dots, a\}, \quad a \geq 2,$$

the *space of the two-sided sequences of symbols* from the alphabet is the metric space

$$\Sigma_A = \{s = (s_j)_{j\in\mathbb{Z}} \mid s_j \in A\}$$

equipped with the metric

$$d(s,t) = \sum_{j\in\mathbb{Z}} \frac{1}{2^{|j|}} \frac{|s_j - t_j|}{1 + |s_j - t_j|}, \quad s,t \in \Sigma_A.$$

The metric has the following *significance.* Two sequences $s,t \in \Sigma_A$ are close, if they agree over a long, central string: $s_j = t_j$ for all $|j| \le N$ with N large. More precisely, the following lemma applies.

Lemma III.13. *If $s,t \in \Sigma_A$, then*

(i) $d(s,t) < \frac{1}{2^{N+1}} \Longrightarrow s_j = t_j,\ |j| \le N,$

(ii) $s_j = t_j, \quad |j| \le N \Longrightarrow d(s,t) \le \frac{1}{2^{N-1}}.$

Proof. (i) Assume $s_j \neq t_j$ for an integer $|j| \le N$, then $\frac{|s_j - t_j|}{1+|s_j-t_j|} \ge \frac{1}{2}$ and therefore

$$d(s,t) \ge \frac{1}{2^{|j|}} \frac{|s_j - t_j|}{1 + |s_j - t_j|} \ge \frac{1}{2^{|j|+1}} \ge \frac{1}{2^{N+1}}.$$

(ii) If $s_j = t_j$ for $|j| \le N$ then, due to $\frac{|s_j - t_j|}{1+|s_j-t_j|} \le 1$,

$$d(s,t) = \sum_{|j|>N} \frac{1}{2^{|j|}} \frac{|s_j - t_j|}{1 + |s_j - t_j|} \le 2 \sum_{j=N+1}^{\infty} \frac{1}{2^j} = \frac{1}{2^{N-1}},$$

as claimed in the lemma □

Lemma III.14 (Properties of (Σ_A, d)). *The metric space (Σ_A, d) is* ***compact*** *and* ***perfect****, i.e., every point is a cluster point.*

Proof [Finiteness of the alphabet]. If $(s^n)_{n\ge1}$ is a sequence in Σ_A, hence $s^n = (s^n_j)_{j\in\mathbb{Z}}$, we shall construct a convergent subsequence in the metric space (Σ_A, d). Since the alphabet is *finite*, there exists for every index $j \in \mathbb{Z}$ a symbol $a_j \in A$ such that $s^n_j = a_j$ for infinitely many n. We now choose a subsequence of s^n whose sequences all have the value a_0 at the index 0. From this subsequence, we choose again a subsequence whose sequences have the values a_{-1} and a_1 at the indices -1 and 1. Iterating the procedure, we finally arrive at a subsequence which, in view of Lemma III.13, converges to the sequence $a := (a_j)_{j\in\mathbb{Z}}$ in Σ_A. Hence, Σ_A is compact.

If $s \in \Sigma_A$ we choose a symbol $a \in A$ such that $s \neq (\dots, a, a, a, \dots)$ and define the sequence s^n by $s^n_j = s_j$ for $|j| \le n$ and $s^j_n = a$ for $|j| > n$. Then $s^n \to s$ and (s^n) is not constant. Therefore, Σ_A is perfect. □

Definition. If (M, d) is a metric space, the mapping $\varphi\colon M \to M$ is called *expansive*, if there exists a universal constant $\alpha > 0$ such that for all $x \neq y$ in M there exists an integer $N \geq 0$ for which $d(\varphi^N(x), \varphi^N(y)) \geq \alpha$.

In case that φ is bijective, one merely requires the existence of an integer $N \in \mathbb{Z}$ having the above property.

If the mapping is expansive, the iterates of two different points visibly separate from each other in the course of time, regardless of how close to each other they start. The dynamical system (M, φ) therefore shows a *sensitive dependence* on the initial conditions.

Equivalently, the map is expansive if there exists a constant $\alpha > 0$, having the property

$$d(\varphi^j(x), \varphi^j(y)) < \alpha \text{ for all } j \Longrightarrow x = y.$$

Hence, if two orbits stay close for all times, then they must be identical.

Proposition III.15. *The dynamical system (Λ, φ) on a* hyperbolic *set Λ of a diffeomorphism φ is expansive.*

Proof [Shadowing lemma]. We assume that δ_0 and $\varepsilon_0 = \varepsilon_0(\delta_0)$ are as in the shadowing lemma and let $p = (p_j)_{j\in\mathbb{Z}}$ and $q = (q_j)_{j\in\mathbb{Z}}$ be two orbits on Λ satisfying

$$|p_j - q_j| = d(\varphi^j(p_0), \varphi^j(q_0)) \leq \delta_0, \quad j \in \mathbb{Z}.$$

Then q is an ε-pseudo orbit (with $\varepsilon = 0$) which is shadowed by the orbit p. Since the shadowing orbit is *unique* and since q is also an orbit, we conclude that $p = q$. □

Definition. The *shift map* σ on the space Σ_A is the mapping $\sigma\colon \Sigma_A \to \Sigma_A$, defined by

$$s \mapsto \sigma(s) = (\sigma(s)_j)_{j\in\mathbb{Z}}, \quad \sigma(s)_j := s_{j+1}.$$

The dynamical system (Σ_A, σ) is called a *Bernoulli system*.

Lemma III.16 (Properties of (Σ_A, σ)). *The Bernoulli system (Σ_A, σ) has the following properties.*

(i) *$\sigma\colon \Sigma_A \to \Sigma_A$ is a homeomorphism.*

(ii) *There exists a countable and* ***dense*** *set of* ***periodic*** *points of σ. All periods exist.*

(iii) *The system is* ***transitive***.

(iv) *The system is* ***expansive***.

(v) *If s, t are two periodic points, then the set of points $r \in \Sigma_A$ satisfying $\sigma^j(r) \to \mathcal{O}(s)$ as $j \to +\infty$ and $\sigma^j(r) \to \mathcal{O}(t)$ as $j \to -\infty$ is* ***dense*** *in Σ_A. These points are therefore heteroclinic to the orbits $\mathcal{O}(s)$ and $\mathcal{O}(t)$.*

Proof. (i) The bijectivity is obvious, the continuity of σ (and hence of σ^{-1}) immediately follows from the definition of the metric.

(ii) The periodic points are precisely the periodic sequences in Σ_A. Obviously, the set of such sequences is countable and dense.

(iii) We construct a dense orbit by taking a symbol sequence which sticks together *all* the possible symbol sequences of finite length $1, 2, 3, \ldots$. According to Lemma III.13, the iterates of this sequence come arbitrarily close to every element of Σ_A proving that the system is transitive.

(iv) We define for $b \in A$ the set

$$\Sigma_A^b = \{s \in \Sigma_A \mid s_0 = b\}.$$

Then

$$\alpha := \inf_{b \neq c} d(\Sigma_A^b, \Sigma_A^c) \geq \frac{1}{2}.$$

If $s \neq t$, then $s_j \neq t_j$ for a some integer $j \in \mathbb{Z}$. Due to $\sigma^j(s)_0 = s_j$, and similarly for t, we have

$$\sigma^j(s) \in \Sigma_A^{s_j}, \quad \text{and} \quad \sigma^j(t) \in \Sigma_A^{t_j},$$

so that $d(\sigma^j(s), \sigma^j(t)) \geq d(\Sigma_A^{s_j}, \Sigma_A^{t_j}) \geq \alpha$.

(v) We assume that s, t are periodic points, then the sequences (s_j), (t_j) are periodic. Let m and n be the periods of the sequences (s_j) and (t_j), and set

$$S := (s_0, \ldots, s_m), \quad T := (t_0, \ldots, t_n),$$

then every symbol sequence

$$r = (\ldots, T, T, X, S, S, \ldots)$$

having an arbitrarily chosen finite central block X converges under the iterates of the shift map to $\mathcal{O}(s)$ and under the iterates of the inverse shift map to $\mathcal{O}(t)$. According to Lemma III.13 these sequences are dense in Σ_A. □

We come to the central result of this chapter.

Theorem III.17 (S. Smale). *Let ν be a transversal homoclinic point of the diffeomorphism φ belonging to the hyperbolic fixed point 0. Let U be an open neighborhood of the closure of the homoclinic orbit $\Lambda = \overline{\mathcal{O}(\nu)} = \mathcal{O}(\nu) \cup \mathcal{O}(0)$ and let A be a finite alphabet. Then there exists an integer $K \geq 1$ and a homeomorphism*

$$\psi \colon \Sigma_A \to \psi(\Sigma_A) =: M \subset U$$

having the following properties.

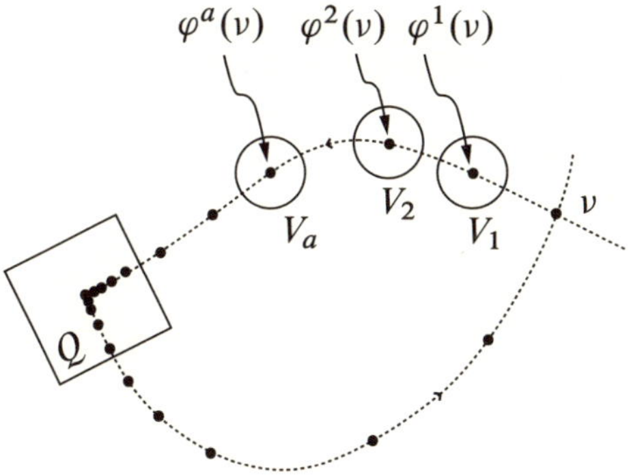

Figure III.9. The geometric construction in the proof of Theorem III.17.

(i) *The compact set* M *is invariant under* φ^K, *so that* $\varphi^K(M) = M$.
(ii) *For every point* $m \in M$ *the orbit* $\mathcal{O}(m) \subset U$ *under* φ *lies in the open set* U.
(iii) $\psi \circ \sigma = \varphi^K \circ \psi$, *so that the following diagram is commutative.*

$$\begin{array}{ccc} M & \xrightarrow{\varphi^K} & M \\ \uparrow \psi & & \uparrow \psi \\ \Sigma_A & \xrightarrow{\sigma} & \Sigma_A \end{array}$$

The theorem guarantees a *continuous embedding* of the Bernoulli system (Σ_A, σ) into $\mathbb{R}^n$ as a *subsystem* of the dynamical system $(\mathbb{R}^n, \varphi^K)$.

Proof [Proposition III.5, Theorem III.7]. Since Σ_A is compact, every injective and continuous mapping $\psi: \Sigma_A \to \psi(\Sigma_A) \subset V$ is a homeomorphism onto its image.

(1) *Strategy of the geometric construction.* If $A = \{1, 2, \ldots, a\}$ is the alphabet, we choose *open neighborhoods* V_j of the finitely many homoclinic points $\varphi^j(\nu)$ for $1 \leq j \leq a$, satisfying

$$\overline{V_i} \cap \overline{V_j} = \emptyset, \quad i \neq j,$$

and set $V = \bigcup_{1 \leq j \leq a} V_j$. We shall construct for every sequence $s = (s_j)_{j \in \mathbb{Z}} \in \Sigma_A$ a point $p_0 = \psi(s) \in V$ having the following property. If $\Phi := \varphi^K$ is the iterated map for a suitable integer $K \geq 1$, then,

$$\Phi^j(p_0) \in V_{s_j} \quad \text{for all } j \in \mathbb{Z}.$$

Thus for *every random sequence* $s \in \Sigma_A$ there exists an orbit $(\Phi^j(p_0))_{j \in \mathbb{Z}}$ of Φ satisfying $\Phi^j(p_0) \in V_{s_j}$. Hence this orbit visits all *randomly* chosen sets V_b where $b \in A$.

(2) *Construction of the ε-pseudo orbit.* We choose $\delta > 0$ so small that the δ neighborhood of Λ is contained in U and the δ neighborhood of the homoclinic point $\varphi^j(v)$ is contained in V_j for every $j \in A$. Let $\varepsilon > 0$ be the real number ε which corresponds to δ in the shadowing lemma. Since v is *a homoclinic point*, there *exists* an ε neighborhood Q of 0 and an integer $N \geq 1$ such that

$$\varphi^{s_j+N-1}(v) \in Q \quad \text{and} \quad \varphi^{s_j-N+1}(v) \in Q \quad \text{for all } s_j \in A.$$

For $s \in \Sigma_A$ the ε-pseudo orbit $q(s) = (q(s)_j)_{j\in\mathbb{Z}} \subset \Lambda$ is constructed in the following way. If $s = (\dots, s_{-1}, s_0, s_1, \dots) \in \Sigma_A$ is given, the points $q(s)_j$ are defined by identifying the points of the first scheme below with the corresponding points of the second scheme

$$\begin{array}{cccccccc}
\vdots & & \vdots & \vdots & \vdots & & \vdots & \vdots \\
q_{-N+1} & \cdots & q_{-1} & q_0 & q_1 & \cdots & q_{N-1} & q_N \\
q_{N+1} & \cdots & q_{2N-1} & q_{2N} & q_{2N+1} & \cdots & q_{3N-1} & q_{3N} \\
q_{3N+1} & \cdots & q_{4N-1} & q_{4N} & q_{4N+1} & \cdots & q_{5N-1} & q_{5N} \\
\vdots & & \vdots & \vdots & \vdots & & \vdots & \vdots
\end{array}$$

and

$$\begin{array}{cccccccc}
\vdots & & \vdots & \vdots & \vdots & & \vdots & \vdots \\
\varphi^{s_0-N+1} & \cdots & \varphi^{s_0-1} & \varphi^{s_0} & \varphi^{s_0+1} & \cdots & \varphi^{s_0+N-1} & 0 \\
\varphi^{s_1-N+1} & \cdots & \varphi^{s_1-1} & \varphi^{s_1} & \varphi^{s_1+1} & \cdots & \varphi^{s_1+N-1} & 0 \\
\varphi^{s_2-N+1} & \cdots & \varphi^{s_2-1} & \varphi^{s_2} & \varphi^{s_2+1} & \cdots & \varphi^{s_2+N-1} & 0 \\
\vdots & & \vdots & \vdots & \vdots & & \vdots & \vdots
\end{array}$$

where, to save space we have abbreviated $q_j = q(s)_j$ and $\varphi^j = \varphi^j(v)$. According to our construction the sequence $q(s)$ is indeed an ε-pseudo orbit on the hyperbolic set Λ. Moreover, it has the following crucial properties,

$$\begin{aligned}
q(s)_{j+2N} &= q(\sigma(s))_j, \qquad && j \in \mathbb{Z}, \\
q(s)_{j2N} &= \varphi^{s_j}(v) \in V_{s_j}, && j \in \mathbb{Z}.
\end{aligned}$$

(3) *Definition of the map* $\psi\colon \Sigma_A \to \mathbb{R}^n$. In view of the shadowing lemma, there exists an orbit $p(s)$ in the neighborhood U which is a δ shadowing orbit of the ε-pseudo orbit $q(s)$, so that $p(s)_j = \varphi^j(p(s)_0)$ and $|p(s)_j - q(s)_j| \leq \delta$ for all $j \in \mathbb{Z}$. From $q(s)_0 = \varphi^{s_0}(v) \in V_{s_0}$, we conclude that the point $p(s)_0$ lies in V_{s_0}.

We now define the mapping ψ by setting $\psi(s) = p(s)_0$ for $s \in \Sigma_A$ and introduce the diffeomorphism $\Phi := \varphi^{2N}$ (hence $K = 2N$). We claim that

$$\Phi^j(\psi(s)) \in V_{s_j}, \quad j \in \mathbb{Z}.$$

Indeed, due to our construction, $|p(s)_{j2N} - q(s)_{j2N}| \leq \delta$ and we conclude from $q(s)_{j2N} = \varphi^{s_j}(v) \in V_{s_j}$ that the point $p(s)_{j2N}$ lies in V_{s_j}, so that

$$p(s)_{j2N} = \varphi^{j2N}(p(s)_0) = \Phi^j(\psi(s)) \in V_{s_j}, \quad j \in \mathbb{Z}.$$

(4) *In order to prove the equation* $\psi \circ \sigma = \Phi \circ \psi$ we first fix $s \in \Sigma_A$ and abbreviate $p_0 := p(s)_0 = \psi(s)$, so that $|\varphi^j(p_0) - q(s)_j| \leq \delta$ for all $j \in \mathbb{Z}$. In particular, $|\varphi^{j+2N}(p_0) - q(s)_{j+2N}| \leq \delta$ and recalling $q(s)_{j+2N} = q(\sigma(s))_j$ and $\Phi = \varphi^{2N}$, we obtain the estimates

$$|\varphi^j(\Phi(p_0)) - q(\sigma(s))_j| \leq \delta, \quad j \in \mathbb{Z}.$$

Hence, $(\varphi^j(\Phi(p_0)))_{j \in \mathbb{Z}}$ is a δ-shadowing orbit of the ε-pseudo orbit $q(\sigma(s))$. According to the definition of the map ψ we have

$$|\varphi^j(\psi(\sigma(s))) - q(\sigma(s))_j| \leq \delta, \quad j \in \mathbb{Z},$$

so that $(\varphi^j(\psi(\sigma(s))))_{j \in \mathbb{Z}}$ is also a δ-shadowing orbit of the pseudo orbit $q(\sigma(s))$. From the uniqueness of the δ-shadowing orbit we conclude that

$$\varphi^j(\Phi(p_0)) = \varphi^j(\psi(\sigma(s))), \quad j \in \mathbb{Z}.$$

In particular, setting $j = 0$ and recalling $p_0 = \psi(s)$ we obtain the equation

$$\Phi(\psi(s)) = \psi(\sigma(s)).$$

This equation holds true for every $s \in \Sigma_A$, as we wanted to prove.

(5) *In order to verify the injectivity of the map* ψ, we take two elements $s \neq s'$ in Σ_A and set $p_0 := \psi(s)$ and $p_0' := \psi(s')$, so that, according to our construction,

$$\Phi^j(p_0) \in V_{s_j}, \quad \Phi^j(p_0') \in V_{s_j'}.$$

There exists an integer $j \in \mathbb{Z}$ for which $s_j \neq s_j'$ and hence $V_{s_j} \cap V_{s_j'} = \emptyset$. Consequently, $\Phi^j(p_0) \neq \Phi^j(p_0')$ and since the map Φ can be inverted, we conclude $p_0 \neq p_0'$, and so $\psi(s) \neq \psi(s')$.

(6) In order to show that the mapping $\psi\colon \Sigma_A \to V$ *is continuous* we take the convergent sequence $s^{(n)}$ in Σ_A satisfying $s^{(n)} \to s$ in (Σ_A, d) and show that $\psi(s^{(n)}) \to \psi(s)$ in $\mathbb{R}^n$. Arguing by contradiction we find a subsequence $\psi(s^{(n)})$ such that

$$|\psi(s^{(n)}) - \psi(s)| \geq \varepsilon^* > 0$$

for all n. Since $\overline{V}$ is bounded, there exists a convergent subsequence again denoted by $\psi(s^{(n)})$. Denote its limit by ξ, then $\xi \in \overline{V}$ and $|\xi - \psi(s)| \geq \varepsilon^*$. We claim that $(\varphi^j(\xi))_{j \in \mathbb{Z}}$ is a δ-shadowing orbit of the pseudo orbit $q(s)$. In view of the uniqueness of the shadowing orbit, it then follows that $\xi = \psi(s)$ which is the desired contradiction. Fixing j we obtain by the triangle inequality

$$\begin{aligned}|\varphi^j(\xi) - q(s)_j| \le |\varphi^j(\xi) - \varphi^j(\psi(s^{(n)}))| \\ + |\varphi^j(\psi(s^{(n)})) - q(s^{(n)})_j| \\ + |q(s^{(n)})_j - q(s)_j|.\end{aligned}$$

We estimate the terms on the right-hand side. The first term converges to 0 as $n \to \infty$. The second term is estimated by $\le \delta$, as we have already seen in (4) (where $s^{(n)} = s$ and $\psi(s^{(n)}) = p_0$). The third term vanishes for n large enough, in view of the convergence $s^{(n)} \to s$ in the special metric of the Bernoulli system. All in all, $|\varphi^j(\xi) - q(s)_j| \le \delta$ for every $j \in \mathbb{Z}$, as claimed.

(7) The statement (i) of the theorem follows from $\Phi(m) = \psi \circ \sigma \circ \psi^{-1}(m) \in \operatorname{Im}\psi$ if $m \in \operatorname{Im}\psi$. Finally, the points $\varphi^j(\psi(s))$ which, by construction, belong to the δ-shadowing orbit of an ε-pseudo orbit on the set Λ, necessarily lie in a δ neighborhood of Λ and so the statement (ii) holds true by construction. The proof of Theorem III.17 is complete. □

By means of the homeomorphism $\psi\colon \Sigma_A \to \psi(\Sigma_A) =: M \subset \mathbb{R}^n$ the dynamical system $(M, \Phi) := (\psi(\Sigma_A), \varphi^K)$ inherits the properties of the Bernoulli system (Σ_A, σ) listed in Lemma III.16.

Corollary III.18. *The subsystem* $(M, \Phi) := (\psi(\Sigma_A), \varphi^K)$ *introduced in Theorem* III.17 *has the following properties.*

(i) Φ *is transitive on* M.

(ii) *The periodic points of the map* Φ *are countable and dense in* M.

(iii) Φ *is an expansive map on* M.

(iv) *The heteroclinic points of* Φ *(to periodic orbits) are dense in* M.

Proof. We only have to verify the expansiveness. Introducing the positive number

$$\alpha := \min_{b \ne c \in A} d(\overline{V_b}, \overline{V_c}) > 0,$$

we shall show for $x \ne y$ in M that there exists an integer $j \in \mathbb{Z}$ such that $|\Phi^j(x) - \Phi^j(y)| \ge \alpha$. To do so, we take s, s' in Σ_A satisfying $x = \psi(s)$ and $y = \psi(s')$. Then $s_j \ne s'_j$ for some integer $j \in \mathbb{Z}$. According to the geometric construction of Theorem III.17 we know that that $\Phi^j(x) \in V_{s_j}$ and $\Phi^j(y) \in V_{s'_j}$ and so, $|\Phi^j(x) - \Phi^j(y)| \ge \alpha$ as claimed. □

III.4 Existence of transversal homoclinic points

We illustrate the chaotic behavior caused by a homoclinic point with the example of a periodically perturbed mathematical pendulum. Starting with the unperturbed situation and assuming all physical constants to be normalized, the mathematical pendulum is determined by the differential equation

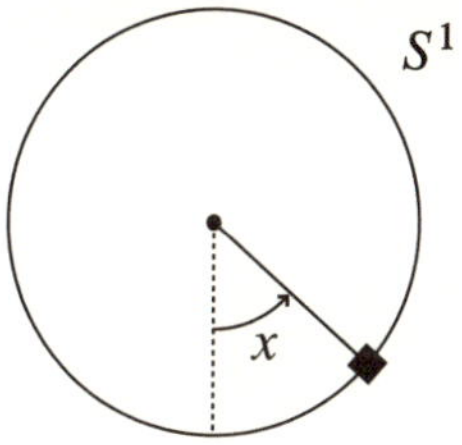

Figure III.10. The pendulum.

$$\ddot{x} + \sin x = 0$$

of second order, where $x \pmod{2\pi}$ is the angle of the swing of the pendulum. Written as an equivalent system of differential equations of first order, the pendulum is described by

$$\begin{cases} \dot{x} = y, \\ \dot{y} = -\frac{dV}{dx}(x), \quad V(x) := -\cos x. \end{cases}$$

Due to the periodicity, the phase space is equal to $S^1 \times \mathbb{R}$, it is, however, more convenient to work in the *covering space* $\mathbb{R}^2$. We write the system as a vector field in $\mathbb{R}^2$,

$$\dot{z} = X(z) \in \mathbb{R}^2, \quad z = (x, y) \in \mathbb{R}^2.$$

The flow $\varphi^t(z)$ of the vector field X is defined by the unique solutions of the Cauchy initial value problem

$$\begin{cases} \frac{d}{dt}\varphi^t(z) = X(\varphi^t(z)), \quad t \in \mathbb{R}, \\ \varphi^0(z) = z. \end{cases}$$

For fixed z the curve $t \mapsto \varphi^t(z) \in \mathbb{R}^2$ is the solution of the initial value problem having the initial conditions z at the time $t = 0$. For fixed time t the mapping $z \mapsto \varphi^t(z)$ is a diffeomorphism of $\mathbb{R}^2$.

The orbits are quickly sketched, since there exists an *integral* $H : \mathbb{R}^2 \to \mathbb{R}$ given by

$$H(x, y) = \frac{1}{2}y^2 + V(x) = \frac{1}{2}y^2 - \cos x.$$

We recall that an integral of the vector field X is a function $H\colon \mathbb{R}^2 \to \mathbb{R}$ satisfying $dH(X) = 0$. Equivalently, the flow of X leaves the function H invariant, so that $H(\varphi^t(z)) = H(z)$ for all t and z. Therefore, the orbits lie on the level lines

$$E_c := \{(x, y) \in \mathbb{R}^2 \mid H(x, y) = c\},$$

consisting of the two branches $y = \pm\sqrt{2(c + \cos x)}$. Figure III.11 shows that the mathematical pendulum possesses the following *orbit types*.

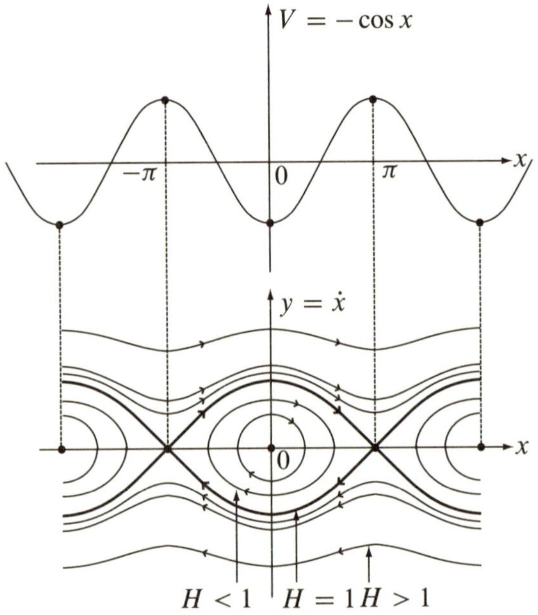

Figure III.11. Level sets of the integral H. The separatrix is marked.

Equilibrium points. The equilibrium points are, on one hand, the constant orbits in the level set $\{H = -1\}$, these are the so-called elliptic equilibrium points located in $(x, y) = (2\pi n, 0)$ (on the left figure). On the other hand, the hyperbolic equilibrium points located in $(x, y) = ((2n + 1)\pi, 0)$ are on the level set $\{H = 1\}$.

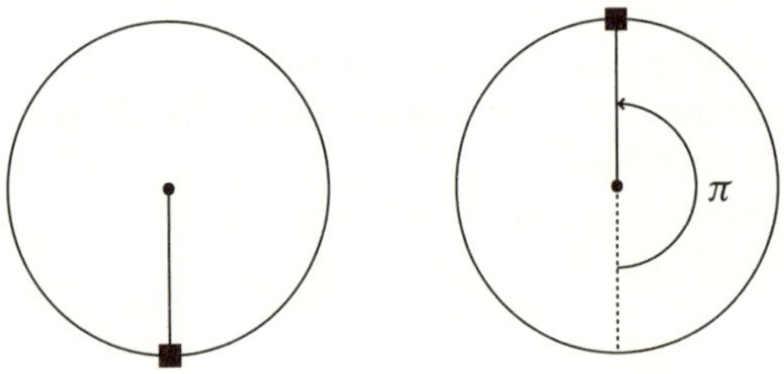

Oscillation. The oscillations around the lowest point are on the level sets $\{-1 < H < 1\}$, in Figure III.12 described by the closed curves.

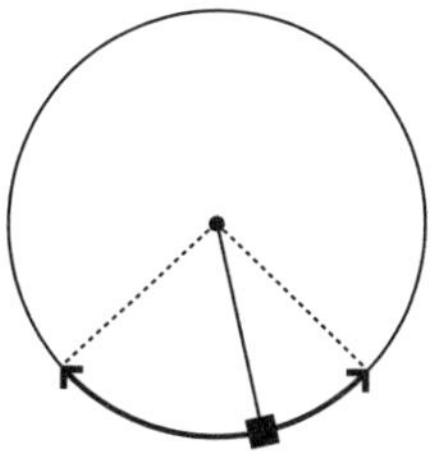

Rotations. The rotational solutions lie on the level sets $\{H > 1\}$. The angle is either strictly increasing (left) or else strictly decreasing (right).

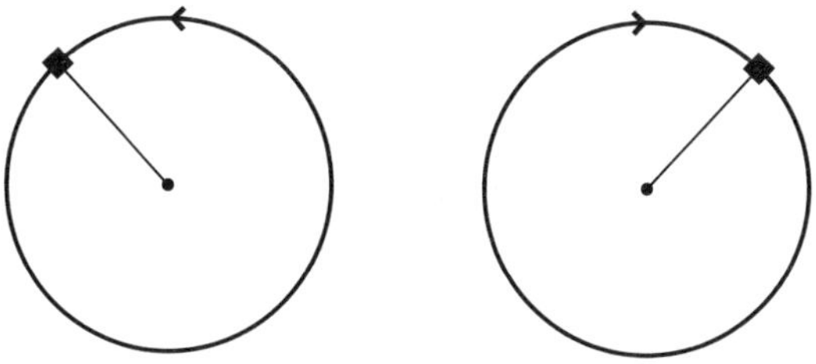

Heteroclinic orbits. The level set $\{H = 1\}$ carries the homoclinic orbits (in $S^1 \times \mathbb{R}$) and the heteroclinic orbits (in $\mathbb{R}^2$) respectively, and the hyperbolic equilibrium points. This level set is called a *separatrix* because it separates the oscillations from the rotations.

Keeping the time $T > 0$ fixed, the flow in time T,

$$\varphi^T : \mathbb{R}^2 \to \mathbb{R}^2,$$

is a diffeomorphism possessing the hyperbolic fixed points $P_n = ((2n+1)\pi, 0)$ for $n \in \mathbb{Z}$. This is easily verified using Lemma II.12. They are 2π-periodically distributed (in the projection on $S^1 \times \mathbb{R}$ they all correspond to the same point). Their stable and unstable invariant manifolds coincide in the sense that $W_+(P_n) = W_-(P_{n+1})$ for all $n \in \mathbb{Z}$. We now *perturb* the pendulum by means of a *time T-periodic excitation* and consider the equation

$$\ddot{x} + \sin x = \mu \sin \omega t, \quad T = \frac{2\pi}{\omega}.$$

The energy function H is *no longer* an integral of the system and the orbit structure changes drastically. The new vector field

$$\dot{z} = X(t, \mu, z) \in \mathbb{R}^2$$

is now *time dependent* and T*-periodic* in time t, so that $X(t+T,\mu,z) = X(t,\mu,z)$ for all t,μ,z. The flow solves the initial value problem

$$\begin{cases} \frac{d}{dt}\varphi^t(z,\mu) = X(t,\mu,\varphi^t(z,\mu)), & t \in \mathbb{R}, \\ \varphi^0(z,\mu) = z. \end{cases}$$

Due to the *uniqueness* of the Cauchy initial value problem, it follows from the 2π periodicity of the vector field X in the variable x that

$$\varphi^t(z + 2\pi j e_1, \mu) = \varphi^t(z,\mu) + 2\pi j e_1,$$

for all $t,\mu \in \mathbb{R}$ and $j \in \mathbb{Z}$, where $e_1 = (1,0)$. Moreover, it follows from the T-periodicity of the vector field in time t that

$$\varphi^{t+T}(z,\mu) = \varphi^t(\varphi^T(z,\mu))$$

for every $t \in \mathbb{R}$ and $z \in \mathbb{R}^2$ (recall that the relation $\varphi^t \circ \varphi^s = \varphi^{t+s}$ is only valid for the flow of a time independent vector field). Keeping the parameter μ fixed, the mapping

$$\psi(z) := \varphi^T(z,\mu) \colon \mathbb{R}^2 \to \mathbb{R}^2$$

is a diffeomorphism satisfying $\psi^j(z) = \varphi^{jT}(z,\mu)$ for every $z \in \mathbb{R}^2$.

Let us assume that there exists a solution $x(t)$ of the equation $\ddot{x} + \sin x = \mu \sin \omega t$ possessing *infinitely many zeros* (mod 2π), at the times $(t_k)_{k\in\mathbb{Z}}$ which are all nondegenerate. The times are ordered according to $t_k < t_l$ if $k < l$, so that

$$x(t_k) = 0 \mod 2\pi, \quad \dot{x}(t_k) \neq 0, \quad k \in \mathbb{Z}.$$

In other words, the pendulum passes the lowest point infinitely often with a non-vanishing velocity. We associate with this solution a two-sided sequence $\sigma(x(t)) = (\sigma_k(x(t)))_{k\in\mathbb{Z}}$, defined by

$$\sigma_k(x(t)) = \operatorname{sign}(\dot{x}(t_k)) = \begin{cases} +1, & \dot{x}(t_k) > 0, \\ -1, & \dot{x}(t_k) < 0. \end{cases}$$

In the *unperturbed* case $\mu = 0$ there exist precisely three types of such sequences, namely

(a) constant $+1$, i.e., $\sigma_k(x(t)) = +1$ for all $k \in \mathbb{Z}$,

(b) constant -1, i.e., $\sigma_k(x(t)) = -1$ for all $k \in \mathbb{Z}$,

(c) alternating, i.e., $\sigma(x(t)) = (\dots, +1, -1, +1, -1, +1, \dots)$.

In sharp contrast to this unperturbed situation, the perturbed mathematical pendulum possesses a solution for every prescribed random sequence as the following theorem shows.

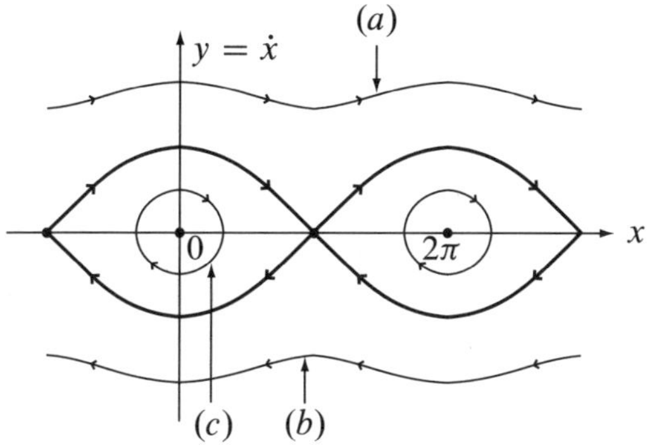

Figure III.12. Types in the unperturbed case.

Theorem III.19. *Let $U \subset \mathbb{R}^2$ be an open neighborhood of the separatrix (in the case $\mu = 0$). If $|\mu| > 0$ is sufficiently small, then there exists for every two-sided sequence $(s_k)_{k\in\mathbb{Z}}$ of integers $s_k \in \{-1, 1\}$ a solution $x(t)$ of the perturbed pendulum equation $\ddot{x} + \sin x = \mu \sin \omega t$ such that $(x(t), \dot{x}(t)) \in U$ which possesses infinitely many nondegenerate zeros (mod 2π) satisfying*

$$\sigma_k(x(t)) = s_k, \quad k \in \mathbb{Z}.$$

In addition, for every finite sequence $s_k \in \{-1, 1\}$ where $-N \leq k \leq M$ there exists a solution $x(t)$ possessing only finitely many nondegenerate zeros (mod 2π) and solving the equations $\sigma_k(x(t)) = s_k$ for $-N \leq k \leq M$. The same applies to half finite sequences s_k for $-\infty < k \leq M$ or for $-N \leq k < \infty$.

In short, one can prescribe any sequence of directions with which the pendulum should consecutively pass through the lowest point and there exists a solution doing precisely that.

Proof [Transversal heteroclinic point, shadowing lemma]. Assuming $\mu \neq 0$, we consider the diffeomorphism ψ of $\mathbb{R}^2$, defined by the time T flow map

$$\psi(z, \mu) := \psi_\mu(z) := \varphi^T(z, \mu),$$

at the time $T = 2\pi/\omega > 0$. In the case $\mu = 0$, the map ψ has the hyperbolic fixed point $P := P_{-1} = (-\pi, 0)$. We shall show that also the diffeomorphism ψ_μ has a *unique hyperbolic fixed point $P(\mu)$ near $P = P(0)$*, which depends differentiably on μ, if μ is small enough. For this, we define the mapping $F: \mathbb{R}^2 \times \mathbb{R} \to \mathbb{R}^2$ by

$$F(z, \mu) = \varphi^T(z, \mu) - z.$$

If $\mu = 0$ then $F(P,0) = \phi^T(P,0) - P = 0$ and the partial derivative in the variable z is given by

$$D_1(P,0) = D_1\varphi^T(P,0) - \mathbb{1} \in \mathcal{L}(\mathbb{R}^2).$$

The linear mapping $D_1F(P,0)$ is an isomorphism, since the hyperbolic matrix $D_1\varphi^T(P,0)$ does not have an eigenvalue equal to 1. In a neighborhood of $\mu = 0$ there exists, by the implicit function theorem, a unique continuously differentiable function $\mu \mapsto P(\mu) \in \mathbb{R}^2$, solving $F(P(\mu),\mu) = 0$ and $P(0) = P$. In other words,

$$P(\mu) = \psi_\mu(P(\mu))$$

is a fixed point of the mapping ψ_μ. The eigenvalues of the derivative $d\psi_\mu(P(\mu))$ depend continuously on μ, hence the linear map $d\psi_\mu(P(\mu))$ possesses for small μ an eigenvalue whose absolute value is > 1 and an eigenvalue whose absolute value is < 1. Consequently, $P(\mu)$ is a hyperbolic fixed point of ψ_μ, if μ is small. Also the points $P(\mu) + 2n\pi$ are hyperbolic fixed points and $P(\mu) + 2n\pi = P_{-1}(\mu) + 2n\pi = P_{n-1}(\mu)$.

From the proof of Theorem II.8 (construction of h) we know that the *local invariant manifolds* issuing from the hyperbolic fixed point $P(\mu)$, denoted by

$$W^{\pm}_{\text{loc}}(P(\mu)),$$

depend differentiably on μ (by the implicit function theorem). For small μ they can, therefore, be represented locally as graphs over the invariant manifolds of the unperturbed system (the branches of the separatrix). If $t \mapsto \gamma(t)$ is a *heteroclinic solution* in the unperturbed case $\mu = 0$ having the x-coordinate at time $t = 0$ equal to $(\gamma(0))_1 = 0$, then $\gamma(t)$ lies on the separatrix. In formulas,

$$\frac{d}{dt}\gamma(t) = X(0,\gamma(t)), \quad t \in \mathbb{R},$$

and $\gamma(t) \to P = P_{-1}$ as $t \to -\infty$ and $\gamma(t) \to P + 2\pi e_1 = P_0$ as $t \to +\infty$.

Denoting by $n(\gamma(t))$ the unit normal vector of the homoclinic orbit γ in the point $\gamma(t)$ as depicted in Figure III.14, we can represent the relevant pieces of the invariant manifolds as follows:

$$\{\gamma(r) + u^-(r,\mu)\cdot n(\gamma(r)) \mid -\infty < r \le M\} \subset W_-(P_{-1}(\mu))$$

and

$$\{\gamma(r) + u^+(r,\mu)\cdot n(\gamma(r)) \mid -M \le r < \infty\} \subset W_+(P_0(\mu)),$$

with a sufficiently large constant $M > 0$ and where in the case $\mu = 0$ the functions $u^+(r,0) = u^-(r,0) = 0$ vanish.

If for a parameter value $r \in \mathbb{R}$,

$$u^-(r,\mu) = u^+(r,\mu) \quad \text{and} \quad \frac{\partial}{\partial r}u^-(r,\mu) \neq \frac{\partial}{\partial r}u^+(r,\mu),$$

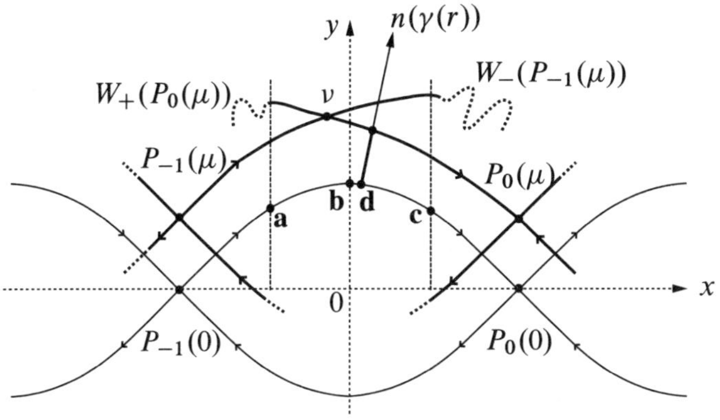

Legend: **a**: $\gamma(-M)$ **b**: $\gamma(0)$ **c**: $\gamma(M)$ **d**: $\gamma(r)$

Figure III.13. The perturbed invariant manifolds possessing a transversal intersection.

then we have found the *transversal* intersection point

$$\nu := \gamma(r) + u^-(r,\mu) \cdot n(\gamma(r)) \in W_-(P_{-1}(\mu)) \cap W_+(P_0(\mu)).$$

In order to study the first-order term in μ of the function $(u^- - u^+)$ we introduce the so-called Melnikov function

$$d(r) := \frac{\partial}{\partial \mu}(u^- - u^+)|_{\mu=0}(r).$$

If $d(r_0) = 0$ and $\frac{d}{dr}d(r_0) \neq 0$, then there exists a transversal intersection point near $\gamma(r_0)$, for small $\mu \neq 0$. This follows from

$$(u^- - u^+)(r,\mu) = \mu\big(d(r) + O(\mu)\big)$$

in view of the implicit function theorem. The first approximation $d(r)$ can be explicitly calculated by means of the following *Melnikov formula.*

Theorem III.20 (Melnikov). *Let*

$$\dot{z} = f(z) + \mu g(t,z) \in \mathbb{R}^2,$$

$z \in \mathbb{R}^2$ *be a smooth vector field satisfying* div $f = 0$, *where* g *is a* T*-periodic vector field for some* $T > 0$, *so that* $g(t+T,z) = g(t,z)$. *We assume that for* $\mu = 0$ *there exists a homoclinic (resp. heteroclinic) orbit* γ *of the vector field* f, *hence satisfying* $\frac{d}{dt}\gamma(t) = f(\gamma(t))$ *for all* $t \in \mathbb{R}$ *and*

$$\gamma(t) \to P, \quad t \to -\infty,$$
$$\gamma(t) \to Q, \quad t \to +\infty$$

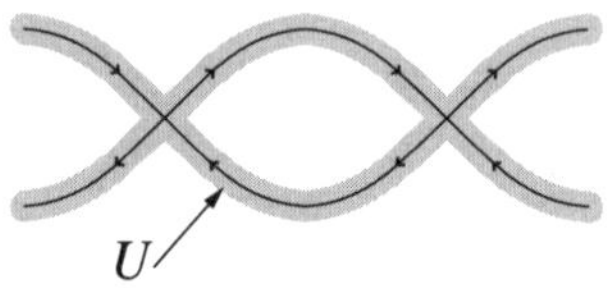

Figure III.14. A neighborhood U of the separatrix.

for two hyperbolic fixed points P and Q of f. Then, setting $f = (f_1, f_2)$ and $g = (g_1, g_2)$ the following formula holds true:

$$d(r) = \frac{1}{|f(\gamma(r))|} \int_{-\infty}^{\infty} (f_1 g_2 - f_2 g_1)(s, \gamma(r+s))\, ds.$$

For a *proof* we refer to C. Robinson in [91, S. 304]. In order to apply the formula to our pendulum, we consider the upper branch of the unperturbed separatrix,

$$\begin{cases} -\pi \le x \le \pi, \\ y = +\sqrt{2(1+\cos x)} = 2\cos(\frac{x}{2}) \end{cases}$$

where $y = \dot{x}$. The solution of the equation $\dot{x} = 2\cos(\frac{x}{2})$ is given by

$$x(t) = 2\arcsin(\tanh(t)), \quad t \in \mathbb{R},$$

and differentiating we obtain

$$\dot{x}(t) = y(t) = \frac{2}{\cosh(t)}.$$

Hence $\gamma(t) = (x(t), y(t))$ is the heteroclinic orbit. Inserting the curve γ into the Melnikov formula results in

$$d(r) = \frac{1}{|X(\gamma(r))|} \frac{2\pi \sin(\omega r)}{\cosh(\frac{\omega\pi}{2})}.$$

The function $d(r)$ has the *nondegenerate* zeros $r = \frac{\pi}{\omega} j$ for all $j \in \mathbb{Z}$. Therefore, there exists a transversal heteroclinic point ν. In the same way, there exists near the lower branch of the separatrix a transversal heteroclinic point η. The closure of the heteroclinic orbits is the *hyperbolic* set

$$\Lambda = \bigcup_{j,k\in\mathbb{Z}} \psi^j(\nu + 2\pi k e_1) \cup \{P_k(\mu)\} \cup \psi^j(\eta + 2\pi k e_1).$$

In order to finish the proof of Theorem III.19 we choose a neighborhood U of the separatrix and we choose the parameters ε, δ as in the shadowing lemma. For the given sequence $s = (s_k)_{k\in\mathbb{Z}}$ we construct the following ε-pseudo orbit q, described by Figure III.15. If $s_0 = 1$, we start in the heteroclinic point $q_0 = \nu$ and if $s_0 = -1$ we start in the heteroclinic point $q_0 = \eta$. Then we follow the heteroclinic orbit $\psi^j(\nu)$, resp. $\psi^j(\eta)$ into the $(\varepsilon/2)$-neighborhood of the next hyperbolic fixed point. There one has again two possibilities. If $s_1 = 1$ we jump onto the heteroclinic orbit of the upper branch to the *right* while if $s_1 = -1$ we jump onto the heteroclinic orbit of the lower branch to the *left*, and so on.

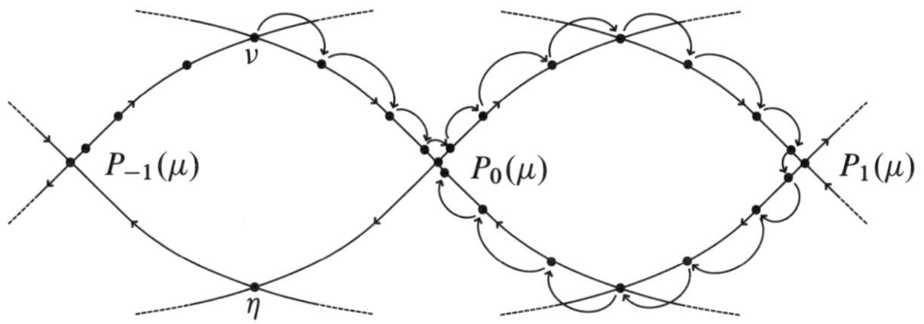

Figure III.15. The ε-pseudo orbit associated with the sequence $(s_k) = (\ldots, s_0, s_1, s_2, \ldots) = (\ldots, 1, 1, -1, \ldots)$.

The associated δ-shadowing orbit $p = (p_j)_{j\in\mathbb{Z}}$ guaranteed by the shadowing lemma,

$$p_j = \psi^j(p_0) = \varphi^{jT}(p_0, \mu), \quad j \in \mathbb{Z},$$

lies on the desired *solution* $t \mapsto \varphi^t(p_0, \mu)$ of the perturbed vector field $X(t, \mu, z)$ starting at the point $\varphi^0(p_0, \mu) = p_0$ at the time $t = 0$ and remaining in the open neighborhood U of the separatrix. Of course, this solution loses a lot of time near the hyperbolic equilibrium points, away from these neighborhoods it moves quite fast. We point out that all the solutions found this way start in a small neighborhood of the homoclinic points ν resp. η!

The passages near the transversal heteroclinic points ν resp. η correspond to the passages of the pendulum through the point $x = 0 \mod 2\pi$, which is the lowest position of the pendulum. In order to obtain a solution defined by a finite sequence (s_k), one constructs an ε-pseudo orbit q as before which, however, at the ends is equal to the orbits of hyperbolic fixed points. Then, the corresponding solution of the pendulum equation makes finitely many swings back and forth and then remains almost immobile near the highest position of the pendulum! This completes the proof of Theorem III.19. □

For a detailed study of the chaotic behavior of the periodically perturbed pendulum we refer to U. Kirchgraber and D. Stoffer in [59].

III.5 Torus automorphisms

A diffeomorphism of a compact Riemann manifold that is a hyperbolic set for the diffeomorphism is called an *Anosov diffeomorphism*. In contrast to our previous studies of transversal homoclinic orbits of a hyperbolic fixed point, the hyperbolic structure is postulated for an Anosov diffeomorphism. Hyperbolic torus automorphisms are a standard example of a hyperbolic system on a compact manifold.

We describe a simple but typical example on the 2-torus $T^2 = S^1 \times S^1$ whose covering space is $\mathbb{R}^2$. The covering map

$$p \colon \mathbb{R}^2 \to T^2, \quad (x_1, x_2) \mapsto (e^{2\pi i x_1}, e^{2\pi i x_2}),$$

satisfies $p(x+\omega) = p(x)$ for all $\omega \in \mathbb{Z}^2$. Clearly, $T^2 \cong \mathbb{R}^2/\mathbb{Z}^2$. We consider the linear mapping

$$A = \begin{pmatrix} 2 & 1 \\ 1 & 1 \end{pmatrix} \in \mathcal{L}(\mathbb{R}^2).$$

Then $\det A = 1$, and the matrix elements of the inverse map

$$A^{-1} = \begin{pmatrix} 1 & -1 \\ -1 & 2 \end{pmatrix}$$

are also integers. In particular, $A(\mathbb{Z}^2) = \mathbb{Z}^2$ and $A^{-1}(\mathbb{Z}^2) = \mathbb{Z}^2$. Consequently, A *induces* the mapping φ_A on the torus given by

$$\varphi_A(e^{2\pi i x_1}, e^{2\pi i x_2}) = \big(e^{2\pi i (2x_1 + x_2)}, e^{2\pi i (x_1 + x_2)}\big).$$

This map is a diffeomorphism of T^2, having the inverse mapping

$$\varphi_A^{-1} = \varphi_{A^{-1}} \colon (e^{2\pi i x_1}, e^{2\pi i x_2}) \mapsto \big(e^{2\pi i (x_1 - x_2)}, e^{2\pi i (-x_1 + 2x_2)}\big).$$

In view of $\varphi_A \circ p = p \circ A$ the diagram

$$\begin{array}{ccc} T^2 & \xrightarrow{\varphi_A} & T^2 \\ {\scriptstyle p}\big\uparrow & & \big\uparrow{\scriptstyle p} \\ \mathbb{R}^2 & \xrightarrow{A} & \mathbb{R}^2 \end{array}$$

is commutative.

We note that if A, B belong to the *group* $\mathrm{SL}_2(\mathbb{Z})$ of the (2×2) matrices having determinants equal to 1 and integer matrix elements, then

$$\varphi_A \circ \varphi_B = \varphi_{AB}, \quad \varphi_A^{-1} = \varphi_{A^{-1}}, \quad \varphi_{\mathbb{1}} = \mathrm{Id}_{T^2}.$$

We next consider the *linear* map in the covering space $\mathbb{R}^2$. The matrix A is symmetric and has the eigenvalues $0 < \lambda_1 < 1 < \lambda_2$,

$$\lambda_1 = \frac{3 - \sqrt{5}}{2}, \quad \lambda_2 = \lambda_1^{-1} = \frac{3 + \sqrt{5}}{2}.$$

The corresponding *orthogonal eigenvectors* are the vectors

$$v_1 = \begin{pmatrix} 1 \\ \frac{-\sqrt{5}+1}{2} \end{pmatrix}, \quad v_2 = \begin{pmatrix} 1 \\ \frac{\sqrt{5}+1}{2} \end{pmatrix},$$

satisfying $Av_1 = \lambda_1 v_1$ and $Av_2 = \lambda_2 v_2$. Therefore, the origin 0 is a *hyperbolic* fixed point of the map $A\colon \mathbb{R}^2 \to \mathbb{R}^2$ possessing the invariant splitting $\mathbb{R}^2 = E_+ \oplus E_-$, defined by the eigenspaces

$$E_+ = \{tv_1 \mid t \in \mathbb{R}\} \quad \text{and} \quad E_- = \{tv_2 \mid, t \in \mathbb{R}\}.$$

Clearly,

$$A^j x_+ = \lambda_1^j x_+, \quad x_+ \in E_+,$$
$$A^j x_- = \lambda_2^j x_-, \quad x_- \in E_-.$$

The image of the fixed point 0 under the projection mapping,

$$p(0) = 0^* \in T^2,$$

is a hyperbolic fixed point of the diffeomorphism $\varphi_A\colon T^2 \to T^2$, since in suitable local coordinates the derivative is equal to $d\varphi_A(0^*) = A$.

From $\varphi_A \circ p = p \circ A$ it follows that $\varphi_A^j \circ p = p \circ A^j$. Hence $\varphi_A^j(p(x)) \to 0^*$ as $j \to \pm\infty$, precisely if $A^j x \to 0 \mod \mathbb{Z}^2$ as $j \to \pm\infty$, hence precisely if $x \in E_+$ resp. $x \in E_-$. The stable and unstable manifolds issuing from the hyperbolic fixed point 0^* are therefore the curves

$$W_+(0^*, \varphi_A) = p(E_+),$$
$$W_-(0^*, \varphi_A) = p(E_-).$$

These immersed curves are *dense* on T^2. Indeed, since $\langle v_1, j\rangle \notin \mathbb{Z}$ for all $0 \neq j \in \mathbb{Z}^2$, already the set $\{p(nv_1) \mid n \in \mathbb{N}\}$ is dense in T^2 in view of Corollary I.7 (Kronecker). The same argument applies to the unstable manifold $p(E_-)$. The *homoclinic points* $W_+(0^*, \varphi_A) \cap W_-(0^*, \varphi_A) \setminus 0^*$ are also *dense* in T^2 and *transversal*. This is obvious, if we look at the invariant manifolds in the fundamental domain $[0, 1]^2$ of T^2 (Figure III.16).

Proposition III.21. *The diffeomorphism $\varphi_A\colon T^2 \to T^2$ has the following properties.*

(i) *φ_A is hyperbolic.*

(ii) *φ_A is ergodic in relation to the Lebesgue measure m.*

(iii) *The set $R^+ = \{x \in T^2 \mid \overline{\mathcal{O}^+(x)} = T^2\}$ of points with dense positive semi orbit has full measure $m(R^+) = m(T^2)$, is dense on T^2 and of second Baire category.*

(iv) *The periodic points are countable and dense. They are the points* $p(\frac{1}{k}j)$ *for* $k \in \mathbb{N}$ *and* $j \in \mathbb{Z}^2$.

(v) *The transversal homoclinic points belonging to the hyperbolic fixed point* 0^* *are dense.*

(vi) *The diffeomorphism* φ_A *is structurally stable.*

We see that every open subset of T^2 *contains points whose orbits behave in the long run completely differently under the diffeomorphism* φ_A.

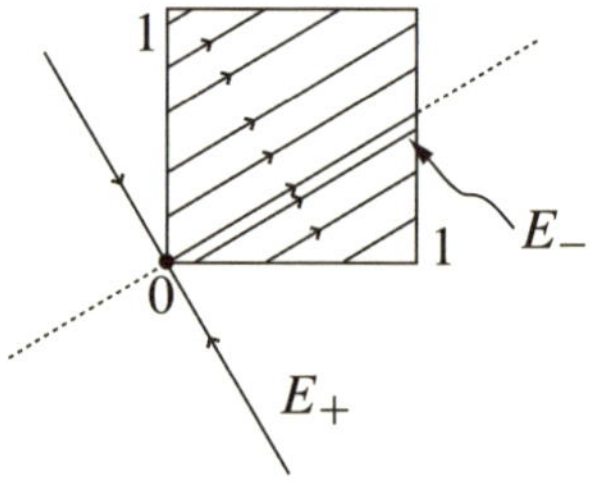

Figure III.16. E_- in the fundamental domain of the torus.

Proof. (i) The torus T^2 is compact. At the point $p \in T^2$ we have $d\varphi_A(p) = A$ in suitable local coordinates so that the desired splitting is defined by the eigenspaces of A.

(ii) We shall show that every invariant function $f \in L^1(T^2)$ is constant almost everywhere. The lift $F : \mathbb{R}^2 \to \mathbb{C}$ of the function $f : T^2 \to \mathbb{C}$ is defined by

$$F(x_1, x_2) = f(e^{2\pi i x_1}, e^{2\pi i x_2}).$$

If the function f is invariant, that is, $f = f \circ \varphi_A$, then also $F = F \circ A$ is an invariant function. We calculate the *Fourier coefficients* f_k for $k \in \mathbb{Z}^2$ of F. Let A^T be the transposed of A and $j \in \mathbb{Z}^2$. For $k = A^T j$ we obtain, using the invariance of the function F and $|\det A| = 1$,

$$\begin{aligned} f_{A^T j} &= \int F(x) e^{-2\pi i \langle A^T j, x\rangle}\, dx \\ &= \int F(x) e^{-2\pi i \langle j, Ax\rangle}\, dx \\ &= \int F(Ax) e^{-2\pi i \langle j, Ax\rangle}\, dx \\ &= \int F(y) e^{-2\pi i \langle j, y\rangle}\, dy \\ &= f_j. \end{aligned}$$

This holds true for every $j \in \mathbb{Z}^2$ and setting $B := A^T$, we therefore deduce

$$f_j = f_{Bj} = f_{B^2 j} = \cdots = f_{B^n j} = \cdots, \quad j \in \mathbb{Z}^2.$$

For f belonging to L^1 we know that

$$f_k \to 0, \quad |k| \to \infty,$$

where $|k| = \sqrt{k_1^2 + k_2^2}$. Hence, if the sequence $|B^n j|$ for every fixed $0 \neq j \in \mathbb{Z}^2$ is unbounded, then $f_j = 0$ for all $j \neq 0$ and consequently the function f is constant almost everywhere.

The sequence $|B^n j|$ is indeed unbounded, because the set $\{B^n j \mid n \in \mathbb{N}\}$ is an infinite set, since for $j \neq 0$ all these points are different. If they were not, we would find an integer $m \in \mathbb{N}$ satisfying $B^m j = j$, so that B^m would have an eigenvalue 1. This, however, is not possible, since B and hence B^m is a hyperbolic map. We have demonstrated that every invariant integrable function is constant almost everywhere. Consequently, due to Proposition I.17 the diffeomorphism φ_A is ergodic.

(iii) The statement (iii) follows from the ergodicity of φ_A (Proposition I.21).

(iv) Since A is an invertible matrix over the ring $\mathbb{Z}$, we have $A(\mathbb{Z}^2) = \mathbb{Z}^2$, hence, due to the linearity $A(\frac{1}{k}\mathbb{Z}^2) = \frac{1}{k}\mathbb{Z}^2$. On the torus we conclude $\varphi_A(p(\frac{1}{k}\mathbb{Z}^2)) = p(A(\frac{1}{k}\mathbb{Z}^2)) = p(\frac{1}{k}\mathbb{Z}^2)$. This set is *finite* (since it corresponds bijectively to the set $[0,1) \times [0,1) \cap \frac{1}{k}\mathbb{Z}^2$), hence φ_A is a permutation of this set and the orbits in $\frac{1}{k}\mathbb{Z}^2$ are all periodic. The points $p(\frac{1}{k}\mathbb{Z}^2)$ with $k \in \mathbb{N}$ lie densely on the torus. We claim that we have found already all the periodic points. Indeed, if $p(z)$ is a periodic point of φ_A, where $z = (x, y) \in \mathbb{R}^2$, then there exist an integer n and an integer vector $k = (k_1, k_2) \in \mathbb{Z}^2$ such that $A^n z = z + k$ and hence $(A^n - \mathbb{1})z = k$. Since A^n does not have an eigenvalue 1, this equation has a unique solution $z = (x, y)$, given by a rational function of the integer matrix elements of A^n and of k_1, k_2 and therefore $x, y \in \mathbb{Q}$ are rational, as claimed.

(v) The statement (v) has been proved previously.

(vi) We assume that g is a C^1 perturbation of φ_A, so that $g = \varphi_A + \hat{g}$ where the map $\hat{g}$ is small in the C^1-sense. We consider the homotopy $g_t = \varphi_A + t\hat{g}$, connecting $g_0 = \varphi_A$ with $g_1 = g$ and we choose a lift G of the end point $g = g_1$, so that $G(0)$ is near 0. We then lift the homotopy to a homotopy G_t. We may assume that $G_0 = A$. From

$$G_0(x + j) - G_0(x) = A(x + j) - Ax = Aj \in \mathbb{Z}^2, \quad z \in \mathbb{Z}^2$$

it follows, since $\mathbb{Z}^2 \subset \mathbb{R}^2$ is discrete, that

$$G_t(x + j) - G_t(x) = Aj \in \mathbb{Z}^2$$

for every $t \in [0,1]$. In particular also for $t = 1$ and we set $G_1 = G$. The mapping $\widehat{G} := G - A$ satisfies

$$\widehat{G}(x+j) = \widehat{G}(x), \quad j \in \mathbb{Z}^2$$

and therefore

$$|\widehat{G}|_{C^0} = \sup_{x \in [0,1]^2} |G(x)| < \infty.$$

Assuming $\|d\hat{g}\| \leq \varepsilon$ we obtain the Lipschitz estimate

$$|\widehat{G}(x) - \widehat{G}(y)| \leq \varepsilon |x-y|.$$

We can therefore apply the arguments of the proof of Theorem II.4 to the mapping $G = A + \widehat{G}$, replacing the space $X = C_b(\mathbb{R}^n, \mathbb{R}^n)$ by the space of periodic mappings $X' = C_{\text{per}}(\mathbb{R}^n, \mathbb{R}^n)$,

$$C_{\text{per}}(\mathbb{R}^n, \mathbb{R}^n) := C(\mathbb{R}^n, \mathbb{R}^n) \cap \{f \colon \mathbb{R}^n \to \mathbb{R}^n \mid f(x+j) = f(x),\ j \in \mathbb{Z}^n\}.$$

Arguing as in Theorem II.4 we find a homeomorphism H of $\mathbb{R}^2$ of the form $H = \text{Id} + v$ with $v \in C_{\text{per}}(\mathbb{R}^2, \mathbb{R}^2)$ which solves the equation

$$G \circ H = H \circ A.$$

In view of $H = \text{Id} + v$ we have for $j \in \mathbb{Z}^2$,

$$H(x+j) = x + j + v(x+j) = x + j + v(x) = H(x) + j,$$

so that H induces a continuous and surjective map $h \colon T^2 \to T^2$ solving

$$g \circ h = h \circ \varphi_A.$$

In order to prove the injectivity of the map h we assume that $h(p(x)) = h(p(y))$. Then $H(x) = H(y) + j$ for an integer vector $j \in \mathbb{Z}^2$ and recalling $H(y) + j = H(y+j)$ we obtain $H(x) = H(y+j)$. Due to the injectivity of H we conclude $x = y + j$ and hence $p(x) = p(y+j) = p(y)$. Finally, in view of the compactness of T^2 the map h is a homeomorphism and we have verified that the diffeomorphism φ_A is indeed structurally stable. □

We would like to remark that, more generally, a hyperbolic $(n \times n)$-matrix A with integer matrix elements and satisfying $\det A = \pm 1$ induces a diffeomorphism φ_A on the torus $T^n = \mathbb{R}^n / \mathbb{Z}^n$ which also has the properties of Proposition III.21. Such a mapping φ_A is called a *hyperbolic torus automorphism*. It turns out that *every* Anosov diffeomorphism of the torus T^n is topologically conjugated to such a hyperbolic torus automorphism φ_A, for details we refer to the book [49, p. 587] by B. Hasselblatt and A. Katok.

Remark. In the ergodic theorem criterion (Theorem I.18) we have seen the following. If φ is an ergodic transformation of a finite measure space, then for every integrable function f the mean value over an orbit is equal to the mean value over the whole space, so that $f^*(x) = \bar{f}$, or explicitly,

$$\lim_{n\to\infty} \frac{1}{n} \sum_{j=0}^{n-1} f(\varphi^j x) = \frac{1}{m(X)} \int_X f,$$

for *almost every* x. In general, this equation does not hold true for all points x, as for example a fixed point of φ shows.

We will show next that $f^*(x) = \bar{f}$ can fail even on a *dense* subset of a topological measure space.Looking at the above torus automorphism $\varphi = \varphi_A$ on T^2 equipped with the (normalized) Lebesgue measure and taking the continuous map $f\colon T^2 \to \mathbb{R}$, defined by

$$f(e^{2\pi i x_1}, e^{2\pi i x_2}) := e^{2\pi i x_1},$$

we see that the mean value over the space vanishes,

$$\bar{f} = \int_{T^2} f = \int_0^1 \int_0^1 e^{2\pi i x_1}\, dx_1 dx_2 = 0.$$

We claim that for every $x \in W_+(0^*, \varphi_A) \setminus \{0^*\}$,

$$\lim_{n\to\infty} \frac{1}{n} \sum_{j=0}^{n-1} f(\varphi_A^j x) = 1,$$

so that $f^*(x) = 1$. In order to prove this claim we first observe that, in view of the definition of the stable invariant manifold W_+ and using the continuity of f, we have $f(\varphi_A^j x) \to f(0^*) = 1$. The claim now follows from the fact that for a converging sequence of numbers $a_j \to 0$, also the sequence of means $\frac{1}{n}\sum_{j=0}^{n-1} a_j$ converges to 0.

III.6 Invariant manifolds of Λ

We would like to describe, without giving proofs, the generalization of the invariant manifolds issuing from a hyperbolic fixed point to the above situation of a hyperbolic set $\Lambda \subset \mathbb{R}^n$. It turns out that the subset of points in $\mathbb{R}^n$, which under the forward iterates lie near an orbit q on Λ, form a local manifold called the *local stable manifold of the orbit* q. Similarly, the set of points that remain close to the orbit q on Λ under the backward iterates form the *local unstable manifolds of the orbit* q. Actually, this holds true even for an ε-pseudo orbit q on the hyperbolic set Λ.

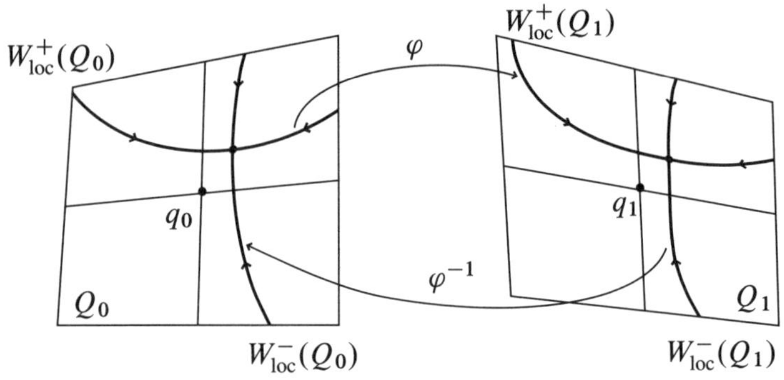

Figure III.17. Invariance of the local invariant manifolds.

We focus on the ε-pseudo orbit q of the diffeomorphism φ on Λ and define for $\delta > 0$ the sequence of neighborhoods

$$Q_j = \{x \in \mathbb{R}^n \mid |x - q_j| \le \delta\}, \quad j \in \mathbb{Z}$$

containing the *subsets* of $\mathbb{R}^n$

$$W^+_{\mathrm{loc}}(Q_j) = \{x \in Q^j \mid \varphi^k(x) \in Q_{j+k} \text{ for all } k \ge 0\},$$
$$W^-_{\mathrm{loc}}(Q_j) = \{x \in Q^j \mid \varphi^{-k}(x) \in Q_{j-k} \text{ for all } k \ge 0\},$$

for $j \in \mathbb{Z}$. Clearly, these sets are $\pm$ invariant in the sense that

$$\varphi(W^+_{\mathrm{loc}}(Q_j)) \subset W^+_{\mathrm{loc}}(Q_{j+1}),$$
$$\varphi^{-1}(W^-_{\mathrm{loc}}(Q_j)) \subset W^-_{\mathrm{loc}}(Q_{j-1}).$$

We abbreviate the splittings in $T\Lambda = T\mathbb{R}^n|_\Lambda$ and simply write

$$E^+_j \oplus E^-_j := E^+_{q_j} \oplus E^-_{q_j} = T_{q_j}\Lambda = \mathbb{R}^n.$$

If $\delta > 0$ is sufficiently small, then there exists an $\varepsilon = \varepsilon(\delta) > 0$ such that for *every ε-pseudo orbit* $q = (q_j)$ of the diffeomorphism φ on Λ the following holds true:

$$W^+_{\mathrm{loc}}(Q_j) = \{(x_+, h^+_j(x_+)) \in (E^+_j \oplus E^-_j) \cap Q_j\},$$
$$W^-_{\mathrm{loc}}(Q_j) = \{(h^-_j(x_-), x_-) \in (E^+_j \oplus E^-_j) \cap Q_j\},$$

for every $j \in \mathbb{Z}$, where the maps $h^+_j : E^+_j \to E^-_j$ are of class of C^k, if φ is of class C^k. The analogous statement holds true for the maps $h^-_j : E^-_j \to E^+_j$. Moreover,

$$W^+_{\mathrm{loc}}(Q_j) \cap W^-_{\mathrm{loc}}(Q_j) =: \{p_j\} \quad \text{for all integers } j \in \mathbb{Z},$$

and $p_j = \varphi^j(p_0)$, $j \in \mathbb{Z}$ is the δ-shadowing orbit of the ε-pseudo orbit q. If, in particular, $q = p$ is an *orbit* of the diffeomorphism φ on the hyperbolic set Λ, we have, in addition,

(i) $W^+_{\text{loc}}(Q_j) \cap W^-_{\text{loc}}(Q_j) = \{p_j\}$,

(ii) $T_{p_j} W^{\pm}_{\text{loc}}(Q_j) = E^{\pm}_j$ for the tangent spaces,

(iii) $W^+_{\text{loc}}(Q_j) = \{x \in Q_j \mid |\varphi^n(x) - p_{j+n}| \to 0,\ n \to \infty\}$,

(iv) $W^-_{\text{loc}}(Q_j) = \{x \in Q_j \mid |\varphi^{-n}(x) - p_{j-n}| \to 0,\ n \to \infty\}$.

We see that the bounds imply the convergence, as before.

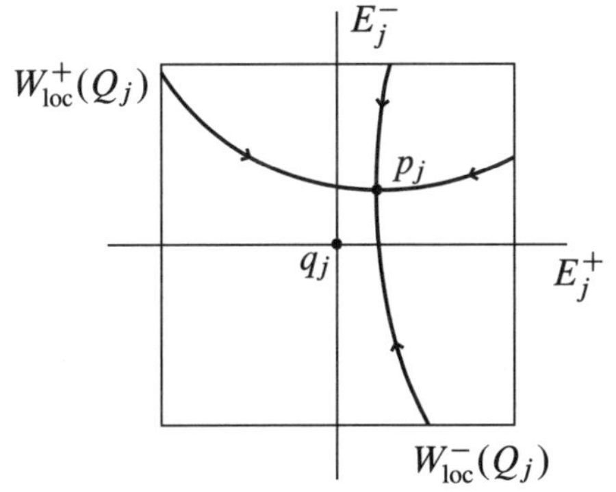

Figure III.18. Intersection of $W^+_{\text{loc}}(Q_j)$ and $W^-_{\text{loc}}(Q_j)$.

We leave the proofs of these statements as an exercise to the reader with the hint that one combines the proof of Theorem II.8 with the shadowing lemma, the contraction principle, and the implicit function theorem. For proofs of these statements, known under the names *stable manifold theorem* and *Hadamard–Perron theorem* we refer, for example, to the books [49, S. 242, S. 267] by B. Hasselblatt and A. Katok, [103, S. 71] by M. Shub, [91, S. 411] by C. Robinson.

The *global* invariant manifolds of the points $p_j = \varphi^j(p_0)$ of an orbit on Λ are the following subsets of $\mathbb{R}^n$:

$$W_+(p_j) = \bigcup_{n\geq 0} \varphi^{-n}(W^+_{\text{loc}}(Q_j)),$$

$$W_-(p_j) = \bigcup_{n\geq 0} \varphi^{n}(W^-_{\text{loc}}(Q_j)).$$

The unions $W_+(p) = \bigcup_{j\in\mathbb{Z}} W_+(p_j)$ and $W_-(p) = \bigcup_{j\in\mathbb{Z}} W_-(p_j)$ are the *global* invariant manifolds of the *orbit* $p = (p_j)$. These sets are characterized by the following properties,

$$W_+(p) = \{x \in \mathbb{R}^n \mid d(p_j, \varphi^j(x)) \to 0,\ j \to \infty\},$$
$$W_-(p) = \{x \in \mathbb{R}^n \mid d(p_j, \varphi^{-j}(x)) \to 0,\ j \to \infty\}.$$

III.7 Structural stability on hyperbolic sets

We conclude the chapter by a sketch of the famous structural stability theorem of D. Anosov, using the shadowing lemma.

Lemma III.22. *Let Λ be a hyperbolic set of the diffeomorphism φ and assume ψ to be another diffeomorphism leaving Λ invariant, $\psi(\Lambda) = \Lambda$, and being near φ in the sense of*

$$|\varphi - \psi|_{C^1} < \varepsilon.$$

Then, for ε small enough, Λ is also a hyperbolic set of ψ.

Proof. Exercise. Hint: To find the hyperbolic splitting of the tangent space in $q_0 \in \Lambda$ consider the orbit of ψ,

$$q_j = \psi^j(q_0), \quad j \in \mathbb{Z},$$

which is an ε-pseudo orbit on Λ for the diffeomorphism φ, since

$$d(q_{j+1}, \varphi(q_j)) = d(\psi(q_j), \varphi(q_j)) \leq |\psi - \varphi|_{C^1} < \varepsilon, \quad j \in \mathbb{Z}.$$

Using the assumption $|d\varphi(q_j) - d\psi(q_j)| < \varepsilon$, the existence of the invariant splitting follows by means of the arguments in the proof of the shadowing lemma. □

Theorem III.23 (Structural stability $\varphi|_\Lambda$). *Let Λ be a hyperbolic set of the diffeomorphism φ and let ψ be another diffeomorphism satisfying $\psi(\Lambda) = \Lambda$. Then, there exists for every $\delta > 0$ an $\varepsilon > 0$ having the following property. If*

$$|\varphi - \psi|_{C^1} < \varepsilon,$$

then there exists a ***homeomorphism*** *$h\colon \Lambda \to \Lambda$ satisfying*

$$\psi = h^{-1} \circ \varphi \circ h$$

and $|h - \mathrm{Id}|_{C^0} + |h^{-1} - \mathrm{Id}|_{C^0} \leq 2\delta$. Moreover, if $\delta > 0$ is small enough, there exists a unique homeomorphism h having these properties.

Proof [Shadowing lemma]. (1) *Definition of the homeomorphism h.* In view of the remark in the shadowing lemma, we find for every $\delta < \delta_0'$ two constants $\varepsilon' = \varepsilon'(\delta) > 0$ and $\mu = \mu(\delta) > 0$ such that for every diffeomorphism ψ satisfying $|\varphi - \psi|_{C^1} \leq \mu$ and for every ε'-pseudo orbit $q = (q_j)_{j \in \mathbb{Z}}$ of φ there exists a unique δ-shadowing orbit for the diffeomorphism ψ. Setting $\varepsilon = \min\{\varepsilon', \mu\}$ we define for ψ satisfying $|\varphi - \psi|_{C^1} < \varepsilon$ the map $h\colon \Lambda \to \Lambda$ in the following way. Consider the point $q_0 \in \Lambda$ and let $q_j = \psi^j(q_0)$ for $j \in \mathbb{Z}$ be the orbit of q_0 under the map ψ. Due to

$$d(q_{j+1}, \varphi(q_j)) = d(\psi(q_j), \varphi(q_j)) \le |\varphi - \psi|_{C^1} < \varepsilon,$$

the sequence $q = (q_j)$ is an ε-pseudo orbit *for the diffeomorphism* φ. If $p = (p_j)$ is the corresponding unique δ-shadowing orbit of q for the diffeomorphism φ, we define

$$h(q_0) = p_0.$$

This map h satisfies the estimate $d(q, h(q)) \le \delta$ for all q in Λ.

(2) *Continuity of h.* In order to prove the continuity of the map h we take a sequence $x_m \in \Lambda$ satisfying $x_m \to x$. Since Λ is compact, the sequence $y_m := h(x_m)$ has a convergent subsequence in Λ. Denoting by y its limit, we claim that $y = h(x)$. In view of the definition of h,

$$d(\varphi^j(y_m), \psi^j(x_m)) \le \delta$$

for every $m \ge 0$ and consequently,

$$d(\varphi^j(y), \psi^j(x)) \le \delta \le \delta_0', \quad j \in \mathbb{Z}.$$

Therefore, the orbit $(\varphi^j(y))_{j\in\mathbb{Z}}$ is a δ_0'-shadowing orbit of the sequence $(\psi^j(x))_{j\in\mathbb{Z}}$ on Λ. In view of the definition of the map h, the orbit $(\varphi^j(h(x)))_{j\in\mathbb{Z}}$ is another δ_0'-shadowing orbit (even δ-shadowing) of the same sequence. From the uniqueness of the shadowing orbit, we conclude $\varphi^j(h(x)) = \varphi^j(y)$ and hence $h(x) = y$, as claimed.

(3) *h is a homeomorphism.* According to Lemma III.22 we can exchange the roles of ψ and φ. By the same construction we obtain a continuous mapping h' and conclude from the uniqueness statement in the shadowing lemma that $h' = h^{-1}$.

(4) *Conjugation and estimate.* In view of the construction,

$$h^{-1} \circ \varphi \circ h(q_0) = h^{-1} \circ \varphi(p_0) = h^{-1}(p_1) = q_1 = \psi(q_0),$$

hence h is the desired conjugation.

(5) *Uniqueness.* Let $\delta > 0$ be so small that the uniqueness in the shadowing lemma holds true and assume that the homeomorphism h satisfies the estimate $|h - \mathrm{Id}|_{C^0} < \delta$, where h is a conjugation, so that

$$h(\psi^j(x)) = \varphi^j(h(x))$$

for all $x \in \Lambda$ and $j \in \mathbb{Z}$. Due to $d(h(\psi^j(x)), \psi^j(x)) \le |h - \mathrm{Id}|_{C^0} < \delta$, the orbit $\varphi^j(h(x)) = h(\psi^j(x))$ is therefore the unique δ-shadowing orbit of the sequence $\psi^j(x)$. In particular, the initial point $h(x)$ of the orbit is uniquely determined by the point x. Hence, the homeomorphism h is unique and the proof of Theorem III.23 is complete. □

Corollary III.24. *An Anosov diffeomorphism of a compact manifold is structurally stable. In particular, every hyperbolic torus automorphism is structurally stable.*

We would like to add that in case the hyperbolic set is a compact Riemannian manifold, J. Moser gave in [71] a direct analytical proof of the structural stability theorem of D. Anosov avoiding the shadowing lemma. The proof is also presented in the book by W. Szlenk [113].

Literature. The proof of Smale's theorem about embeddings of Bernoulli systems is due to K. Palmer in [80]. Useful references about hyperbolicity theory are the monograph [49] by B. Hasselblatt and A. Katok, the monograph [85] by S. Y. Pilyugin, the survey articles [48] by B. Hasselblatt and [83] by Y. Pesin. An efficient introduction to hyperbolic dynamics are the lectures [119] by J.-C. Yoccoz. Classics are A. Katok's article [58] and D. Anosov's work [7]. The partial hyperbolicity is treated in the lectures [84] by Y. Pesin.

Chapter IV
Gradientlike flows

So far we have dealt with discrete dynamical systems whose dynamics is determined by iterations of mappings. We now consider systems whose dynamics are described by flows generated by smooth vector fields and start the chapter by recalling some facts about ordinary differential equations. In order to describe the asymptotic behavior of orbits of flows, the concepts of limit sets and attractors are introduced. In the example of the gradient flow on a compact manifold, every orbit approaches, in forward and in backward time, the set of equilibrium points. Here all the movements slow down and come to a halt at the equilibrium points. Equilibrium points are found by mini-max principles based on index functions such as the Lusternik–Schnirelman category. There is an intimate and fruitful relation between the dynamics of gradient flows on compact manifolds and the topology of the underlying manifold, described by the Morse inequalities presented at the end of the chapter. The Morse theory will only be sketched.

IV.1 Flow of a vector field, recollections from ODE

In order to recall the concept of the flow of a vector field, we consider the open set $\Omega \subset \mathbb{R}^n$ and the vector field $X : \Omega \to \mathbb{R}^n$ of class C^r for $r \geq 1$. The vector field X is, in particular, locally Lipschitz-continuous.

If $x^* \in \Omega$, then there exists an open ball $B_a(x^*) \subset \Omega$ around x^* of radius $a > 0$, an open interval $I = (-\varepsilon, \varepsilon)$ around 0 and a C^r map

$$\varphi\colon I \times B_a(x^*) \to \Omega \subset \mathbb{R}^n, \quad (t,x) \mapsto \varphi(t,x) = \varphi^t(x),$$

such that for every $x \in B_a(x^*)$ the curve $t \mapsto \varphi^t(x)$, $t \in I$ is the *unique* solution of the Cauchy initial value problem

$$\begin{cases} \frac{d}{dt}\varphi^t(x) = X(\varphi^t(x)), & t \in \mathbb{R}, \\ \varphi^0(x) = x. \end{cases}$$

The map φ is called the *local flow* of X. For the existence, uniqueness and regularity of the flow, we refer to the book [5, Chapter II] by H. Amann. In view of its uniqueness, every solution $\varphi^t(x)$ with the initial condition $\varphi^0(x) = x$ can be *continued* onto the *maximal* interval

$$I_x = \big(-t^-(x), t^+(x)\big) \subset \mathbb{R}$$

where the upper bound of the interval (or the exit time of the solution) is defined by

$$t^+(x) := \sup\big\{\tau \geq 0 \mid \dot{x} = X(x),\ x(0) = x \text{ has a solution on } [0,\tau]\big\}$$

and analogously the lower bound (respectively the entrance time) $t^-(x)$ is defined.

We point out that, in case the vector field is not linear, the maximal existence interval I_x is not necessarily the real line $\mathbb{R}$, as the following example shows.

Example. Let $\Omega = \mathbb{R}$ and $X(x) = x^2$. Then the flow of the Cauchy initial value problem

$$\begin{cases} \dot{x} = x^2, \\ x(0) = x \end{cases}$$

is equal to $\varphi^t(x) = \frac{x}{1-tx}$ and so the maximal existence intervals are the intervals

$$\begin{aligned} I_x &= (-\infty, \tfrac{1}{x}), & x > 0, \\ I_x &= \mathbb{R}, & x = 0, \\ I_x &= (\tfrac{1}{x}, \infty), & x < 0. \end{aligned}$$

It follows from the uniqueness of the solution of the Cauchy initial values problem that the flow of a time independent vector field satisfies

$$\varphi^t(\varphi^s(x)) = \varphi^{t+s}(x) = \varphi^s(\varphi^t(x))$$

on the domains of definition. Indeed, the two curves $t \mapsto \varphi^t(\varphi^s(x))$ and $t \mapsto \varphi^{t+s}(x)$ solve the same Cauchy initial value problem with the same initial condition $x(0) = \varphi^s(x)$ at the time $t = 0$.

In the special case of $\Omega = \mathbb{R}^n$ and $I_x = \mathbb{R}$ for all $x \in \mathbb{R}^n$ the flow defines a one-parameter group of diffeomorphisms $\{\varphi^t \mid t \in \mathbb{R}\}$,

$$\varphi^t : \mathbb{R}^n \to \mathbb{R}^n, \; x \mapsto \varphi^t(x)$$

satisfying, for all $t, s \in \mathbb{R}$,

$$\varphi^t(\varphi^s(x)) = \varphi^{t+s}(x) = \varphi^s(\varphi^t(x)), \quad (\varphi^t)^{-1} = \varphi^{-1}, \quad \varphi^0 = \mathrm{Id}\,.$$

If $\Omega \subset \mathbb{R}^n$ is an open subset, the flow defines local groups φ^t of diffeomorphisms on their domains of definition. A flow is called a *global flow (or a complete flow)*, if it is defined on $\mathbb{R} \times \Omega$. Next, we describe conditions which guarantee that the flow is complete. We first remark that if a solution does not exist for all times, then it must leave every compact set in positive or in negative time.

Proposition IV.1. *We assume that $X : \Omega \to \mathbb{R}^n$ is a C^1 vector field and that the point $x \in \Omega$ satisfies $t^+(x) < \infty$. Then, there exists for every compact set K containing $x \in K \subset \Omega$ a time $\tau_K < t^+(x)$ such that*

$$\varphi^t(x) \notin K, \quad \tau_K < t < t^+(x).$$

Hence, if the solution $\varphi^t(x)$ remains in a compact set $K \subset \Omega$ for all $0 \leq t < t^+(x)$, then $t^+(x) = \infty$.

An analogous result holds true for $t^-(x)$.

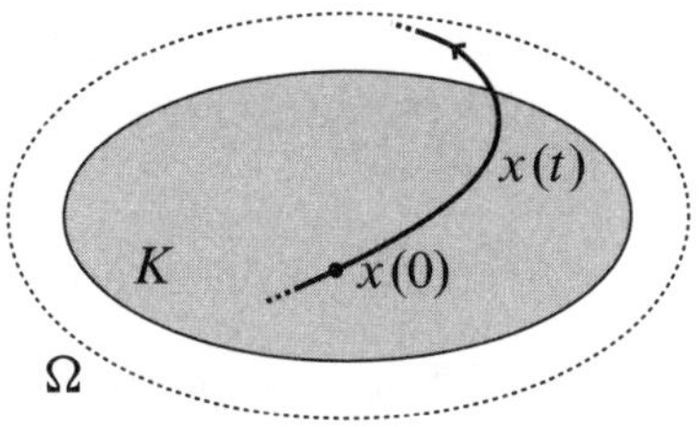

Figure IV.1. If $t^+(x) < \infty$, then the solution $x(t) = \varphi^t(x)$ leaves every compact set K.

Proof [Local flow]. Arguing by contradiction we assume the existence of a compact set K containing $x \in K \subset \Omega$ and the existence of a sequence $t_n \nearrow t^+(x) < \infty$ satisfying

$$x_n := \varphi^{t_n}(x) \in K, \quad n \geq 1.$$

Since K is compact, we find a convergent subsequence

$$x_n = \varphi^{t_n} \to x^* \in K.$$

In view of the existence of the local flow, we find a ball $B_a(x^*) \subset \Omega$ and an interval $I = (-\varepsilon, \varepsilon)$ such that the solution $\varphi^t(\xi)$ exists for every $\xi \in B_a(x^*)$ and every $t \in I$. For n large we conclude from the uniqueness of the flow that

$$\varphi^t(x_n) = \varphi^t(\varphi^{t_n}(x)) = \varphi^{t+t_n}(x), \quad t \in I,$$

so that the solution exists for all $0 \leq t < t_n + \varepsilon$. However, if n is large, then $\varepsilon + t_n > t^+(x)$, contradicting the definition of $t^+(x)$ and completing the proof. □

Proposition IV.2. *Let $X \colon \mathbb{R}^n \to \mathbb{R}^n$ be a bounded C^1 vector field, satisfying*

$$\sup_{x \in \mathbb{R}^n} |X(x)| =: M < \infty.$$

Then, $I_x = \mathbb{R}$ for all $x \in \mathbb{R}^n$.

Proof. Arguing again by contradiction, we find a point x_0 satisfying $I_{x_0} \neq \mathbb{R}$ and we assume that $t^+(x_0) < \infty$. The local flow satisfies

$$\varphi^t(x_0) - x_0 = \int_0^t X(\varphi^s(x_0))\, ds,$$

so that

$$|\varphi^t(x_0) - x_0| \leq \int_0^t |X(\varphi^s(x_0))|\, ds \leq tM \leq t^+(x_0) M < \infty$$

for all $0 \le t < t^+(x_0)$. Consequently, the solution remains in the compact set $\overline{B_R(x_0)}$, where $R = t^+(x_0)M$, in contradiction to Proposition IV.1. An analogous argument applies in case $t^-(x_0)$ is finite. □

More generally, the flow of a linearly bounded vector field is complete.

Proposition IV.3. *Let $X\colon \mathbb{R}^n \to \mathbb{R}^n$ be a linear bounded C^1 vector field, hence satisfying*

$$|X(x)| \le a + b|x|, \quad x \in \mathbb{R}^n$$

with non-negative constants $a, b \in \mathbb{R}$. Then, $I_x = \mathbb{R}$ for all $x \in \mathbb{R}^n$.

Proof. Assume that $t^+(x_0) < \infty$ and consider the time interval $0 \le t < t^+(x_0)$. From

$$\varphi^t(x_0) - x_0 = \int_0^t X(\varphi^s(x_0))\, ds$$

we deduce the estimate

$$|\varphi^t(x_0)| \le a(t) + b \int_0^t |\varphi^s(x_0)|\, ds$$

for all $0 \le t < t^+(x_0)$, where $a(t) := |x_0| + at$.

According to the Lemma of Gronwall below, the solution $\varphi^t(x_0)$ is therefore bounded, in contradiction to Proposition IV.1. Analogous arguments lead to a contradiction under the assumption that $t^-(x_0)$ is finite, and the proof of the proposition is finished. □

The Lemma of Gronwall is a useful tool to obtain estimates in ordinary differential equations. We only need a special case and refer to Chapter II in the book [5] by H. Amann for more general versions.

Lemma IV.4 (Gronwall). *We consider an interval $I \subset \mathbb{R}$, a point $t_0 \in I$ and non-negative functions $u, a, b\colon I \to \mathbb{R}$ and we assume that the function a is increasing. If*

$$u(t) \le a(t) + \left| \int_{t_0}^t b(s)u(s)\, ds \right| \quad \text{for all } t \in I,$$

then

$$u(t) \le a(t) e^{\left|\int_{t_0}^t b(s)\, ds\right|} \quad \text{for all } t \in I.$$

Proof. We carry out the proof for $t_0 = 0$ and $t \ge 0$. Introducing the functions

$$v(t) := \int_0^t b(s)u(s)\, ds,$$

$$\gamma(t) := e^{-\int_0^t b(s)\, ds},$$

it follows from the assumption that $u(t) \le a(t) + v(t)$. Therefore,

$$\dot{v}(t) \le b(t)[a(t) + v(t)].$$

Hence, using $\dot{\gamma} = -b(t)\gamma(t)$ one obtains

$$\frac{d}{dt}\big(\gamma(t)v(t)\big) \le a(t)b(t)\gamma(t).$$

Integrating this inequality from 0 to t and observing that $v(0) = 0$ one arrives at the estimate

$$\gamma(t)v(t) \le \int_0^t a(s)b(s)\gamma(s)\,ds.$$

In view of the monotonicity of the function a we get $\gamma(t)v(t) \le a(t)\int_0^t b(s)\gamma(s)\,ds$ and hence,

$$\begin{aligned} u(t)\gamma(t) &\le [a(t) + v(t)]\gamma(t) \\ &\le a(t)\Big[\gamma(t) + \int_0^t b(s)\gamma(s)\,ds\Big]. \end{aligned}$$

The integration of $\dot{\gamma}(t) = -b(t)\gamma(t)$ together with $\gamma(0) = 1$ results in

$$\int_0^t b(s)\gamma(s)\,ds = 1 - \gamma(t).$$

Hence, it follows from the above estimate that, indeed, $u(t)\gamma(t) \le a(t)$ and the lemma is proved. □

We next recall the concept of *equivalent vector fields*. We consider a diffeomorphism $u\colon \Omega_1 \to \Omega$ between the open sets $\Omega_1, \Omega \subset \mathbb{R}^n$. If X is a vector field on Ω and $x(t) = \varphi^t(x)$ a solution of the equation $\dot{x} = X(x)$, then we can associate with the differentiable curve $x(t)$ in Ω the differentiable curve $y(t)$ in Ω_1 defined by $x(t) = u(y(t))$ or

$$y(t) = u^{-1}(x(t)).$$

The curve $y(t)$ is a solution of a vector field on Ω_1. Indeed, from

$$X(u(y(t))) = X(x(t)) = \dot{x}(t) = du(y(t))\,\dot{y}(t),$$

it follows that the curve $y(t)$ solves the equation $\dot{y}(t) = [du(y(t))]^{-1}X(u(y(t)))$. Therefore, the curve $y(t)$ is a solution of the vector field Y defined by

$$Y(y) = [du(y)]^{-1}X(u(y)), \quad y \in \Omega_1.$$

The vector field

$$Y = (du)^{-1}X \circ u =: u^*X$$

is called the *transformed vector field or the pull-back of the vector field X under the diffeomorphism u.*

Definition. The vector field X on Ω and the vector field Y on Ω_1 are called *equivalent*, if there exists a diffeomorphism $u\colon \Omega_1 \to \Omega$ satisfying

$$du(y)Y(y) = X(u(y)), \quad y \in \Omega_1,$$

or equivalently,

$$Y(y) = (du(y))^{-1}X(u(y)) = (u^*X)(y), \quad y \in \Omega_1.$$

Two vector fields are equivalent, if and only if their local flows are conjugated.

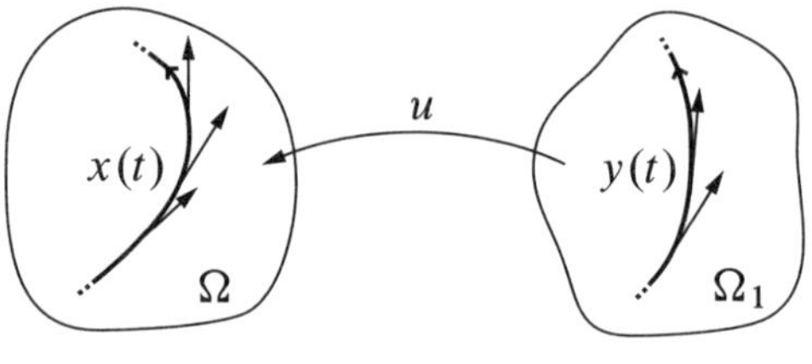

Figure IV.2. Two equivalent vector fields with flows.

Proposition IV.5. *Let $u\colon \Omega_1 \to \Omega$ be a diffeomorphism. If X is a vector field on Ω having the flow φ^t and Y a vector field on Ω_1 having the flow ψ^t, then the following two statements are equivalent.*

(i) *$du(y)Y(y) = X(u(y))$ for all $y \in \Omega_1$,*

(ii) *$u \circ \psi^t(y) = \varphi^t \circ u(y)$ on the domains of definition.*

Proof [Definition and uniqueness of the flow]. (ii)$\Longrightarrow$(i): Differentiating the conjugacy identity in t at the time $t = 0$, one finds, using the chain rule,

$$du(\psi^t(y))\frac{d}{dt}\psi^t(y) = du(\psi^t(y))Y(\psi^t(y)) = \frac{d}{dt}\varphi^t(u(y)) = X(\varphi^t(u(y))).$$

Due to $\psi^0(y) = y$ and $\varphi^0(u(y)) = u(y)$, the identity $du(y)Y(y) = X(u(y))$ follows for $t = 0$.

(i)$\Longrightarrow$(ii): Introducing the family of diffeomorphisms $\psi^t(y) := u^{-1}\circ\varphi^t\circ u(y)$, we obtain the identity

$$(*) \quad u \circ \psi^t(y) = \varphi^t \circ u(y).$$

We want to show that ψ^t is the flow of Y. In view of the uniqueness it suffices to verify that $\psi^0(y) = y$ (this follows immediately from $\varphi^0 = \mathrm{Id}$) and that

$\frac{d}{dt}\psi^t(y) = Y(\psi^t(y))$ for $t \in I_y$. By differentiating the equation $(*)$ in t we obtain by the chain rule

$$\begin{aligned} du(\psi^t(y))\frac{d}{dt}\psi^t(y) &= \frac{d}{dt}\varphi^t(u(y)) \\ &= X(\varphi^t(u(y))) \\ &= X(u(\psi^t(y))), \end{aligned}$$

and using the assumption (i) the desired equation $\frac{d}{dt}\psi^t(y) = Y(\psi^t(y))$ follows. □

Instead of going immediately after the solutions of a given vector field, it is often more advisable to first transform the vector field into a simple form from which the solutions can be read off easily. This is illustrated by the following local result.

Theorem IV.6 (Local straightening out theorem). *Let x^* be a* ***regular*** *point of the C^1 vector field X, that is, $X(x^*) \neq 0$. Then, there exists a neighborhood U of x^* on which the vector field X is equivalent to the* ***constant*** *vector field $Y(y) = e_1 = (1, 0, \ldots, 0)$.*

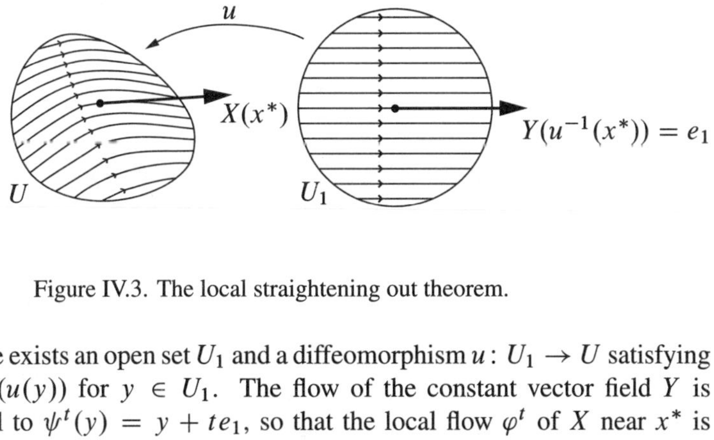

Figure IV.3. The local straightening out theorem.

Hence, there exists an open set U_1 and a diffeomorphism $u \colon U_1 \to U$ satisfying $du(y)e_1 = X(u(y))$ for $y \in U_1$. The flow of the constant vector field Y is obviously equal to $\psi^t(y) = y + te_1$, so that the local flow φ^t of X near x^* is represented by

$$\varphi^t(u(y)) = u(y + te_1).$$

Proof of the theorem [Inverse function theorem]. By translation, we can assume that $x^* = 0$ is the origin. Since $X(0) \neq 0$, we can, after renumbering of the coordinates, assume that

$$\langle X(0), e_1 \rangle = X_1(0) \neq 0.$$

Using the local flow φ^t of the vector field X near $x^* = 0$ we define the map $u\colon \mathbb{R}^n \to \mathbb{R}^n$ in the neighborhood of 0 by

$$y = (y_1, \dots, y_n) \mapsto u(y) := \varphi^{y_1}(0, y_2, \dots, y_n).$$

Then $u(0) = \varphi^0(0) = 0$ and

$$du(0) = \left(\begin{array}{c|ccc} X_1(0) & 0 & \dots & 0 \\ \hline X_2(0) & 1 & & \\ \vdots & & \ddots & \\ X_n(0) & & & 1 \end{array}\right).$$

This follows directly from the definition of the flow of the vector field X. Hence, $\det du(0) = X_1(0) \neq 0$ and in view of the inverse function theorem the map u is

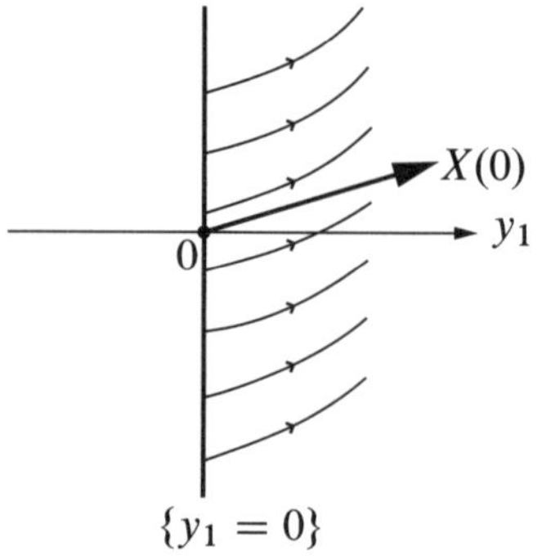

Figure IV.4. Geometrically, the plane $\{y_1 = 0\}$ is locally a transversal section of X.

a local diffeomorphism $u\colon U_1 \to u(U_1) =: U$ leaving the origin fixed, where U_1 is an open neighborhood of the origin. We define the constant vector field Y on U_1 by $Y(y) = e_1$. Its flow is equal to $\psi^t(y) = y + te_1$ and one computes,

$$\begin{aligned} u \circ \psi^t(y) &= u(y_1 + t, y_2, \dots, y_n) \\ &= \varphi^{y_1+t}(0, y_2, \dots, y_n) \\ &= \varphi^t \circ \varphi^{y_1}(0, y_2, \dots, y_n) \\ &= \varphi^t \circ u(y). \end{aligned}$$

Consequently, the flows are conjugated and the theorem follows from Proposition IV.5. □

Corollary IV.7. *All the vector fields of the same dimension are equivalent* **locally** *in the neighborhood of regular points. Near a regular point, there exist no local invariants.*

The dynamics near a regular point is clearly not interesting and we introduce next the *singular points* where the vector field vanishes.

Definition. A point x^* is called a *singular point* (or *equilibrium point* or *rest point*) of a vector field X, if $X(x^*) = 0$.

The solution of a vector field through the singular point x^* satisfies $\varphi^t(x^*) = x^*$ for all t. At a singular point we shall encounter *local* invariants of the vector field. We assume that X and Y are equivalent vector fields, so that

$$du(y)Y(y) = X(u(y))$$

with a local diffeomorphism u. If $x^* = u(y^*)$ is a singular point of X, then

$$X(x^*) = 0 \Longleftrightarrow Y(y^*) = 0.$$

We linearize the vector fields at the corresponding singularities. Differentiating the previous identity in the variable y at the point $y = y^*$ one obtains the equation

$$du(y^*)\, dY(y^*) = dX(x^*)\, du(y^*)$$

or equivalently,

$$dY(y^*) = [du(y^*)]^{-1} dX(x^*) du(y^*).$$

Hence, the two linear maps $dY(y^*)$ and $dX(x^*)$ belonging to $\mathcal{L}(\mathbb{R}^n)$ are *equivalent*. Consequently, they possess the same eigenvalues, and, more generally, they have the same *Jordan normal forms*. These data are the linear invariants of the vector field at the singular point.

Remark (Reparametrization of the time). We consider the vector field X and denote its flow by φ^t. The curve $t \mapsto \varphi^t(x)$ is a solution, hence a parameterized orbit. We can reparametrize this orbit as a solution of the rescaled vector field Y defined by

$$Y(x) = \varrho(x)X(x), \quad x \in \mathbb{R}^n,$$

where $\varrho\colon \mathbb{R}^n \to \mathbb{R}$ is a positive C^∞ function. We claim that *the flow ψ^t of this vector field Y satisfies*

$$\psi^{s(t,x)}(x) = \varphi^t(x),$$

where $t \mapsto s(t,x)$ is the strictly increasing function from $\mathbb{R}$ into $\mathbb{R}$, defined as the unique solution of the Cauchy initial value problem

$$\begin{cases} \frac{ds}{dt} = \frac{1}{\varrho(\psi^s(x))} > 0, \\ s(0,x) = 0, \end{cases}$$

in which x enters as a parameter.

Proof. In order to prove the claim, we verify that the curve $t \mapsto \psi^{s(t,x)}(x)$ solves the Cauchy initial value problem for the vector field X with the initial condition

$$\psi^{s(0,x)}(x) = \psi^0(x) = x.$$

Differentiating in time t and using the chain rule, we compute

$$\begin{aligned}
\frac{d}{dt}\psi^{s(t,x)}(x) &= \frac{d}{ds}\psi^s(x)\frac{ds}{dt}(t,x) \\
&= Y(\psi^s(x))\frac{ds}{dt}(t,x) \\
&= \varrho(\psi^s(x))X(\psi^s(x))\frac{ds}{dt}(t,x) \\
&= \varrho(\psi^s(x))X(\psi^s(x))\frac{1}{\varrho(\psi^s(x))} \\
&= X(\psi^{s(t,x)}(x)).
\end{aligned}$$

The uniqueness of the flow implies indeed that $\varphi^t(x) = \psi^s(x)$ with $s = s(t,x)$ as claimed. □

From

$$\frac{ds}{dt}(t,x) = \frac{1}{\varrho(\psi^s(x))} = \frac{1}{\varrho(\varphi^t(x))}$$

we obtain in view of $s(0,x) = 0$, the following transformation formula for the time parameter,

$$s(t,x) = \int_0^t \frac{d\tau}{\varrho(\varphi^\tau(x))}.$$

Example. If $\varrho(x) = \lambda > 0$, then $s(t,x) = \frac{1}{\lambda}t$.

IV.2 Limit sets, attractors and Lyapunov functions

Where do the solutions come from and where do they go to? In order to study the *asymptotic* behavior as $|t| \to \infty$, it is useful to first introduce the concept of a limit set of an orbit. To do so, we consider the flow φ^t of the vector field X in $\mathbb{R}^n$ and abbreviate it as

$$\varphi^t(x) = x \cdot t, \quad t \in I_x.$$

We recall that the differentiable mapping $(t,x) \mapsto x \cdot t$ on its domain of definition $U \subset \mathbb{R} \times \mathbb{R}^n$ into $\mathbb{R}^n$ satisfies the properties

$$\begin{aligned}
x \cdot 0 &= x, \quad x \in \mathbb{R}^n, \\
(x \cdot t) \cdot s &= x \cdot (t+s), \quad s, t, t+s \in I_x,
\end{aligned}$$

where

$$I_x = \big(-t^-(x), t^+(x)\big) \subset \mathbb{R}$$

is the maximal interval of definition of the orbit with initial condition x. We shall denote the *unparametrized orbits* of the flow as the sets

$$\begin{aligned}\mathcal{O}^+(x) &= \{x \cdot t \mid 0 \le t < t^+(x)\},\\ \mathcal{O}^-(x) &= \{x \cdot t \mid t^-(x) < t \le 0\},\\ \mathcal{O}(x) &= \{x \cdot t \mid t^-(x) < t < t^+(x)\}.\end{aligned}$$

Definition. If $t^+(x) = \infty$, the *ω-limit set* of the point x is the set

$$\omega(x) := \{y \in \mathbb{R}^n \mid \text{there exists a sequence } t_j \to \infty \text{ satisfying } x \cdot t_j \to y\}.$$

If $t^-(x) = -\infty$, we define

$$\omega^*(x) := \{y \in \mathbb{R}^n \mid \text{there exists a sequence } t_j \to \infty \text{ satisfying } x \cdot (-t_j) \to y\}.$$

The limit sets can be *empty*. The limit sets can equivalently be defined by

$$\begin{aligned}\omega(x) &= \bigcap_{s>0} \overline{\mathcal{O}^+(x \cdot s)},\\ \omega^*(x) &= \bigcap_{s<0} \overline{\mathcal{O}^-(x \cdot s)}\end{aligned}$$

from which one concludes that $\omega(x)$ and $\omega^*(x)$ are *closed* sets. In order to verify the above representations of the limit sets we assume that $y \in \overline{\mathcal{O}^+(x \cdot s)}$ for every $s > 0$. Therefore, there exists for every positive integer n a real number $t(n) > n$, such that the distance between $x \cdot t(n)$ and y is smaller than n^{-1}. The sequence $t(n)$ satisfies $\lim t(n) = \infty$ and $\lim x \cdot t(n) = y$, so that y is an element of $\omega(x)$. Conversely, if y is an element of $\omega(x)$, then there exists a sequence $t_n \to \infty$ satisfying $x \cdot t_n \to y$. If $s > 0$ is given, then $x \cdot t_n \in \mathcal{O}^+(x \cdot s)$ for sufficiently large integers n and hence $y = \lim x \cdot t_n \in \overline{\mathcal{O}^+(x \cdot s)}$. This holds true for every $s > 0$.

Example. On $\mathbb{R}^2$ we study the differential equations

$$\begin{aligned}\dot{x}_1 &= x_2 + x_1(1 - r^2),\\ \dot{x}_2 &= -x_1 + x_2(1 - r^2),\end{aligned}$$

where $r^2 = x_1^2 + x_2^2$. In order to decouple this coupled system of differential equations we introduce the polar coordinates $(x_1, x_2) = (r \cos\vartheta, r \sin\vartheta)$ and obtain by means of the transformation formula for vector fields the two decoupled equations

$$\begin{aligned}\dot{r} &= r(1 - r^2),\\ \dot{\vartheta} &= -1.\end{aligned}$$

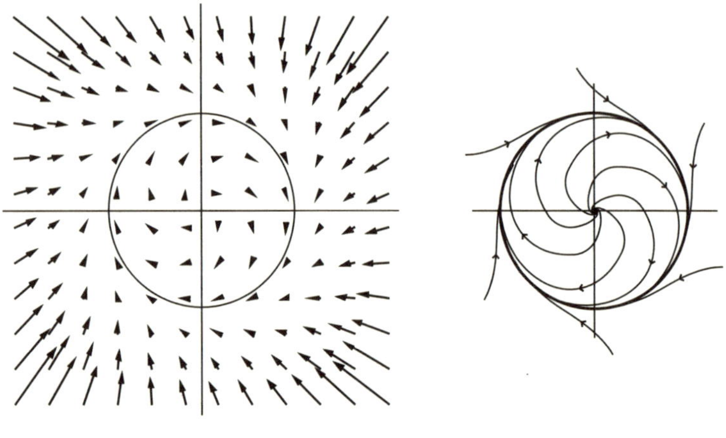

Figure IV.5. The vector field $(\dot{x}_1, \dot{x}_2)$ from the example, shown with the circle S^1 (left), and some solutions (right).

Clearly, the solutions of the second equation are the functions $\vartheta(t) = -t + \vartheta(0)$. The first equation can, of course, also be solved explicitly by a formula for the solutions $r(t)$. This, however, is not necessary because the qualitative behavior of the solutions on the positive real axis are obvious. Namely, the points $r = 0$ and $r = 1$ are the rest points and all the solutions starting in the complements of these rest points have to converge to the rest point $r = 1$ as $t \to \infty$. In backward time the solutions starting at points r smaller than 1 converge to the rest point $r = 0$ while those starting at points larger than 1 diverge to infinity.

Therefore the limit sets of the solutions $x(t)$ in $\mathbb{R}^2$ are the sets

$$\begin{aligned} \omega(0) &= \{0\}, & & \\ \omega(x) &= S^1, & & x \in \mathbb{R}^2 \setminus \{0\}, \\ \omega^*(x) &= \emptyset, & & |x| > 1, \\ \omega^*(x) &= S^1, & & |x| = 1, \\ \omega^*(x) &= \{0\}, & & |x| < 1. \end{aligned}$$

Here S^1 is the unit circle centered at the origin.

Proposition IV.8 (Properties of the limit sets). *The ω-limit set $\omega(x)$ is invariant under the flow so that $\omega(x) \cdot t \subset \omega(x)$ for every t.*

If the closure $\overline{\mathcal{O}^+(x)}$ of the orbit is compact (*and hence $t^+(x) = \infty$ by Proposition* IV.1), *the limit sets have the following properties.*

(i) *$\omega(x)$ is compact and not empty,*

(ii) *$x \cdot t \to \omega(x)$ as $t \to \infty$,*

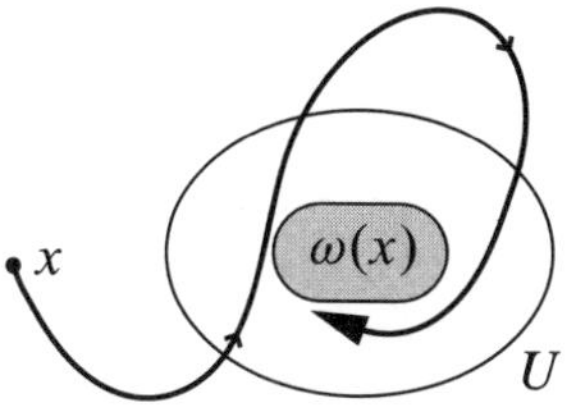

Figure IV.6. Proposition IV.8 (ii).

(iii) $\omega(x)$ *is a connected set.*

The limit set $\omega^*(x)$ has analogous properties as the time tends to minus infinity.

Proof. If $y \in \omega(x)$, there exists a sequence $t_n \to \infty$ satisfying $x \cdot t_n \to y$ and using the continuity of the flow we conclude for every real number t,

$$y \cdot t = \big(\lim_{n\to\infty} x \cdot t_n\big) \cdot t = \lim_{n\to\infty} x \cdot (t_n + t).$$

Since $t_n + t \to \infty$ as $n \to \infty$ we find $y \cdot t \in \omega(x)$, so that the limit set is indeed invariant under the flow.

(i) We take a sequence $t_n \to \infty$. Then the sequence $(x \cdot t_n)_{n\geq 1}$ lies in the compact set $\overline{\mathcal{O}^+(x)}$, so that there exists a convergent subsequence whose limit belongs to $\omega(x)$, in view of the definition of a limit set. Consequently, the limit set $\omega(x)$ is not empty. Since $\omega(x)$ is closed and contained in the compact set $\overline{\mathcal{O}^+(x)}$, the limit set $\omega(x)$ is compact.

(ii) We shall prove that for every sequence $t_n \to \infty$ and for every open neighborhood U of $\omega(x)$ there exists an integer n_U, satisfying $x \cdot t_n \in U$ for all $n > n_U$. Arguing by contradiction we assume that $x \cdot t$ does not converge to $\omega(x)$. Then there exists an open neighborhood U of $\omega(x)$ and a sequence $t_n \to \infty$ satisfying $x \cdot t_n \notin U$ for all n. Since $\overline{\mathcal{O}^+(x)} \cap U^c$ is *compact*, there exists a convergent subsequence $x \cdot t_j \to y \in U^c$. Its limit satisfies $y \in \omega(x) \cap U^c$, in contradiction to the assumption $U \supset \omega(x)$.

(iii) In order to prove the statement in (iii), we argue again by contradiction and assume the existence of two closed sets $\omega_1, \omega_2 \neq \emptyset$ satisfying $\omega(x) = \omega_1 \cup \omega_2$ and $\omega_1 \cap \omega_2 = \emptyset$. With $\omega(x)$ also the sets ω_1 and ω_2 are compact. Therefore, there exist open sets $U_1, U_2 \subset \mathbb{R}^n$ satisfying

$$\omega_1 \subset U_1, \quad \omega_2 \subset U_2, \quad U_1 \cap U_2 = \emptyset.$$

The set $U := U_1 \cup U_2$ is an open neighborhood of $\omega(x)$. In view of (ii),

$$x \cdot t \in U, \quad t \geq t_U$$

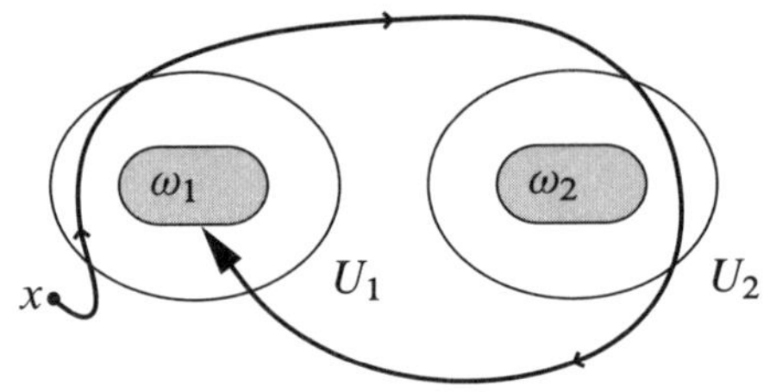

Figure IV.7. Proposition IV.8 (iii).

for some time $t_U \in \mathbb{R}$. Since the interval $[t_U, \infty)$ is connected and the flow continuous, the image set $\{x \cdot t \mid t_U \leq t < \infty\}$ under the flow is *connected.* In view of the definition of U this set must therefore lie completely in one of the two open sets, let us say in U_1. Consequently, there exists *no* sequence $t_n \to \infty$, for which $x \cdot t_n$ converges to a point in $\omega_2 \subset U_2$ and since $\omega_2 \subset \omega(x)$ we conclude that $\omega_2 = \emptyset$, in contradiction to the assumption. We have verified that the limit set $\omega(x)$ is connected and the proposition is proved. □

Definition. A point $x \in \mathbb{R}^n$ is *attracted* by a subset $M \subset \mathbb{R}^n$, if $t^+(x) = \infty$ and

$$x \cdot t \to M \quad \text{for } t \to \infty.$$

The *domain of attraction* of M is the set of all points attracted by M and is denoted by $A(M)$. A closed set M is called an *attractor*, if $A(M)$ is a neighborhood of M.

Lyapunov functions can be a useful tool to localize limit sets and to find attractors.

Definition. A continuous function $V : \mathbb{R}^n \to \mathbb{R}$ is called a *Lyapunov function* of the flow, if it decreases along every orbit, so that for every $x \in \mathbb{R}^n$,

$$V(x \cdot t) \leq V(x \cdot s) \quad \text{for all } t \geq s.$$

The *derivative of a Lyapunov function along the flow* is, by definition, the function

$$\dot{V}(x) := \varliminf_{t \searrow 0} \frac{1}{t}[V(x \cdot t) - V(x)].$$

Here we made use of the well-known definition for a function $\varphi : (a, b) \to \mathbb{R}$,

$$\varliminf_{h \searrow a} \varphi(h) := \sup_{a<t\leq b} \Big[\inf_{a<h<t} \varphi(h)\Big] = \lim_{t \searrow a} \Big[\inf_{a<h<t} \varphi(h)\Big].$$

Clearly, a Lyapunov function satisfies $-\infty \leq \dot{V}(x) \leq 0$.

Example. If $V\colon \mathbb{R}^n \to \mathbb{R}$ is any *differentiable* function and $x \cdot t = \varphi^t(x)$ is the flow of the vector field X, then by the chain rule and the definition of the flow of a vector field,

$$\begin{aligned}\frac{d}{dt}V(\varphi^t(x)) &= dV(\varphi^t(x))\frac{d}{dt}\varphi^t(x)\\ &= dV(\varphi^t(x)\,X(\varphi^t(x))\\ &= \langle \nabla V(\varphi^t(x)), X(\varphi^t(x))\rangle.\end{aligned}$$

The vector field ∇V is the gradient of the function V with respect to the Euclidean scalar product. Therefore,

$$\dot V(x) = \frac{d}{dt}V(\varphi^t(x))\Big|_{t=0} = \langle \nabla V(x), X(x)\rangle.$$

We will frequently make use of this calculation.

Lemma IV.9. *Let X be a vector field with flow φ^t and let $V\colon \mathbb{R}^n \to \mathbb{R}$ be a differentiable function. Then V is a Lyapunov function of the flow φ^t if and only if $\langle \nabla V(x), X(x)\rangle \le 0$ for all $x \in \mathbb{R}^n$.*

Proof. If $\langle \nabla V(x), X(x)\rangle \le 0$ for all $x \in \mathbb{R}^n$, then it follows for $t \ge s$ that

$$\begin{aligned}V(\varphi^t(x)) - V(\varphi^s(x)) &= \int_s^t \frac{d}{d\tau}V(\varphi^\tau(x))\,d\tau\\ &= \int_s^t \langle \nabla V(\varphi^\tau(x)), X(\varphi^\tau(x))\rangle\,d\tau\\ &\le 0,\end{aligned}$$

so that V is a Lyapunov function. Conversely if V is a Lyapunov function, then $\dot V(x) \le 0$ and in view of the above calculation, $\langle \nabla V(x), X(x)\rangle = \dot V(x) \le 0$, as claimed. □

Theorem IV.10 (La Salle's principle). *We assume that V is a Lyapunov function on the closed set M. Let $\mathcal{O}^+(x) \subset M$ be an orbit whose limit set $\omega(x) \ne \emptyset$ is not empty. Then*

(i) *$\omega(x) \subset V^{-1}(\alpha)$ for an $\alpha \in \mathbb{R}$, i.e., $V|_{\omega(x)} = \text{const.}$,*
(ii) *$\omega(x) \subset \{y \in M \mid \dot V(y) = 0\}$, i.e., $\dot V|_{\omega(x)} = 0$.*

Proof. Clearly, $\omega(x) \subset \overline{\mathcal{O}^+(x)} \subset \overline{M} = M$. Since $V(x \cdot t)$ is decreasing, there exists the limit $\lim_{t\to\infty} V(x\cdot t) = \inf_{t\ge 0} V(x\cdot t) =: \alpha$ in $\mathbb{R} \cup \{-\infty\}$. If $y \in \omega(x)$, there exists a sequence $t_n \to \infty$ satisfying $x \cdot t_n \to y$ and the *continuity* of the function V implies

$$\lim_{n\to\infty} V(x \cdot t_n) = V(y) = \alpha.$$

In particular, $\alpha \in \mathbb{R}$. This holds true for every $y \in \omega(x)$, so that $V(y) = \alpha$ for all $y \in \omega(x)$ and (i) is proved. Since $\omega(x)$ is *invariant* under the flow in view of Proposition IV.8, it follows for $y \in \omega(x)$, that $V(y \cdot t) = V(y) = \alpha$ for all t and consequently $\dot{V}(y) = 0$, as claimed in the theorem. □

Definition. The *maximal invariant set* of the subset $M \subset \mathbb{R}^n$ is the set

$$I(M) = \{x \in M \mid x \cdot t \in M \text{ for all } t \text{ in } \mathbb{R}\}.$$

Example. If 0 is a *hyperbolic* equilibrium point of a vector field then there exists an open neighborhood M whose maximal invariant subset consists of the equilibrium point, so that $I(M) = \{0\}$.

Proposition IV.11. *Let V be a Lyapunov function on a closed set M and assume that the orbit $\mathcal{O}^+(x) \subset M$ has a compact closure $\overline{\mathcal{O}^+(x)}$. Then for the set C defined by*

$$C := \{y \in M \mid \dot{V}(y) = 0\},$$

it holds true that

$$x \cdot t \to I(C), \quad t \to \infty.$$

If, in addition, the set M is positively invariant, so that $M \cdot t \subset M$ for all $t \geq 0$, and if $\overline{\mathcal{O}^+(x)}$ is compact for every $x \in M$, then

$$M \subset A(I(C)).$$

Proof [Theorem IV.10]. Take $x \in M$, then $\omega(x) \neq \emptyset$ due to Proposition IV.8. In view of Theorem IV.10,

$$\omega(x) \subset \{y \in M \mid \dot{V}(y) = 0\} = C.$$

Since $\omega(x)$ is invariant under the flow (Proposition IV.8), we have $\omega(x) \subset I(C)$. From $x \cdot t \to \omega(x)$ as $t \to \infty$ we conclude $x \cdot t \to I(C)$.

Finally, if the additional assumptions of Proposition IV.11 are satisfied, then $x \cdot t \to I(C)$ for every $x \in M$ so that $M \subset A(I(C))$, as we wanted to prove. □

Proposition IV.12. *Let V be a coercive Lyapunov function on $\mathbb{R}^n$, i.e., $V(x) \to \infty$ for $|x| \to \infty$. Then the closure $\overline{\mathcal{O}^+(x)}$ of every orbit is compact, the set $C := \{y \in M \mid \dot{V}(y) = 0\}$ is not empty and the maximal invariant set $I(C)$ is a* ***global attractor****,*

$$A(I(C)) = \mathbb{R}^n.$$

Proof. Since V is continuous, the sublevel set $V^a := \{x \in \mathbb{R}^n \mid V(x) \leq a\}$ is closed for every $a \in \mathbb{R}$. If a is sufficiently large, the set $V^a \neq \emptyset$ is not empty. Due to the coerciveness of the function V, the set V^a is bounded and hence *compact*. If $x \in V^a$, then $V(x \cdot t) \leq V(x) \leq a$ for every $0 \leq t < t^+(x)$ so that $t^+(x) = \infty$, in view of Proposition IV.1, and hence the sublevel set V^a is positively invariant. Therefore, for every $x \in V^a$ the set $\overline{\mathcal{O}^+(x)}$ is compact and $\omega(x) \neq \emptyset$ and, in view of Proposition IV.11, we have the convergence $x \cdot t \to I(C)$. Due to the coerciveness of the function V there exists for every $x \in \mathbb{R}^n$ a real number a satisfying $x \in V^a$. The proof is complete. □

Definition. A singular point x^* of the vector field X is called *Lyapunov-stable* or, in short, *L-stable* (or $+$-stable), if every open neighborhood O of x^* contains an open neighborhood $x^* \in O_1 \subset O$ such that

$$O_1 \cdot t \subset O \quad \text{for all } t \geq 0.$$

The point x^* is called *asymptotically stable*, if the open neighborhood O_1 can, in addition, be chosen such that $O_1 \subset A(\{x^*\})$ and hence

$$\lim_{t\to\infty} x \cdot t = x^* \quad \text{for all } x \in O_1.$$

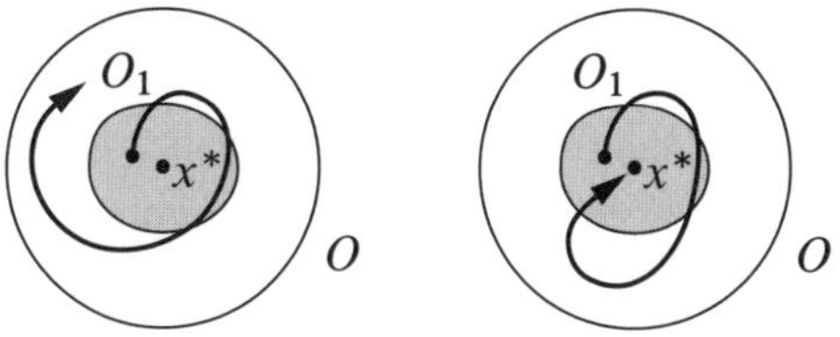

Figure IV.8. Lyapunov stability (left) and asymptotic stability (right).

Theorem IV.13 (Lyapunov). *Let x^* be a singular point of the vector field X and let U be an open neighborhood of x^* and let $V : U \to \mathbb{R}$ be a continuous function that is differentiable on $U \setminus \{x^*\}$. If*

(1) *$V(x^*) = 0$ and $V(x) > 0$ for $x \in U \setminus \{x^*\}$,*

(2) *$\dot{V}(x) \leq 0$ for $x \in U \setminus \{x^*\}$,*

then, x^* is L-stable.

If, in addition, the inequality in (2) *is strict, then x^* is asymptotically stable.*

Proof. If O is an open neighborhood of x^*, we choose a closed ball $\overline{B_\delta(x^*)} \subset U \cap O$, where $\delta > 0$. Since the function V is continuous and since the boundary $\partial B_\delta(x^*)$ is compact, it follows from (1) that

$$\alpha := \inf\{V(x) \mid x \in \partial B_\delta(x^*)\} > 0.$$

We define the open neighborhood $O_1 := \{x \in B_\delta(x^*) \mid V(x) < \alpha\}$ of x^* and choose a point $x \in O_1$. In view of (2) we have $V(x \cdot t) \leq V(x) < \alpha$, and since

$V \geq \alpha$ on $\partial B_\delta(x^*)$, the continuous curve $x \cdot t$ cannot leave the ball $B_\delta(x^*)$ for $t \geq 0$. Therefore, the open set O_1 is positively invariant under the flow and the point x^* is L-stable.

We assume now that $\dot{V}(x) < 0$ for all $x \in U \setminus \{x^*\}$ and show for every $x^* \neq x \in O_1$ that $x \cdot t_n \to x^*$ for every sequence $t_n \to \infty$. Arguing by contradiction we find, due to the compactness of $\overline{B_\delta(x^*)}$, a convergent sequence

$$x \cdot t_n \to y^* \neq x^*,$$

where $y^* \in \overline{B_\delta(x^*)}$. In view of the assumptions, $V(y^*) > 0$. Moreover, the function $V(z \cdot t)$ is strictly decreasing for every $z \in U \setminus \{x^*\}$. It follows that

$$(*) \quad V(x \cdot t) \geq V(y^*) > 0 \quad \text{for all } t \geq 0, \quad \text{and} \quad V(y^* \cdot s) < V(y^*) \quad \text{if } s > 0.$$

Since the function V is continuous, there exists an open neighborhood U_2 of y^* within which

$$V(y \cdot s) < V(y^*) \quad \text{if } s \geq 1 \text{ and } y \in U_2.$$

However, for n large, the point $x \cdot t_n$ is in U_2 and therefore $V(x \cdot (t_n + s)) < V(y^*)$, in contradiction to $(*)$. The proof of Theorem IV.13 is complete. □

Remark. In physics, one can often take the energy as a Lyapunov function V. In this case the L-stable point x^* is a local *minimum of the energy*.

Example. We study the *pendulum with linear friction* which is determined by the ordinary differential equation of second order

$$\ddot{x} + \sin x = -\mu \dot{x}$$

with a positive constant $\mu > 0$. The equation is equivalent to the system of equations

$$\begin{pmatrix} \dot{x} \\ \dot{y} \end{pmatrix} = \begin{pmatrix} y \\ -\sin x - \mu y \end{pmatrix} =: X(x, y) \in \mathbb{R}^2.$$

As usual, we work in the covering space $\mathbb{R}^2$ of the phase space $S^1 \times \mathbb{R}$. Since the vector field X is linearly bounded, the flow φ^t exists for all times, in view of Proposition IV.3.

The *energy function* $H\colon \mathbb{R}^2 \to \mathbb{R}$, defined by

$$H(x, y) := \frac{1}{2} y^2 - \cos x \geq -1,$$

is a *Lyapunov function*, since

$$\dot{H}(x, y) = \langle \nabla H(x, y), X(x, y) \rangle = -\mu y^2 \leq 0.$$

In particular, $\dot{H}(x, y) = 0$ if and only if $y = 0$. Therefore, the function $H(\varphi^t(x, y))$ is strictly decreasing outside of the set $\{y = 0\}$. Since the gradient

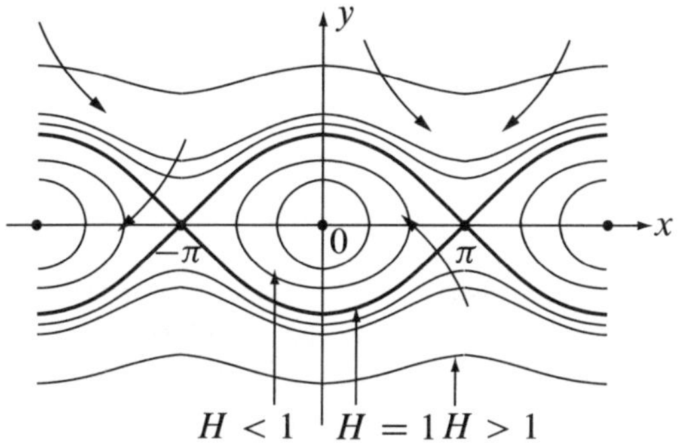

Figure IV.9. Level sets of the energy H and a few solutions.

∇H is orthogonal to the level lines of the function H, the vector field X is transversal to the level lines in all the points (x, y) satisfying $y \neq 0$.

We consider the set

$$C := \{(x, y) \in \mathbb{R}^2 \mid \dot{H}(x, y) = 0\} = \{y = 0\}.$$

In order to determine the maximal invariant set $I(C)$ in C we assume that the solution $(x(t), y(t)) \in C$ remains in C for all $t \in \mathbb{R}$. Then $\sin x(t) \equiv 0$ and consequently the function $x(t) = k\pi$ is constant, where $k \in \mathbb{Z}$.

Conversely, these points lie in $I(C)$, since they are equilibrium points, so that the maximal invariant set in C, given by

$$I(C) = \{(k\pi, 0) \mid k \in \mathbb{Z}\},$$

is the set of the singular points of the vector field X.

If the solution $z(t) = (x(t), y(t)) \in \mathbb{R}^2$ is *bounded* for $t \geq 0$, then the closure $\overline{\mathcal{O}^+(z(0))}$ of the orbit is compact and by Proposition IV.11,

$$z(t) \to I(C) \quad \text{as } t \to \infty.$$

Because the set $I(C)$ is discrete, there exists a unique integer $k \in \mathbb{Z}$ such that

$$z(t) \to (k\pi, 0), \quad t \to \infty,$$

and the pendulum converges to the highest or to the lowest position. We shall prove that, since $\mu > 0$, *every* solution is bounded in $\mathbb{R}^2$ for $t \geq 0$.

Proof. (1) We first verify that the function $y(t)$ *is bounded for* $t \geq 0$. Since the function $H(\varphi^t(z))$ is decreasing along every solution,

$$\begin{aligned}\frac{1}{2}y(t)^2 &= H(z(t)) + \cos x(t)\\ &\le H(z(0)) + 1\\ &=: M(z(0)),\end{aligned}$$

where we have defined the positive constant $M = M(z(0)) > 0$. In terms of physics, the kinetic energy is bounded by the initial energy.

(2) We next show that the function $x(t)$ *is bounded for* $t \ge 0$.

Every eye A_j in Figure IV.10 is a bounded set which is positively invariant, since the function H is decreasing along the solutions and is, on the boundary ∂A_j, equal to $H = 1$.

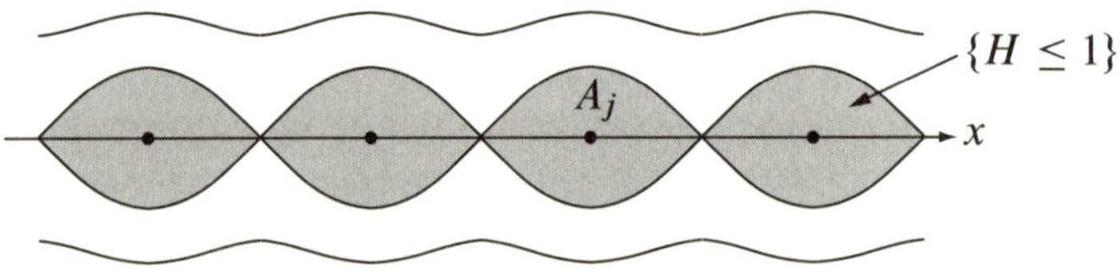

Figure IV.10. An eye A_j.

Every solution $z(t)$, which is not constant and which hits the separatrix $\{H = 1\}$, remains in one of the eyes A_j for all future times, since the function $H(z(t))$ is *strictly* decreasing outside of the set $\{y = 0\}$.

We assume that there exists a solution $z(t) = (x(t), y(t))$ whose function $x(t)$ is *not* bounded. Then, in view of the above discussion, $\varphi^t(z(0)) \notin \{H \le 1\}$ for all $t \ge 0$, and so,

$$\varphi^t(z(0)) \in \{H > 1\}, \quad t \ge 0.$$

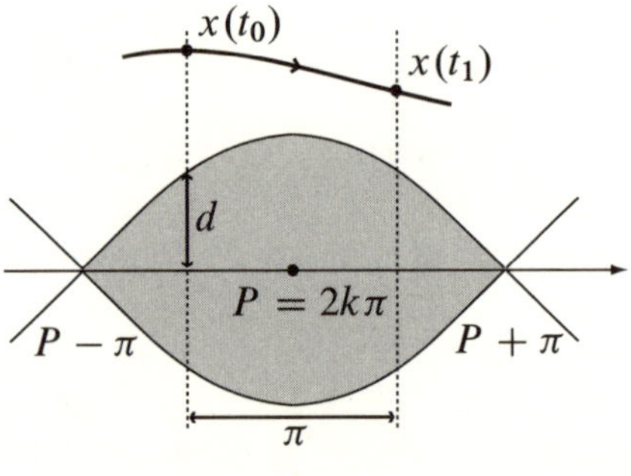

Figure IV.11. To prove that $x(t)$ is bounded.

In particular, sign $y(t)$ is constant (no oscillation of the pendulum), so that $z(t)$ lies completely above or below the branches of the separatrix. We assume that sign $y(t) = +1$. Because $x(t)$ is continuous and not bounded, there exists, in view of the mean value theorem, for every integer $k \in \mathbb{Z}$ sufficiently large, a time $t_0 \in \mathbb{R}$ at which $x(t_0) = 2k\pi - \frac{1}{2}\pi$ and a time $t_1 \in \mathbb{R}$ at which $x(t_1) = 2k\pi + \frac{1}{2}\pi$.

The branches of the separatrix are given by $|y| = \sqrt{2(\cos x + 1)} = |2\cos\frac{x}{2}|$, so that the function $y(t)$ satisfies

$$|y(t)| \geq |2\cos(k\pi - \frac{\pi}{4})| = 2^{-\frac{1}{2}} =: d$$

on the interval $t_0 \leq t \leq t_1$. From

$$\pi = x(t_1) - x(t_0) = \int_{t_0}^{t_1} \dot{x}(\tau)\, d\tau = \int_{t_0}^{t_1} y(\tau)\, d\tau \leq (t_1 - t_0)\sqrt{2M}$$

we conclude that $t_1 - t_0 \geq \frac{\pi}{\sqrt{2M}}$ with the previously defined constant $M = M(z(0)) > 0$. Consequently,

$$\begin{aligned}
H(\varphi^{t_1}(z)) - H(\varphi^{t_0}(z)) &= \int_{t_0}^{t_1} \frac{d}{d\tau} H(\varphi^{\tau}(z))\, d\tau \\
&= \int_{t_0}^{t_1} \dot{H}(\varphi^{\tau}(z))\, d\tau \\
&= -\int_{t_0}^{t_1} \mu y(\tau)^2\, d\tau \\
&< -(t_1 - t_0)\mu d^2 \\
&\leq -\frac{\pi}{\sqrt{2M}}\mu d^2 \\
&=: K.
\end{aligned}$$

□

Hence, the function $H(\varphi^t(z(0)))$ decreases on the interval $[t_0, t_1]$ at least by the fixed amount $K \neq 0$. Since this is true for every k sufficiently large, it follows that

$$H(\varphi^t(z(0))) \to -\infty, \quad t \to \infty.$$

This, however, is impossible, because $H \geq -1$ on all of $\mathbb{R}^2$. The case sign $y(t) = -1$ is completely analogous and we see that every solution $z(t)$ is bounded for $t \geq 0$ and consequently approaches an equilibrium point in the future.

There are three types of solutions of the pendulum with friction. Either, a solution

(1) is constant and sits in one of the equilibrium points, or

(2) rotates at most finitely many times, then slows down and converges, possibly with oscillations, to the lowest position of the pendulum, or

(3) rotates at the most finitely many times, slows down and converges without oscillations to the highest position of the pendulum.

In any case, due to the friction, every movement of the pendulum eventually comes to a standstill. Clearly, most of the solutions (in the sense of Baire category and in the sense of Lebesque measure) are of type (2).

Remark. We consider the linearized vector field

$$DX(k\pi, 0) = \begin{pmatrix} 0 & 1 \\ \cos k\pi & -\mu \end{pmatrix}$$

at the singular points. In the point $P = (2k\pi, 0)$ the matrix has the eigenvalues

$$\lambda_{1,2} = \frac{1}{2}(-\mu \pm i\sqrt{4 - \mu^2})$$

which both have a negative real part. These points are attractive equilibrium points. In the points $P = ((2k+1)\pi, 0)$, the eigenvalues

$$\lambda_{1,2} = \frac{1}{2}(-\mu \pm \sqrt{\mu^2 + 4})$$

are real and satisfy $\operatorname{Re}\lambda_1 < 0 < \operatorname{Re}\lambda_2$.

Hence, the points $P = ((2k+1)\pi, 0)$ are *hyperbolic fixed points* of the flow maps φ^T for $T > 0$. The initial conditions of the solutions of type (3) constitute the stable manifold $W_+(P)$ of the hyperbolic fixed point P. It is a *one-dimensional* submanifold of $\mathbb{R}^2$. The stable manifold converges without rotations, as illustrated in Figure IV.12, to an attracting equilibrium point.

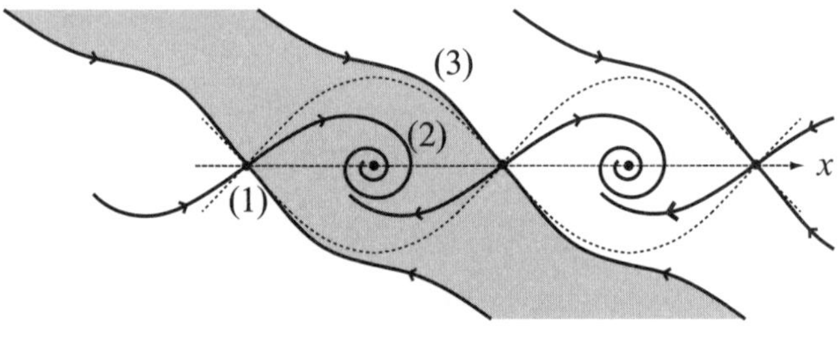

Figure IV.12. Every solution of the pendulum with friction approaches an equilibrium point in the future.

Literature. For the qualitative theory of ordinary differential equations we recommend the book [5] by H. Amann.

IV.3 Gradient systems

Let us recall first the definition of the *gradient vector field* $\nabla V\colon \mathbb{R}^n \to \mathbb{R}^n$ of a function
$$V\colon \mathbb{R}^n \to \mathbb{R}$$
of class C^{r+1}, where $r \geq 1$, with respect to the Euclidean scalar product $\langle\cdot,\cdot\rangle$. At every point $x \in \mathbb{R}^n$, the derivative of the function V can be represented by means of the Euclidean scalar product with the unique vector $\nabla V(x)$ satisfying
$$\langle \nabla V(x), u\rangle = dV(x)u \quad \text{for all } u \in \mathbb{R}^n.$$
The vector field $\nabla V\colon \mathbb{R}^n \to \mathbb{R}^n$ obtained this way belongs to the class C^r. The function V is a Lyapunov function for the flow φ^t of the (negative) gradient vector field
$$\dot{x} = -\nabla V(x), \quad x \in \mathbb{R}^n.$$
Indeed, using the chain rule and the definition of the flow one computes
$$\begin{aligned}\frac{d}{dt}V(\varphi^t(x)) &= dV(\varphi^t(x))\frac{d}{dt}\varphi^t(x)\\ &= dV(\varphi^t(x))(-\nabla V(\varphi^t(x)))\\ &= -\langle \nabla V(\varphi^t(x)), \nabla V(\varphi^t(x))\rangle\\ &= -|\nabla V(\varphi^t(x))|^2\end{aligned}$$
so that the derivative of the function V along the flow is given by
$$V(x) = -|\nabla V(x)|^2 \leq 0, \quad x \in \mathbb{R}^n.$$
In this special case, the Lyapunov-function V decreases *strictly* along the non-constant solutions. The ω- (and ω^*-) limit sets of the orbits consist of *critical* points of V and hence in view of Theorem IV.10 are contained in the set
$$C = \{x \in \mathbb{R}^n \mid \dot{V}(x) = 0\} = \{x \in \mathbb{R}^n \mid \nabla V(x) = 0\}.$$
By Proposition IV.8 the limit sets are connected. Hence, if the critical points are *isolated*, there exists for every solution $x \cdot t$ which is bounded in $t \geq 0$ a unique critical point x^* of V such that
$$x \cdot t \to x^*, \quad t \to +\infty.$$
Similarly, there exists for every solution $x \cdot t$ which is bounded in $t \leq 0$ a critical point y^* satisfying $x \cdot t \to y^*$ as $t \to -\infty$.

In short, *bounded solutions come from critical points of the function V and also go to critical points.* From Proposition IV.12 we deduce the following result.

Theorem IV.14. *If the function* $V : \mathbb{R}^n \to \mathbb{R}$ *of class* C^2 *is* coercive, *i.e.,* $V(x) \to \infty$ *as* $|x| \to \infty$, *then every solution of the gradient equation* $\dot{x} = -\nabla V(x)$ *is bounded in* $t \geq 0$, *for every* $x \in \mathbb{R}^n$ *and* $t^+(x) = \infty$. *Moreover,*

$$x \cdot t \to \{y \in \mathbb{R}^n \mid \nabla V(y) = 0\}, \text{ as } \quad t \to \infty.$$

If the critical points of the function V *are isolated, there exists for every* $x \in \mathbb{R}^n$ *a critical point* x^* *of* V *such that* $x \cdot t \to x^*$ *as* $t \to \infty$.

The gradient ∇V is orthogonal to the regular level surfaces

$$N_c = \{x \mid V(x) = c\}$$

of the function V, so that the solutions of $\dot{x} = -\nabla V(x)$ are *orthogonal* to the level surfaces of V. Indeed, in every point $x \in N_c$ of a regular level surface, the tangent space is given by

$$T_x N_c = \{\xi \in \mathbb{R}^n \mid dV(x)\xi = 0\} = \{\xi \in \mathbb{R}^n \mid \langle \nabla V(x), \xi \rangle = 0\}.$$

Clearly, gradient systems do *not* admit non-constant *periodic* solutions, because V decreases strictly along non-constant solutions.

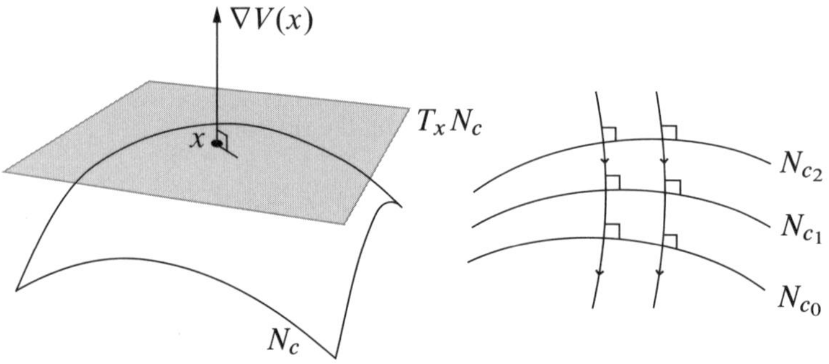

Figure IV.13. The tangent space $T_x N_c$ is orthogonal to $\nabla V(x)$ (left). Orthogonal trajectories to the level surfaces with $c_0 < c_1 < c_2$ (right).

Example. In $\mathbb{R}^2$ we consider the *coercive* function

$$V(x, y) = x^2(x-1)^2 + y^2$$

and the corresponding gradient equation

$$\dot{x} = -\nabla V(x, y) = -(2x(x-1)(2x-1), 2y).$$

The critical points of V are the points in the set $C = I(C) = \{(0,0), (\frac{1}{2}, 0), (1,0)\}$. In view of Theorem IV.14 every solution converges to one of these points,

$$x \cdot t \to x^* \in C \quad \text{as } t \to \infty.$$

This is illustrated in Figure IV.14.

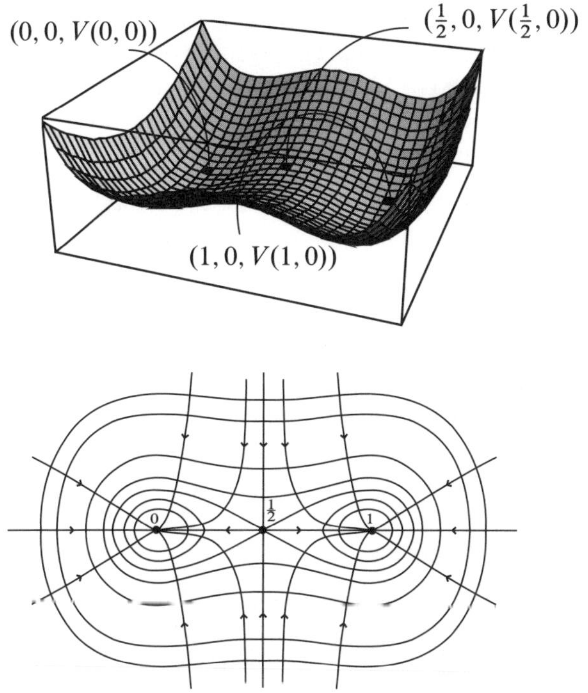

Figure IV.14. Graph of V and level lines with some solutions.

One can easily extend the qualitative concepts and results to the abstract situation of continuous *gradientlike* flows on compact metric spaces (M, d) whose metric is denoted by d.

Definition. A *flow on a compact metric space* M is a continuous mapping $\varphi\colon \mathbb{R} \times M \to M$ satisfying

$$\begin{aligned} \varphi(0, x) &= x, \\ \varphi(t, \varphi(s, x)) &= \varphi(t + s, x) \end{aligned}$$

for all $x \in M$ and for all $t, s \in \mathbb{R}$.

We note that the flow under consideration is, by assumption, complete. We continue to use the notation

$$\varphi(t, x) = \varphi^t(x) = x \cdot t.$$

In view of Proposition IV.1 the flows of vector fields on compact manifolds are complete and hence are examples of flows on compact metric spaces.

Limit sets and orbits of the continuous flow are defined as in the case of smooth flows of vector fields. Again, the orbits of a continuous flow partition the space such that each point M is contained in precisely one orbit. Indeed, if $x \cdot t = y \cdot s$ then it follows that $(x \cdot t) \cdot \tau = (y \cdot s) \cdot \tau$ and so, in view of the group structure of the flow, $x \cdot (t + \tau) = y \cdot (s + \tau)$. This holds true for all τ in $\mathbb{R}$ and consequently the orbits $\mathcal{O}(x)$ and $\mathcal{O}(y)$ must coincide.

Definition. A *rest point* of the flow φ (or *constant orbit*, or *critical point*) is a point $x \in M$ satisfying

$$x \cdot t = x \quad \text{for } all \ t \in \mathbb{R}.$$

The following observation is important for the understanding of a rest point of a continuous flow. If $x \cdot t_j = x$ for a sequence of non-vanishing real numbers $\{t_j \mid j \geq 1\}$ which converges to 0, then $x \cdot t = x$ for all $t \in \mathbb{R}$, so that x is a rest point of the flow. In order to prove this claim we first observe that, if $t = nt_j$ for two integers $n \in \mathbb{Z}$ and j, then $x \cdot t = x$ in view of the group structure of the flow. From this the claim follows, because the set $\{nt_j \mid n \in \mathbb{Z},\ j \in \mathbb{N}\}$ is dense in $\mathbb{R}$ and the flow is continuous. If, for example, $x \cdot t = x$ for t varying in an interval, then x is a rest point of the flow.

Definition. We denote the set of rest points by

$$K := \{x \in M \mid x \text{ is a rest point of the flow } \varphi\}.$$

The flow φ is called *gradientlike*, if there exists a continuous function $V\colon M \to \mathbb{R}$, which decreases *strictly* along the non-constant orbits, i.e.,

$$V(x \cdot t) < V(x \cdot s) \quad \text{for all } t > s, \text{ and for all } x \notin K.$$

Since M is compact, it follows by the arguments in Proposition IV.8, (ii) for every point $x \in M$ that

$$\begin{aligned} x \cdot t &\to \omega(x), & t &\to \infty, \\ x \cdot t &\to \omega^*(x), & t &\to -\infty. \end{aligned}$$

Arguing as in Theorem IV.10 we conclude that

$$\omega(x) \subset K \quad \text{and} \quad \omega^*(x) \subset K,$$

and obtain the following result.

Theorem IV.15. *Let φ be a gradientlike flow (with respect to V) on a compact metric space M. We assume that the set of rest points*

$$K = \{x_1, \dots, x_N\}$$

is ***finite*** *and label the points x_j in such a way that*

$$V(x_j) \le V(x_k) \quad \text{if } j < k.$$

Then there exists for ***every*** *$x \notin K$ a pair $j < k$ of indices satisfying*

$$\omega(x) = x_j \quad \text{and} \quad \omega^*(x) = x_k.$$

The invariant sets $\{x_1\}, \dots, \{x_N\} \subset M$ having the properties as in the theorem are said to constitute a *Morse decomposition* of the dynamical system (M, φ).

Definition. If $a \in \mathbb{R}$ we denote by V^a the *sublevel set*

$$V^a := \{x \in M \mid V(x) \le a\}.$$

The *set of constant orbits on the level a* is denoted by

$$K_a := K \cap V^{-1}(a).$$

The real number a is called a *critical value of the function V* if $K_a \neq \emptyset$.

Theorem IV.16 (Deformation lemma). *Let φ be a gradientlike flow (with respect to the continuous function V) on a compact metric space M and let U be an open neighborhood of the constant orbits K_a in M on the level $a \in \mathbb{R}$. Then there exists an $\varepsilon > 0$ such that the time 1 flow map φ^1 satisfies*

$$\varphi^1(V^{a+\varepsilon} \setminus U) \subset V^{a-\varepsilon}.$$

Special case: if $K_a = \emptyset$ and $U = \emptyset$, then

$$\varphi^1(V^{a+\varepsilon}) \subset V^{a-\varepsilon}.$$

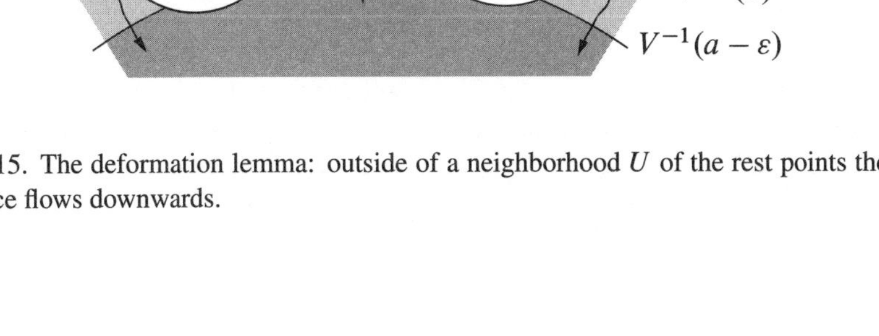

Figure IV.15. The deformation lemma: outside of a neighborhood U of the rest points the whole space flows downwards.

Proof. If $X := V^{-1}(a) \setminus U$, then $X \cap K_a = \emptyset$. Since the function V decreases strictly along non-constant orbits, there exists for every $x \in X$ an open neighborhood U_x in M and a positive real number $\delta_x > 0$ such that

$$\varphi^1(U_x) \subset V^{a-\delta_x}.$$

Because X is a compact set there exists a finite covering by such open sets, so that

$$U_{x_1} \cup \cdots \cup U_{x_m} \supset X.$$

Setting $\delta = \min\{\delta_{x_1}, \ldots \delta_{x_m}\}$, it follows for $\varepsilon \leq \delta$ that

$$\varphi^1(U_{x_1} \cup \cdots \cup U_{x_m}) \subset V^{a-\varepsilon}.$$

The set $O := U \cup U_{x_1} \cup \cdots \cup U_{x_m}$ is an open neighborhood of $V^{-1}(a)$.

Claim. If $\varepsilon > 0$ is small enough, then $V^{-1}([a-\varepsilon, a+\varepsilon]) \subset O$.

Postponing the proof of the claim, we first finish the proof of the theorem. From

$$V^{a+\varepsilon} = V^{a-\varepsilon} \cup V^{-1}([a-\varepsilon, a+\varepsilon]) \quad \subset \quad V^{a-\varepsilon} \cup O$$

together with $O \setminus U \ \subset \ U_{x_1} \cup \cdots \cup U_{x_m}$ it follows that

$$V^{a+\varepsilon} \setminus U \quad \subset \quad V^{a-\varepsilon} \cup U_{x_1} \cup \cdots \cup U_{x_m},$$

and therefore

$$\varphi^1(V^{a+\varepsilon} \setminus U) \quad \subset \quad \varphi^1(V^{a-\varepsilon}) \cup \varphi^1(U_{x_1} \cup \cdots \cup U_{x_m}) \quad \subset \quad V^{a-\varepsilon},$$

as stated in the theorem and it remains to prove the claim.

Proof of the claim Arguing indirectly we assume that for every $\varepsilon > 0$ there exists a point $x_\varepsilon \in V^{-1}([a-\varepsilon, a+\varepsilon]) \cap O^c$. Hence, taking a sequence $\varepsilon_n \to 0$ we find, in view of the compactness of M and the fact that the set O^c is closed, a sequence x_n converging to $x^* \in O^c$. The continuity of V implies $V(x^*) = \lim_{n\to\infty} V(x_n) = a$ and hence $x^* \in V^{-1}(a)$, so that $x^* \in O^c \cap V^{-1}(a)$, which is in contradiction to $V^{-1}(a) \subset O$. The claim and therefore also the theorem are proved. □

Index functions are a very useful tool in the search for rest points of a gradientlike flow on a metric space. In order to define them we first recall the concept of a deformation.

A *deformation of a subset* A of M is a continuous map $f : A \to M$ which is homotopic to the inclusion map $A \hookrightarrow M$. In other words, there exists a continuous map $F : [0,1] \times A \to M$ (a homotopy) satisfying $F(0, x) = x$ and $F(1, x) = f(x)$ for all $x \in A$.

Definition. An *index function* ν on the compact metric space M associates with every subset A of M a non-negative integer in $\mathbb{N} \cup \{0\}$,

$$A \subset M \mapsto \nu(A) \in \mathbb{N} \cup \{0\},$$

such that the following properties for subsets $A, B \subset M$ hold true.

(i) Monotonicity: $A \subset B \Longrightarrow \nu(A) \leq \nu(B)$.

(ii) Subadditivity: $\nu(A \cup B) \leq \nu(A) + \nu(B)$.

(iii) $\nu(\emptyset) = 0$.

(iv) Normalization: $\nu(\{x\}) = 1$ for every point $x \in M$.

(v) Outer regularity : Every subset $A \subset M$ possesses an open neighborhood $U \supset A$ such that $\nu(A) = \nu(U)$.

(vi) Deformation monotonicity: If $A \subset M$ is a closed set and if $f : A \to M$ is a deformation of A, then $\nu(A) \leq \nu(f(A))$.

If ν is an index function on M and if $f : A \to M$ is a deformation of A satisfying $f(A) = B$ and $B \subset A$, then $\nu(B) \leq \nu(A)$ in view of the monotonicity of the index function. On the other hand, by the deformation monotonicity, $\nu(A) \leq \nu(B)$ and therefore $\nu(A) = \nu(B)$. In particular, if $B \subset M$ is a deformation retract of M, then $\nu(B) = \nu(M)$.

Theorem IV.17 (Lusternik–Schnirelman). *Let M be a compact metric space and assume that an index function ν on M* ***exists****. Then the number $\#K$ of rest points of a gradientlike flow φ on M is estimated from below by*

$$\#K \geq \nu(M).$$

Proof. We assume that φ is a gradientlike flow with respect to the continuous function V. If $\#K = \infty$, there is nothing to prove, and hence we assume $\#K < \infty$. We define the real numbers c_k for $1 \leq k \leq \nu(M)$ by

$$c_k := \inf_{\substack{A \subset M \\ \nu(A) \geq k}} \big[\sup_{x \in A} V(x)\big].$$

If we take a smaller k, the infimum is taken over a larger set and hence the numbers c_k are ordered according to

$$c_1 \leq c_2 \leq \cdots \leq c_{\nu(M)}.$$

In view of the definition of an infimum there exists for every $\varepsilon > 0$ a subset $A \subset M$ such that $\nu(A) \geq k$ and

$$\sup_{x \in A} V(x) < c_k + \varepsilon.$$

Therefore,

$$A \subset V^{c_k+\varepsilon} = \{x \in M \mid V(x) \le c_k + \varepsilon\}.$$

From the monotonicity of the index function ν it therefore follows that

$$k \le \nu(A) \le \nu(V^{c_k+\varepsilon}).$$

Moreover, $\nu(V^{c_k-\varepsilon}) < k$, because otherwise we would arrive at the contradiction

$$c_k = \inf_{\substack{A\subset M \\ \nu(A)\ge k}} \Big[\sup_{x\in A} V(x)\Big] \le \sup_{x\in V^{c_k-\varepsilon}} V(x) \le c_k - \varepsilon.$$

Hence,

$$(*) \quad \nu(V^{c_k+\varepsilon}) \ge k > \nu(V^{c_k-\varepsilon}) \quad \text{for all } 1 \le k \le \nu(M),\ \varepsilon > 0.$$

Claim. On every level c_k there exists a rest point, so that

$$V^{-1}(c_k) \cap K \ne \emptyset.$$

Proof of the claim. Arguing by contradiction we assume that $K_{c_k} = V^{-1}(c_k) \cap K = \emptyset$. In view of the special case of the deformation lemma we find a small positive number ε such that

$$\varphi^1(V^{c_k+\varepsilon}) \subset V^{c_k-\varepsilon}.$$

Since φ^1 is a *deformation* (taking the flow φ as a homotopy), the deformation monotonicity of ν leads to the estimate

$$\nu(V^{c_k+\varepsilon}) \le \nu(\varphi^1(V^{c_k+\varepsilon})) \le \nu(V^{c_k-\varepsilon}),$$

where the second inequality follows from the monotonicity of the index ν. This estimate, however, contradicts the estimate $\nu(V^{c_k+\varepsilon}) > \nu(V^{c_k-\varepsilon})$ in $(*)$ and the claim is proved. □

If $c_1 < c_2 < \cdots < c_{\nu(M)}$, there exist, therefore, at least $\nu(M)$ rest points, as claimed in the theorem. If the levels are not different from each other, there still exist sufficiently many rest points, as we will verify next.

Claim. If $c_k = c_{k+1} = \cdots = c_{k+r}$ for a positive integer $r > 0$ and an integer k, then there exist at least $r + 1$ rest points on this level,

$$\#K_{c_k} = \#\big(V^{-1}(c_k) \cap K\big) \ge r + 1.$$

Proof of the claim. Let $K_{c_k} = \{x_1, \dots, x_l\}$ be the rest points in $V^{-1}(c_k)$. From the subadditivity properties of the index function we conclude $\nu(\{x_1, \dots, x_l\}) \leq l$.

(†) Due to the outer regularity of the index ν there exists an open neighborhood U of the set $\{x_1, \dots, x_l\}$ satisfying $\nu(U) \leq l$. According to the deformation lemma (Theorem IV.16) we find a positive ε such that

$$\varphi^1(V^{c_k+\varepsilon} \setminus U) \subset V^{c_k-\varepsilon}.$$

In view of the deformation monotonicity and of the estimate in $(*)$ we therefore obtain

$$(**) \quad \nu(V^{c_k+\varepsilon} \setminus U) \leq \nu(\varphi^1(V^{c_k+\varepsilon} \setminus U)) \leq \nu(V^{c_k-\varepsilon}) < k.$$

From the assumption $c_k = c_{k+r}$ we conclude, using the subadditivity of the index function and the estimate in $(*)$,

$$\begin{aligned} \nu(V^{c_k+\varepsilon} \setminus U) + \nu(U) &= \nu(V^{c_{k+r}+\varepsilon} \setminus U) + \nu(U) \\ &\geq \nu\big([V^{c_{k+r}+\varepsilon} \setminus U] \cup U\big) \\ &\geq \nu(V^{c_{k+r}+\varepsilon}) \\ &\geq k + r. \end{aligned}$$

Using $(**)$ it follows that

$$\nu(U) \geq k + r - \nu(V^{c_k+\varepsilon} \setminus U) > r,$$

hence $l \geq \nu(U) \geq r + 1$, and the claim is proved. The proof of Theorem IV.17 is complete. □

Remark. *The critical levels c_k are also characterized by*

$$c_k = \inf\{c \in \mathbb{R} \mid \nu(V^c) \geq k\}.$$

This is a consequence of the estimate in $(*)$.

The Lusternik–Schnirelman category. An example of an index function on topological spaces is the Lusternik–Schnirelman category which is a strong tool for the search of solutions of variational problems. It is a topological invariant defined as the smallest cardinality of closed contractible subsets covering the spaces. We shall formulate the definition in metric spaces and start with the notion of a contractible set.

Definition. Let M be a metric space. A subset $A \subset M$ is called *contractible in M*, if the inclusion mapping $A \hookrightarrow M$ is nullhomotopic in M. In other words, there exists a continuous map $F\colon [0,1] \times A \to M$ satisfying $F(0,x) = x$ and $F(1,x) \equiv y^* \in M$ for all $x \in A$.

We point out that we do *not* require that the homotopy F is a map of $[0,1] \times A$ into A (instead of into M).

Definition. Let M be a metric space and $A \subset M$ a subset. The *Lusternik–Schnirelman category* of A in M, denoted by

$$\operatorname{cat}_M(A) \in \mathbb{N} \cup \{0\} \cup \{\infty\},$$

is defined as follows. We set $\operatorname{cat}_M(A) = k \in \mathbb{N} \cup \{0\}$ if there exist k closed, in M contractible sets $A_j \subset M$ which cover A, so that

$$A \subset A_1 \cup \cdots \cup A_k,$$

and if k is the smallest integer having this property. If such a finite covering does not exist, we set $\operatorname{cat}_M(A) = \infty$.

Since the properties being closed and contractible are preserved under homeomorphisms, the category is a *topological invariant*.

Proposition IV.18. *Let M be a metric space. The Lusternik–Schnirelman category* cat_M *has the following properties for subsets A and B of M.*

(i) *Monotonicity:* $A \subset B \Longrightarrow \operatorname{cat}_M(A) \leq \operatorname{cat}_M(B)$.

(ii) *Subadditivity:* $\operatorname{cat}_M(A \cup B) \leq \operatorname{cat}_M(A) + \operatorname{cat}_M(B)$.

(iii) $\operatorname{cat}_M(\emptyset) = 0$.

(iv) *Normalization:* $\operatorname{cat}_M(\{x\}) = 1$, $x \in M$.

(v) $\operatorname{cat}_M(A) = \operatorname{cat}_M(\bar{A})$.

(vi) *Deformation monotonicity: If the subset $A \subset M$ is closed and if $f : A \to M$ is a deformation of A, then* $\operatorname{cat}_M(A) \leq \operatorname{cat}_M(f(A))$.

It follows, in particular, that $\operatorname{cat}_M(A) = \operatorname{cat}_M(B)$ *if A is a deformation retract of the subset B. The Lusternik–Schnirelman category* cat_M *has, with the exception of the outer regularity, all the properties of an index function, provided* $\operatorname{cat}_M(M) < \infty$.

Proof. All the statements except the deformation monotonicity follow directly from the definition. In order to verify the deformation monotonicity we consider the deformation $f : A \to M$ of the closed subset A. If $\operatorname{cat}_M(f(A)) = \infty$ there is nothing to prove and we assume that $\operatorname{cat}_M(f(A)) = k < \infty$. Then there exist k closed, in M contractible sets B_j such that

$$f(A) \subset B_1 \cup \cdots \cup B_k.$$

The sets $A_j := f^{-1}(B_j) \subset A$ are, due to the continuity of f, closed in $A = \bar{A}$ and therefore also closed in M. Since $f|_{A_j} : A_j \to B_j$ is a deformation in B_j and since B_j is contractible in M, also the set A_j is contractible in M. From $A \subset A_1 \cup \cdots \cup A_k$ it follows that $\operatorname{cat}_M(A) \leq k = \operatorname{cat}_M(f(A))$ and the proof of the proposition is finished. □

We now restrict the class of metric spaces under consideration and require that the metric space M is *semilocally contractible*. In other words we require that every point $x \in M$ possesses an open neighborhood U whose closure $\bar{U}$ is contractible in M. Every normed vector space, as for example $\mathbb{R}^n$, possesses this property, because we can just take an open ball around x for the neighborhood U. Since it is locally homeomorphic to a normed vector space, a metric manifold is semilocally contractible.

If $A \subset M$ is a *compact* subset, we choose for every point $x \in A$ a neighborhood U_x such that its closure $\bar{U}_x$ is contractible in M. There exists a finite subcovering satisfying

$$A \subset U_{x_1} \cup \cdots \cup U_{x_N} \subset \bar{U}_{x_1} \cup \cdots \cup \bar{U}_{x_N},$$

and we conclude from the definition of cat_M that $\mathrm{cat}_M(A) \leq N$. Therefore, *compact subsets A of a semilocally contractible space M have a finite category,* $\mathrm{cat}_M(A) < \infty$.

In particular, cat_M is finite on a compact manifold M.

The Lusternik–Schnirelman category of a space is in general not easy to compute. Here are some well-known examples.

Examples. (1) If V is a normed vector space, then $\{0\}$ is a deformation retract of V, and hence

$$\mathrm{cat}_V(V) = 1.$$

(2) If $A = \{x_1, \dots, x_n\}$ is a non-empty finite subset of a path-connected space M, then

$$\mathrm{cat}_M(A) = \mathrm{cat}_M(\{x_k\}) = 1.$$

(3) We consider the n-dimensional sphere S^n for $n \geq 1$ equipped with the metric induced from $\mathbb{R}^{n+1}$. We choose a point $x^* \in S^n$ and a small open ball U centered at x^* in the sphere. Then $\bar{U}$ is contractible to the point $\{x^*\}$ and we conclude from property (v) that

$$\mathrm{cat}_{S^n}(U) = \mathrm{cat}_{S^n}(\bar{U}) = \mathrm{cat}_{S^n}(\{x^*\}) = 1.$$

The complement $U^c = S^n \setminus U$ is also contractible on the sphere S^n and hence has also the category equal to 1. In view of the subadditivity, $\mathrm{cat}_{S^n}(S^n) \leq \mathrm{cat}_{S^n}(U) + \mathrm{cat}_{S^n}(U^c) = 2$ and since S^n is not contractible in S^n we find

$$\mathrm{cat}_{S^n}(S^n) = 2, \quad n \geq 1.$$

(4) In contrast, the unit sphere $S^\infty = \{x \in V \mid \|x\| = 1\}$ in an *infinite-dimensional* normed vector space V is a retract of V (for a proof we refer to p. 66 in [26] by K. Deimling), from which it follows that S^∞ is contractible in itself, so that

$$\mathrm{cat}_{S^\infty}(S^\infty) = 1.$$

(5) The Lusternik–Schnirelman category of the n-dimensional torus T^n is equal to

$$\mathrm{cat}_{T^n}(T^n) = n + 1, \quad n \geq 1,$$

The proof can be found in [101] by J. Schwartz and in [22] by O. Cornea et al.

Theorem IV.19 (Lusternik–Schnirelman). *If φ^t is a gradientlike flow on a compact manifold M (or more generally, on a semilocally contractible and compact metric space M), then the number of its rest points K is estimated from below by*

$$\#K \geq \mathrm{cat}_M(M).$$

Proof [Theorem IV.17]. In view of Proposition IV.18 we can take over almost the complete proof of Theorem IV.17 for the function cat_M instead of the index function ν. We have used the outer regularity only in the proof of the second claim in the proof of Theorem IV.17 and have marked this passage with (†). Instead, we now use the semilocal contractibility in order to verify that the finite set of rest points $\{x_1, \dots, x_l\} \subset M$ possesses an open neighborhood U satisfying $\mathrm{cat}_M(U) \leq l$.

We choose open neighborhoods U_i of x_i in such a way that $\overline{U_i}$ is contractible in M. Then the union $U = U_1 \cup \dots \cup U_l$ is an open neighborhood of $\{x_1, \dots, x_l\}$ satisfying $U \subset \overline{U_1} \cup \dots \cup \overline{U_l}$ so that $\mathrm{cat}_M(U) \leq l$. The rest of the proof is identical with the proof of Theorem IV.17. □

Remark. If the metric space M is a manifold then cat_M is regular from the outside. If, moreover, M is a compact manifold, then

$$\mathrm{cat}_M(M) \leq \dim M + 1 < \infty$$

is finite, so that cat_M is an index function. More generally, the inequality

$$\mathrm{cat}_M(A) \leq \dim A + 1$$

holds true for every compact subset of a (Hilbert-)manifold M, where $\dim A$ is the covering dimension from topology. With the differential topological concept of the *cuplength* the following very useful estimate of the Lusternik–Schnirelman category holds true,

$$\mathrm{cuplength}\, M + 1 \leq \mathrm{cat}_M(M).$$

For proofs we refer to the books [22] by O. Cornea et al., [26] by K. Deimling and [101] by J. Schwartz.

Exercise. On the torus $T^2 = \mathbb{R}^2/\mathbb{Z}^2$ the function

$$V\colon T^2 \to \mathbb{R}, \quad V(x,y) = \sin(\pi x)\sin(\pi y)\sin(\pi(x+y))$$

has exactly three critical points, namely the minimum, the maximum and a degenerate saddle point which is a so-called monkey saddle.

In view of $\operatorname{cat}(T^2) = 3$, every function on the 2-torus must have at least three critical points according to the Lusternik–Schnirelman theorem. The above example represents therefore a minimal model having precisely three critical points. One of the critical points (in our example the monkey saddle) is necessarily degenerate. Indeed, we shall see in the next section that the Morse theory guarantees at least four critical points for functions on the 2-torus whose critical points are all nondegenerate.

Literature. For a view of the state of the art in Lusternik–Schnirelman category we point out the monograph [22] by O. Cornea, G. Lupton, J. Oprea and D. Taure. The Lusternik–Schnirelman category is treated in books about nonlinear functional analysis, as for example in [26] by K. Deimling and in [101] by J. Schwartz. The Lusternik–Schnirelman theory on Banach manifolds is developed in [77] by R. S. Palais.

IV.4 Gradient systems on manifolds and Morse theory

We finally describe the relation between the dynamics of gradient flows and the topology of the underlying manifold. We assume M to be a (smooth) *compact manifold* of dimension $\dim M = d$ (as for example a d-dimensional compact submanifold of the $\mathbb{R}^N$). *We choose* a Riemann *metric* g on M, so that there exists at every point $x \in M$ a *scalar product* $g(x)$ (symmetric and positively definite) in the tangent space,

$$g(x)(\xi, \eta) =: \langle \xi, \eta \rangle, \quad \xi, \eta \in T_x M.$$

The induced norm on the tangent space in x is defined by

$$g(x)(\xi, \xi) = \langle \xi, \xi \rangle =: |\xi|^2, \quad \xi \in T_x M.$$

The scalar product depends differentiably on the base point x.

If $V : M \to \mathbb{R}$ is a function of class C^2, its *derivative* $dV(x)$ at the point $x \in M$,

$$dV(x) : T_x M \to \mathbb{R},$$

is a linear function. Hence there exists a *unique vector* $\nabla V(x) \in T_x M$ satisfying

$$g(x)(\nabla V(x), \xi) = dV(x)\xi \quad \text{for all vectors } \xi \in T_x M,$$

or in short,

$$\langle \nabla V(x), \xi \rangle = dV(x)\xi, \quad \xi \in T_x M.$$

The vector $\nabla V(x)$ is called the *gradient vector* of the function V in x. It *depends on* the choice of the metric. The vector field ∇V on M obtained this way is differentiable. The flow of the equation

$$\dot{x} = -\nabla V(x), \quad x \in M$$

is called the *gradient flow* of the function V on M. It is defined by

$$\frac{d}{dt}\varphi^t(x) = -\nabla V(\varphi^t(x)), \quad \varphi^0(x) = x, \quad t \in \mathbb{R},$$

and is, in view of Proposition IV.1, a complete flow.

Again, the function V is decreasing along the parametrized orbits of the flow, as the by now familiar computation shows,

$$\begin{aligned}\frac{d}{dt}V(\varphi^t(x)) &= dV(\varphi^t(x)) \cdot \frac{d}{dt}\varphi^t(x)\\ &= \langle \nabla V(\varphi^t(x)), \frac{d}{dt}\varphi^t(x)\rangle\\ &= -\langle \nabla V(\varphi^t(x)), \nabla V(\varphi^t(x))\rangle\\ &= -|\nabla V(\varphi^t(x))|^2\\ &\leq 0.\end{aligned}$$

The function V decreases strictly along non-constant orbits and all the results about gradientlike flows are applicable.

Definition. A point $x \in M$ is called a *critical point* of the function $V : M \to \mathbb{R}$, if $dV(x) = 0$, or equivalently, if $\nabla V(x) = 0$ in $T_x M$. The *set of critical points* is again denoted by K,

$$K = \{x \in M \mid \nabla V(x) = 0\}.$$

We briefly recall the results about gradientlike flows on compact spaces. In view of Theorem IV.19 the number of critical points is estimated from below by $\#K \geq \mathrm{cat}_M(M)$. Since the solutions are *bounded*, we have for every point $x \in M$ the convergence

$$x \cdot t \to K, \quad t \to \pm\infty.$$

In other words, every solution approaches the critical set K in forward and in backward time. If the critical points are *isolated*, there exist only *finitely* many of them, since M is compact, so that

$$K = \{x_1, x_2, \dots, x_N\}.$$

With every point $x \notin K$ in the manifold M we can associate a pair $k \neq j$ of indices such that

$$\begin{aligned} x \cdot t &\to x_k, \quad t \to +\infty,\\ x \cdot t &\to x_j, \quad t \to -\infty.\end{aligned}$$

In the following, we make *additional assumptions* on the critical points K.

Definition. A critical point $x_k \in K$ is called *nondegenerate*, if, in local coordinates around x_k, the bilinear form

$$d^2V(x_k)(u,v) \in \mathbb{R}, \quad u, v \in \mathbb{R}^d,$$

defined by the *Hesse matrix* of V in x_k, is nondegenerate. In view of $\nabla V(x_k) = 0$ this definition is independent of the choice of the local coordinates.

For the *linearized* gradient vector field at the critical point x_k, we then have (in local coordinates)

$$\langle d\nabla V(x_k)u, v\rangle = d^2V(x_k)(u,v), \quad u, v \in \mathbb{R}^d.$$

The local invariants in x_k are the dimensions of the subspaces of $\mathbb{R}^d$, on which the bilinear form is *positive* respectively *negative*. Hence the tangent space splits, $T_{x_k}M = E^+_{x_k} \oplus E^-_{x_k}$ so that, for $u = u^+ + u^- \in E^+_{x_k} \oplus E^-_{x_k}$,

$$d^2V(x_k)(u^+, u^+) > 0, \quad u^+ \neq 0,$$
$$d^2V(x_k)(u^-, u^-) < 0, \quad u^- \neq 0.$$

Again, this is independent of the choice of the coordinates and the metric.

Definition. The *Morse index* of the function V at the critical point x_k is the integer

$$\mu(x_k) := \dim E^-_{x_k}.$$

Obviously, $0 \leq \mu(x_k) \leq d$. The linearization of the gradient vector field at the critical point x_k is *symmetric* with respect to the scalar product and has therefore d real eigenvalues. The number of negative eigenvalues is equal to $\mu(x_k)$, the number of positive eigenvalues is equal to $d - \mu(x_k)$. Therefore, the equilibrium point x_k of $-\nabla V(x)$ on M is a *hyperbolic* equilibrium point according to Lemma II.12, and its unstable manifold $W^-(x_k) = \{x \in M \mid x \cdot t \to x_k, \text{ as } t \to -\infty\}$ is of dimension

$$\mu(x_k) = \dim W^-(x_k).$$

Here, the concept of hyperbolicity has been somewhat generalized by admitting the cases $\mu(x_k) = 0$ and $\mu(x_k) = d$, in which $T_{x_k}M = E^-_{x_k} \oplus E^+_{x_k} = \{0\} \oplus E^+_{x_k}$ respectively $T_{x_k}M = E^-_{x_k} \oplus \{0\}$. We set $W^{\pm}(x_k) = \{x_k\}$, in the cases $\dim E^{\pm}(x_k) = 0$.

Definition. A function $V: M \to \mathbb{R}$ of class C^2 is called a *Morse function*, if all its critical points are nondegenerate.

In a sufficiently small neighborhood of a nondegenerate critical point, there do not exist other critical points. This follows from the inverse function theorem.

In particular, a Morse function on a compact manifold admits only finitely many critical points.

We should mention that on every compact manifold (without boundary) there exists an abundance of Morse functions. Every smooth function can be uniformly approximated together with all its derivatives by Morse functions, as is demonstrated, for example in the book [21] by L. Conlon.

As already pointed out, $x \cdot t \to K$ as $t \to -\infty$ for every point $x \in M$. In the case of nondegenerate critical points we conclude that the compact manifold M is composed of the unstable manifolds of the critical points,

$$M = \bigcup_{x_k \in K} W^-(x_k).$$

Examples. (1) We consider the sphere $S^2 \subset \mathbb{R}^3$ standing on $\mathbb{R}^2$ and the height function $V: S^2 \to \mathbb{R}$ measuring the levels over $\mathbb{R}^2$, as shown in Figure IV.16. Then, V has two critical points, namely, the maximum at the north pole N satisfying $\mu(N) = 2$ and the minimum in S having the Morse index $\mu(S) = 0$.

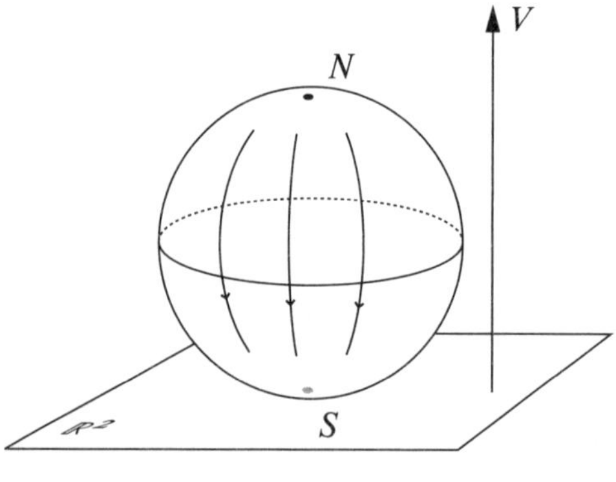

Figure IV.16. The height function on S^2 having the two critical points N and S, the maximum and the minimum. Also indicated are some flow lines.

(2) If $T^2 \subset \mathbb{R}^3$ is the embedded torus as depicted in Figure IV.17 and if the function $V: T^2 \to \mathbb{R}$ is again the height function, then there exist one maximum having Morse index 2 and one minimum having index 0 as before. However, there exist two additional critical points S_1, S_2 which are saddle points whose Morse indices are equal to $\mu(S_i) = 1$.

Since all the points $x \in M$ move under the gradient towards *critical* points, one expects a relation between the *local* invariants $\mu(x_k)$ of the flow at the critical points $x_k \in K$, and the *topological* invariants of the manifold M.

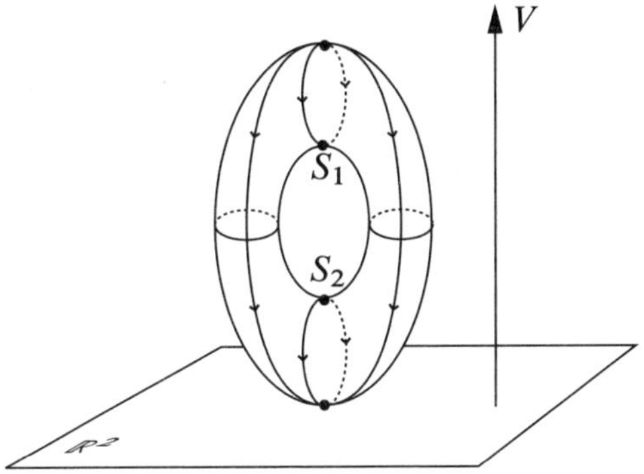

Figure IV.17. The height function on T^2 having four critical points and some flow lines of its gradient flow.

The *dynamical data* of a Morse function V are collected in the *Morse polynomial* of V, defined by

$$m(t, V) := \sum_{\{x \mid \nabla V(x)=0\}} t^{\mu(x)},$$

where $\mu(x)$ is the Morse index of the critical point x. By m_j, we denote the number of critical points whose Morse index is equal

$$m_j = \#\{x \in K \mid \mu(x) = j\}.$$

Then the number of critical points is equal to $\#K = \sum_{j=0}^{\dim M} m_j$ and the Morse polynomial becomes

$$m(t, V) = \sum_{j=0}^{\dim M} m_j t^j.$$

In particular, $\#K = m(1, V)$.

On the other hand, also the *topological data* of M are collected in a polynomial. For this we need the *Betti numbers* $\beta_j = \beta_j(M)$ of M. The j-th Betti number is defined as the dimension of the vector space $H^j(M)$ of the j-th cohomology group of M. In view of the *theorem of de Rham* this vector space is isomorphic to the real vector space of the closed j-forms *modulo* the exact j-forms on M,

$$H^j(M) \cong \frac{\{\omega \in \Omega^j(M) \mid d\omega = 0\}}{\{\omega \in \Omega^j(M) \mid \omega = d\alpha,\ \alpha \in \Omega^{j-1}(M)\}}, \quad 1 \le j \le \dim M,$$

and, for $j = 0$,

$$H^0(M) \cong \{\omega \in \Omega^0(M) \mid d\omega = 0\}.$$

Here $\Omega^j(M)$ denotes the space of the j-forms on M and $d : \Omega^j(M) \to \Omega^{j+1}(M)$ is the *exterior derivative* acting on forms.These concepts will be introduced in the next chapter. The exterior derivative satisfies $d \circ d = 0$, so that every exact form is a closed form. As usual, we make use of the convention that $\Omega^0(M) = C^\infty(M, \mathbb{R})$ denotes the space of smooth functions. On Ω^0 the exterior derivative d coincides with the usual derivative.

The direct sum of the forms $H^j(M)$, denoted by

$$H^*(M) = \bigoplus_j H^j(M),$$

is an algebra (with respect to the exterior product $\wedge$) and is called *de Rham cohomology algebra.*

The *Poincaré polynomial* of the compact manifold M is the polynomial

$$P(t, M) := \sum_{j=0}^{\dim M} \beta_j t^j.$$

The Betti numbers $\beta_j = \beta_j(M)$ contain relevant information about the topology of M.

Example. We consider the space $\Omega^0(M) = C^\infty(M, \mathbb{R})$ of smooth functions. The function $f \in \Omega^0$ is a closed form, if $df(x) = 0$ for all $x \in M$. Hence, these are the locally constant functions on M. If M is connected, then $f =$ constant so that

$$H^0(M) = \{f = \text{ const.}\} \cong \mathbb{R}.$$

Therefore

$$\beta_0 = \dim H^0(M) = 1,$$

if M is connected. More generally, the dimension $\dim H^0(M)$ is the number of connected components of M.

In the following table some Betti numbers $\beta_j = \beta_j(M)$ are listed.

manifold	genus	Betti numbers
S^2	$g = 0$	$(1, 0, 1)$
T^2	$g = 1$	$(1, 2, 1)$
F_g	g	$(1, 2g, 1)$
S^n	–	$(1, 0, \dots, 0, 1)$
T^n	–	$\beta_k = \binom{n}{k}$

The manifold F_g denotes a compact connected 2-dimensional manifold of genus g, hence a closed surface having g holes, as illustrated in Figure IV.18.

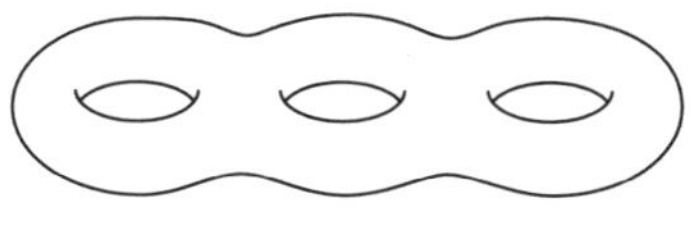

Figure IV.18. A surface of genus $g = 3$.

The following theorem describes the relation between the gradient flow of a Morse function and the topology of the manifold.

Theorem IV.20 (Morse inequalities). *We assume that $V: M \to \mathbb{R}$ is a C^2-function on the compact manifold M whose critical points are all* ***nondegenerate****. Then*

$$m(t, V) = P(t, M) + (1 + t)Q(t),$$

where $Q(t)$ is a polynomial with non-negative integer coefficients. More explicitly,

$$\sum_{j=0}^{\dim M} m_j t^j = \sum_{j=0}^{\dim M} \beta_j t^j + (1 + t)Q(t).$$

The sum on the left-hand side sums up the local invariants of the gradient flow of the vector field ∇V on M defined by the Morse function V. The sum on the right-hand side contains the topological invariants of the manifold M which are independent of the choice of the function V and of the metric used.

We next draw some conclusions from the theorem.

Definition. The *Euler characteristic* of a compact manifold M is the integer

$$\chi(M) := \sum_{j=0}^{\dim M} (-1)^j \beta_j = P(-1, M),$$

where $\beta_j = \beta_j(M)$ are the Betti numbers of M.

Examples. From the above table for the Betti numbers we compute

$$\begin{aligned}
\chi(S^2) &= 1 - 0 + 1 = 2, \\
\chi(T^2) &= 1 - 2 + 1 = 0, \\
\chi(F_g) &= 1 - 2g + 1 = 2 - 2g, \\
\chi(S^n) &= 1 + (-1)^n, \\
\chi(T^n) &= (1 - 1)^n = 0.
\end{aligned}$$

Corollary IV.21. *We assume M to be a compact manifold and $V\colon M \to \mathbb{R}$ a function of class C^2. If all the critical points are nondegenerate, then*

$$\sum_{j=0}^{\dim M} (-1)^j m_j = \chi(M),$$

where m_j is the number of critical points x of the function V whose Morse index is equal to $\mu(x) = j$.

Proof. We choose $t = -1$ in Theorem IV.20. □

The sum on the left-hand side of the equation in the corollary is independent of the choice of the Morse function which at first sight is not evident. Conversely, the topological Euler characteristic on the right-hand side is determined by the dynamics of the gradient vector field $-\nabla V$ on the manifold M.

Example. If the function $V\colon F_g \subset \mathbb{R}^3 \to \mathbb{R}$, defined on the closed surface of genus g, is of class C^2 and possesses only *nondegenerate* critical points, then

$$2 - 2g = m_0 - m_1 + m_2,$$

where m_0 is the number of local minima, m_1 the number of saddle points and m_3 the number of local maxima.

Special case. If $m_0 = m_2 = 1$ (there exists exactly one minimum and exactly one maximum) then the number of saddle points is equal to $m_1 = 2g$. If $g = 0$, that is, if the closed surface $F_g = S^2$ is the sphere, then $m_1 = 0$, so that a Morse function on the sphere S^2 having exactly one minimum and one maximum does not admit any saddle points.

Corollary IV.22. *If M is a compact manifold and if the function $V\colon M \to \mathbb{R}$ of class C^2 has only nondegenerate critical points, then*

$$m_j \geq \beta_j, \quad 0 \leq j \leq \dim M.$$

In particular, the number of critical points K of the function V is estimated by

$$\#K \geq \sum_{j=0}^{\dim M} \beta_j =: \mathrm{SB}(M).$$

Proof. The polynomial $Q(t) = \sum q_j t^j$ in Theorem IV.20 has only integer coefficients $q_j \geq 0$. Therefore the estimates follow if we compare the coefficients of t^j. The second statement then follows recalling that $\#K = \sum_{j=0}^{\dim M} m_j$. □

Examples.

$$\begin{aligned}\mathrm{SB}(S^n) &= 2,\\ \mathrm{SB}(T^n) &= (1+1)^n = 2^n.\end{aligned}$$

In the case of the sphere there is, in comparison to the estimate involving the Lusternik–Schnirelman category, no improvement. In the case of the torus we obtain for the number of critical points $\#K$ the following stronger estimate

$$\#K \geq 2^n = \mathrm{SB}(T^n),$$

which improves the earlier estimate $\#K \geq n+1 = \mathrm{cat}_{T^n}(T^n)$, however under the additional nondegeneracy assumption on the function V.

Corollary IV.23. *If M is a compact manifold and if $V\colon M \to \mathbb{R}$ is of class C^2 and possesses only nondegenerate critical points, then we have, for every $0 \leq k \leq$ $\dim M$, the inequality*

$$m_k - m_{k-1} + m_{k-2} - \cdots \pm m_0 \geq \beta_k - \beta_{k-1} + \beta_{k-2} - \cdots \pm \beta_0.$$

Proof. Setting $Q(t) = \sum t^j q_j$, it follows from Theorem IV.20 by comparing the coefficients of t^j that

$$\sum_{j=0}^{k} t^j m_j = \sum_{j=0}^{k} t^j \beta_j + t^k q_k + (1+t)\sum_{j=0}^{k-1} t^j q_j.$$

Multiplication by t^{-k} leads to

$$\sum_{j=0}^{k} t^{j-k} m_j = \sum_{j=0}^{k} t^{j-k} \beta_j + q_k + t^{-k}(1+t)\sum_{j=0}^{k-1} t^j q_j.$$

Setting $t=-1$ and recalling that $q_k \geq 0$, the corollary follows. □

Next, we develop some preparatory steps for the proof of Theorem IV.20. First, we shall show that the topological type of the sublevel set

$$M^a := \{x \in M \mid V(x) \leq a\}, \quad a \in \mathbb{R}$$

only changes when we pass a critical level, as illustrated in Figure IV.19 for the 2-torus. The compact set M^a is a manifold having the boundary $\partial M^a = V^{-1}(a)$, if a is a regular value of V. For $a < b$ we introduce the subset

$$M_a^b := \{x \in M \mid a \leq V(x) \leq b\}.$$

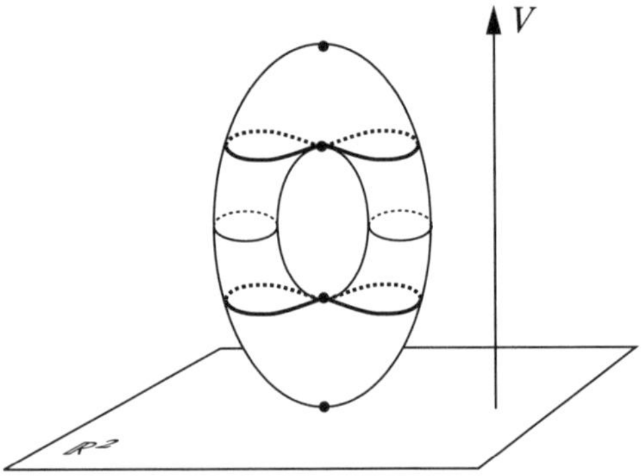

Figure IV.19. The 2-torus with height function V, with the critical levels in bold print.

Proposition IV.24 (Deformation lemma). *If* $V: M \to \mathbb{R}$ *is of class* C^2 *and if* $a < b$ *and if*

$$M_a^b \cap K = \emptyset,$$

then M^b *is* diffeomorphic *to* M^a. *Moreover, the sublevel set* M^a *is a* ***strong deformation retract*** *of* M^b.

Corollary IV.25. *Under the assumptions in the deformation lemma the cohomologies* $H^j(M^b, M^a) = 0$ *of the pair of spaces vanish for all* j.

Proof [Borrowed from topology]. This follows because $M^a \subset M^b$ is a deformation retract of M^b. □

Proof of Proposition IV.24. The diffeomorphism will be constructed by means of the gradient flow. We reparametrize the orbits of the gradient vector field $\dot{x} = -\nabla V(x)$ and define the rescaled vector field

$$X(x) := -\alpha(x)\nabla V(x), \quad x \in M,$$

with a non-negative differentiable function $\alpha: M \to \mathbb{R}$ satisfying on M_a^b the equation

$$\alpha(x) = \frac{1}{|\nabla V(x)|^2}, \quad x \in M_a^b.$$

In view of the compactness there exists an $\varepsilon > 0$ such that $M_{a-\varepsilon}^{b+\varepsilon} \cap K = \emptyset$. On the open neighborhood $V^{-1}(a-\varepsilon, b+\varepsilon)$ of M_a^b, the vector field ∇V does not vanish. The flow ψ^t on M, generated by the vector field X, satisfies

$$\frac{d}{dt}\psi^t(x) = X(\psi^t(x)), \quad \psi^0(x) = x$$

for all $x \in M$ and all $t \in \mathbb{R}$.

(1) *Diffeomorphism.* We shall show that the flow map $\psi^{b-a}: M^b \to M^a$ defines a diffeomorphism. To do this, it suffices to show that $\psi^{b-a}(M^b) = M^a$. If $\psi^t(x) \in M_a^b$, then

$$\begin{aligned}\frac{d}{dt}V(\psi^t(x)) &= dV(\psi^t(x))\, X(\psi^t(x)) \\ &= \langle \nabla V(\psi^t(x)), X(\psi^t(x))\rangle \\ &= -\alpha(\psi^t(x))\, |\nabla V(\psi^t(x))|^2 \\ &= -1.\end{aligned}$$

Therefore, as long as x and $\psi^t(x) \in M_a^b$, we have

$$V(\psi^t(x)) - V(\psi^0(x)) = \int_0^t \frac{d}{dt}V(\psi^t(x))\, dt = -t$$

or

$$(*) \quad V(\psi^t(x)) = V(x) - t.$$

If $t = b-a$ and $V(x) = b$, we obtain $V(\psi^{b-a}(x)) = a$, so that $\psi^{b-a}(M^b) \subset M^a$. If $V(x) = a$, then $V(\psi^{-(b-a)}(x)) = V(x)+b-a = b$, and the surjectivity follows.

(2) *Construction of a retraction* $r_1: M^b \to M^a$. We define for $0 \le t \le 1$ and $x \in M^b$ the mapping r by

$$r(t,x) = r_t(x) = \begin{cases} x, & V(x) \le a, \\ \psi^{t[V(x)-a]}(x), & a \le V(x) \le b. \end{cases}$$

Then the map $r: [0,1] \times M^b \to M^b$ is continuous and satisfies

$$\begin{aligned} r(0,x) &= x, \quad x \in M^b, \\ r(t,x) &= x, \quad x \in M^a \text{ and all } t \in [0,1]. \end{aligned}$$

Using the identity $(*)$ it follows that

$$V\big(\psi^{t[V(x)-a]}(x)\big) = (1-t)V(x) + ta,$$

and setting $t = 1$ one concludes that $r_1(x) = r(1,x) \in M^a$ for all $x \in M^b$. Therefore, M^a is a strong deformation retract of M^b, as claimed. □

In sharp contrast to the above situation, the topological type of M^a changes if we *pass* a *critical* level.

Proposition IV.26. *Let $p \in K \cap V^{-1}(c)$ be a nondegenerate critical point of the C^2 function $V: M \to \mathbb{R}$ having the Morse index $\mu = \mu(p)$. If $a < c < b$ and if*

$$M_a^b \cap K = \{p\},$$

then

$$H^j(M^b, M^a) = \begin{cases} \mathbb{R}, & j = \mu(p), \\ 0, & \textit{otherwise.} \end{cases}$$

In particular, $\dim H^j(M^b, M^a) = 1$ *for* $j = \mu(p)$ *and* 0 *otherwise.*

Figuratively speaking, the sublevel set M^b is homeomorphic to M^a with a $\mu(p)$-handle glued on, as illustrated in Figure IV.20 for the height function V on the torus T^2.

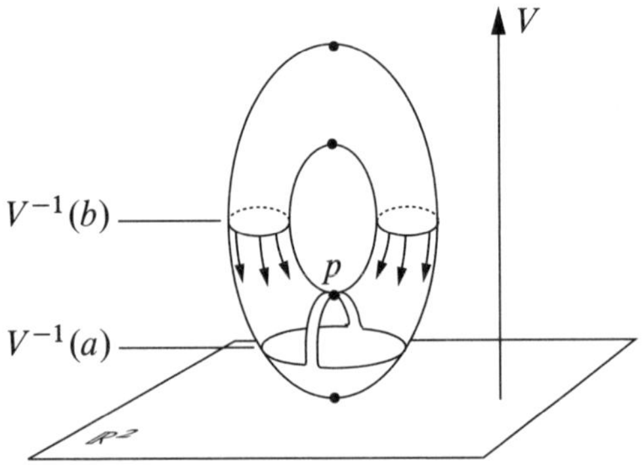

Figure IV.20. Passing the critical level of the saddle point p. The sublevel set M^a with a 1-handle glued on is a deformation retract of M^b.

For the proof of the proposition we first introduce convenient coordinates near the critical point.

Lemma IV.27 (Morse lemma). *We assume* M *to be a smooth manifold of dimension* d *and* $m \in M$ *to be a nondegenerate critical point of the smooth function* $F : M \to \mathbb{R}$, *such that* $dF(m) = 0$. *Moreover, we assume the Hesse matrix* $d^2F(m)$ *to be nondegenerate. Then there exist local smooth coordinates* $\psi : U \subset \mathbb{R}^n \to M$ *near* m *satisfying* $\psi(0) = m$, *in which the function* $f(x) = F \circ \psi(x)$ *is written as*

$$f(x) = f(0) + \frac{1}{2} d^2 f(0)(x, x).$$

Proof [Deformation trick by R. Thom and J. Moser]. Choosing local coordinates we may assume that the function $f : \mathbb{R}^d \to \mathbb{R}$ has the form

$$f(x) = f(0) + \frac{1}{2} d^2 f(0)(x, x) + O_3(x),$$

by using the Taylor formula. Moreover, we may assume that $f(0) = 0$. With the Hesse matrix H we obtain $d^2 f(0)(x, x) = \langle Hx, x \rangle$, so that the function has the form

$$f(x) = \frac{1}{2}\langle Hx, x\rangle + O_3(x).$$

We introduce the function g by $g(x) = \frac{1}{2}\langle Hx, x\rangle$ and look for a local diffeomorphism ψ near $x = 0$, satisfying $\psi(0) = 0$ and

$$(\psi^* f)(x) = g(x)$$

for all x near 0. For this, it suffices to find a local diffeomorphism ψ leaving the origin fixed and satisfying

$$\psi^* df = dg$$

near $x = 0$. We now deform these two 1-forms into each other and define the family ω_t of 1-forms by

$$\omega_t = t\,df + (1-t)dg, \quad 0 \le t \le 1.$$

Then $\omega_1 = df$ and $\omega_0 = dg$, moreover, $d\omega_t = 0$. More ambitiously, we look for a family ψ^t of local diffeomorphisms satisfying $\psi^0 = \mathrm{Id}$ and $\psi^t(0) = 0$ such that

$$(*) \quad (\psi^t)^* \omega_t = dg, \quad 0 \le t \le 1$$

for all x near 0. The desired result then follows for the parameter value $t = 1$.

We now assume that the family ψ^t of local diffeomorphisms is the flow of a time dependent vector field X_t which is defined near $x = 0$ and which has the property that $X_t(0) = 0$ for all $0 \le t \le 1$. The flow satisfies

$$\frac{d}{dt}\psi^t(x) = X_t(\psi^t(x)), \quad 0 \le t \le 1, \quad \text{and} \quad \psi^0(x) = x.$$

In order to find the vector field X_t, we differentiate the equation $(*)$ in the variable t and find, using the formula of Proposition V.14,

$$\begin{aligned} 0 &= \frac{d}{dt}\left[(\psi^t)^*\omega_t\right] \\ &= (\psi^t)^*\Big[d i_{X_t}\omega_t + i_{X_t} d\omega_t + \frac{d}{dt}\omega_t\Big] \\ &= (\psi^t)^* d\left[i_{X_t}\omega_t + f - g\right]. \end{aligned}$$

We therefore look for a vector field X_t which solves the linear equation

$$(**) \quad i_{X_t}\omega_t = g - f, \quad X_t(0) = 0$$

near $x = 0$ for all $0 \le t \le 1$. If we have found such a vector field, then the lemma is proved, because we can make all our steps backwards. Namely, the flow ψ^t exists for all $0 \le t \le 1$ and for all x near 0 and satisfies $\psi^t(0) = 0$ and $\frac{d}{dt}(\psi^t)^*\omega_t = 0$.

Consequently, $(\psi^t)^*\omega_t = (\psi^0)^*\omega_0 = dg$ for all $0 \le t \le 1$ and for $t = 1$ we obtain the desired solution $\psi^1 = \psi$.

The equation $(**)$ for X_t has a smooth solution, since using the Taylor formula and $\omega_t(x)[h] = \langle Hx, h\rangle + tO_2(x)[h]$, the equation $(**)$ to be solved becomes

$$\big\langle X_t(x), [H + tR(x)]h\big\rangle = \langle Q(x), h\rangle$$

for all x near 0, with a smooth matrix function $R(x)$ and with a smooth vector function $Q(x)$ satisfying $R(0) = 0$ and $Q(0) = 0$. The Hesse-matrix H is symmetric and invertible, in view of the assumption of nondegeneracy. Therefore, the desired vector field X_t is given by

$$X_t(x) = \big(H + tR(x)^T\big)^{-1} \cdot Q(x)$$

and satisfies the equation $(**)$ for all x near 0, and, moreover, satisfies $X_t(0) = 0$ for all $0 \le t \le 1$. This proves the lemma. □

Proof of Proposition IV.26 [Sketch]. We abbreviate by $\mu = \mu(p)$ the Morse index of the critical point p and choose by means of the Morse lemma local coordinates near p in which $p = 0$, and in which the function V has the normal form $V(z) = V(0) + \frac{1}{2}d^2V(0)(z,z)$, with the Hesse-matrix

$$H = d^2V(0) = \begin{pmatrix} -\mathbb{1}_\mu & 0 \\ 0 & \mathbb{1}_{d-\mu} \end{pmatrix}$$

in 0. Then, we have the coordinates $z = (x, y) \in D^\mu \times D^{d-\mu}$ with the two small discs $D^\mu = \{x \in \mathbb{R}^\mu \mid |x| \le \varepsilon\}$ and $D^{d-\mu} = \{y \in \mathbb{R}^{d-\mu} \mid |y| \le \varepsilon\}$. Assuming $V(0) = 0$ the function V is of the form

$$V(x, y) = |y|^2 - |x|^2, \quad (x, y) \in D^\mu \times D^{d-\mu}.$$

Locally near 0 the sublevel sets are depicted in Figure IV.21.

Introducing the subset H of the type of a handle connecting the two pieces of the sublevel set V^a as shown in Figure IV.22 one can see by slowing down the gradient flow that the set $V^a \cup H$ is a deformation retract of V^b.

Therefore, borrowing from topology again, we deduce the isomorphism in cohomology

$$H^*(V^b, V^a) \cong H^*(V^a \cup H, V^a).$$

With the two disjoint pieces $H_1 \subset H$ attached to V^a as indicated in Figure IV.23 we see that $V^a \subset V^a \cup H_1 \subset V^a \cup H$ and conclude that the sublevel set V^a is a deformation retract of the set $V^a \cup H_1$.

Consequently, we arrive at the isomorphism

$$H^*(V^a \cup H, V^a) \cong H^*(V^a \cup H, V^a \cup H_1).$$

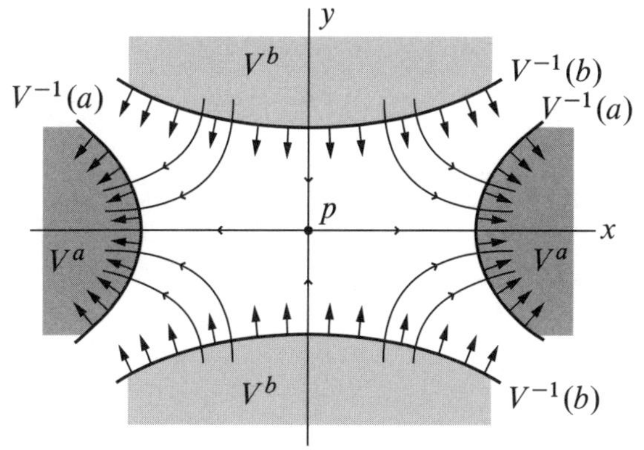

Figure IV.21. Level sets and gradient flow locally near the critical point p.

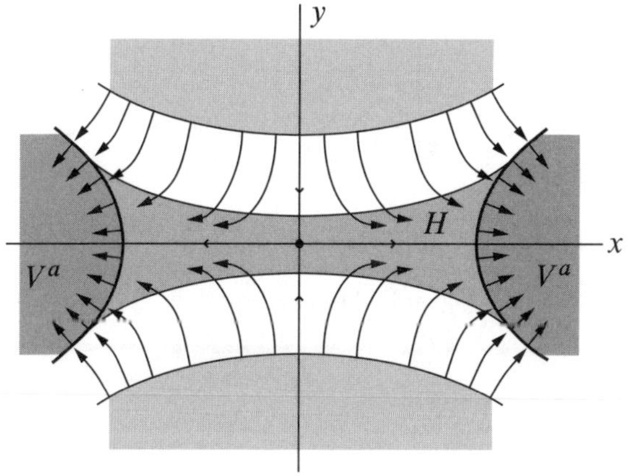

Figure IV.22. The deformation retract $V^a \cup H$ of V^b.

Now cutting off the interior of the parts both spaces of the pair $(V^a \cup H, V^a \cup H_1)$ have in common, we obtain a pair of spaces that is homotopically equivalent to the pair illustrated in Figure IV.24.

Using now the cohomology results about *excisions* and homotopy equivalences one obtains the isomorphism

$$H^*\big(V^a \cup H, V^a \cup H_1\big) \cong H^*\big(D^\mu \times D^{d-\mu}, \partial D^\mu \times D^{d-\mu}\big),$$

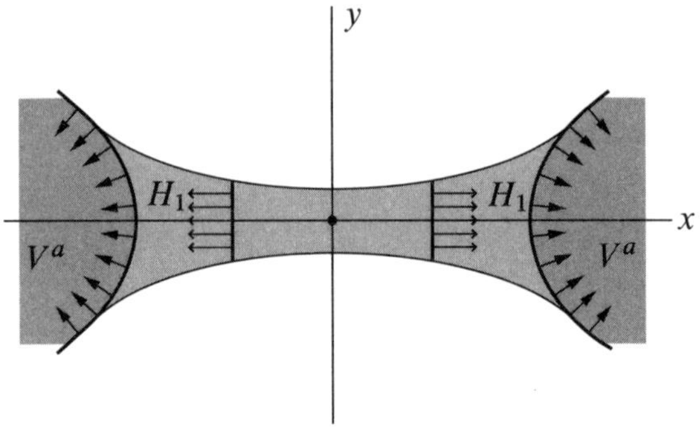

Figure IV.23. Deformation of $V^a \cup H_1$ to V^a.

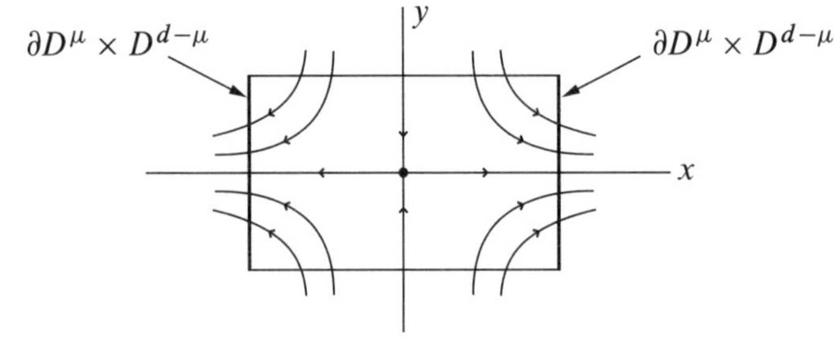

Figure IV.24. After excision.

where the two discs are defined by

$$D^\mu = \{x \mid |x| \le r\} \subset \mathbb{R}^\mu,$$
$$D^{d-\mu} = \{y \mid |y| \le r\} \subset \mathbb{R}^{d-\mu}$$

for a small radius r.

The pair of spaces on the right-hand side of the above isomorphism is of the same homotopy type as the pair $(D^\mu, \partial D^\mu)$, because the disc $D^{d-\mu}$ is contractible. Consequently,

$$H^*\big(D^\mu \times D^{d-\mu}, \partial D^\mu \times D^{d-\mu}\big) \cong H^*(D^\mu, \partial D^\mu).$$

Summing up, we have so far sketched the proof of the isomorphism

$$H^*(V^b, V^a) \cong H^*(D^\mu, \partial D^\mu).$$

Now,

$$H^j(D^\mu, \partial D^\mu) = \begin{cases} \mathbb{R}, & \text{if } j = \mu, \\ 0, & \text{otherwise.} \end{cases}$$

This can be easily verified by collapsing the boundary ∂D^μ into a point $*$ (proceeding as in Figure IV.25). From this we obtain the isomorphism $H^*(D^\mu, \partial D^\mu) = H^*(S^\mu, *)$. Using this, we finally arrive at the desired result,

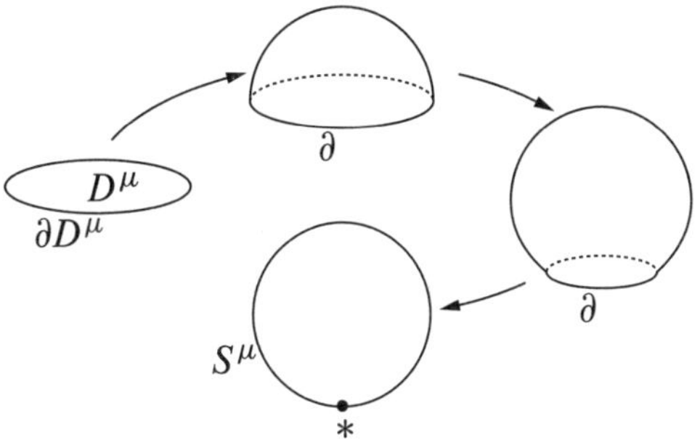

Figure IV.25. Collapsing ∂D^μ into a point.

$$H^j(V^b, V^a) = \begin{cases} \mathbb{R}, & \text{if } j = \mu, \\ 0, & \text{otherwise.} \end{cases}$$

This finishes the sketch of the proof of Proposition IV.26. All the details of this sketch, including the necessary tools from algebraic topology, can be found in the book [55] by J. Jost. □

In Proposition IV.26 we have considered only *one* critical point and thus have obtained only one handle. If several critical points are lying on the same level, we obtain, with the same local arguments applied in disjoint sets, several handles and therefore arrive at the following proposition.

Proposition IV.28. *We assume that* $\{p_1, p_2, \dots, p_m\} \in K \cap V^{-1}(c)$ *are* ***non-degenerate*** *critical points having the Morse indices* $\mu(p_j)$. *If* $a < c < b$ *and if*

$$K \cap M_a^b = \{p_1, p_2, \dots, p_m\},$$

then

$$H^j(V^b, V^a) = \mathbb{R} \oplus \cdots \oplus \mathbb{R} = \mathbb{R}^{s_j},$$

where the integer s_j *is the number of critical points* p *in* $\{p_1, p_2, \dots, p_m\}$ *whose* Morse *index is equal to* $\mu(p) = j$.

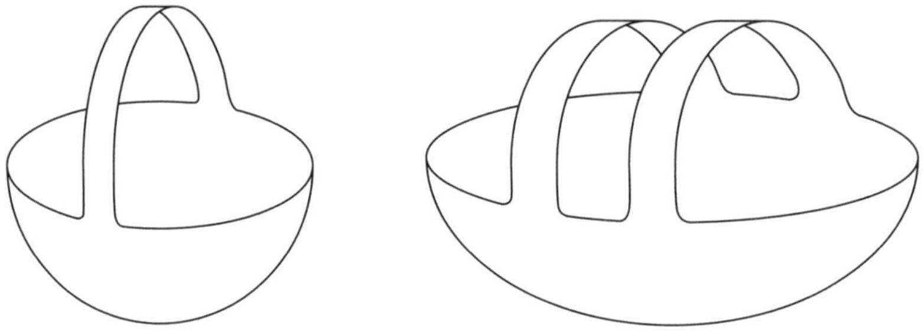

Figure IV.26. One handle (left) corresponds to one critical point, and two handles (right) corresponding to two critical points.

The rest of the proof of the Morse equation is of more algebraic nature. We begin with an easy lemma.

Lemma IV.29. *Given a sequence*

$$0 = A_0 \xrightarrow{a_0} A_1 \xrightarrow{a_1} A_2 \xrightarrow{a_2} A_3 \xrightarrow{a_3} A_4 \xrightarrow{a_4} \cdots \xrightarrow{a_k} A_k$$

*of linear maps a_j between the vector spaces A_j. If the sequence is **exact**, so that* $\ker a_{j+1} = \operatorname{Im} a_j$ *for all j, then*

$$\dim A_1 - \dim A_2 + \cdots + (-1)^{k-1} \dim A_k + (-1)^k \dim (\operatorname{Im} a_k) = 0.$$

Proof. From the exactness assumption we obtain the equations

$$\begin{aligned} \dim A_j &= \dim (\ker a_j) + \dim (\operatorname{Im} a_j) \\ &= \dim (\operatorname{Im} a_{j-1}) + \dim (\operatorname{Im} a_j) \end{aligned}$$

and $\operatorname{Im} a_0 = 0$. The lemma now follows by induction. □

We borrow again from algebraic topology. If $X \supset Y \supset Z$ is a triple of manifolds, then there exists the so-called *long exact sequence*

$$\begin{aligned} 0 \to H^0(X,Y) &\to H^0(X,Z) \to H^0(Y,Z) \\ &\xrightarrow{\delta^0} H^1(X,Y) \to H^1(X,Z) \to H^1(Y,Z) \\ &\qquad \xrightarrow{\delta^1} H^2(X,Y) \to H^2(X,Z) \to H^2(Y,Z) \xrightarrow{\delta^2} \cdots. \end{aligned}$$

Introducing the non-negative integers

$$\begin{aligned} r_j(X,Y) &:= \dim H^j(X,Y), \\ \nu_j(X,Y,Z) &:= \dim (\operatorname{Im} \delta^j), \end{aligned}$$

it follows from Lemma IV.29 that for every m,

$$\sum_{j=0}^{m}(-1)^j[r_j(X,Y)-r_j(X,Z)+r_j(Y,Z)]-(-1)^m\nu_m(X,Y,Z)=0,$$

and so, by subtraction,

$$(*)\quad (-1)^{m-1}\nu_{m-1}(X,Y,Z)=(-1)^m\nu_m(X,Y,Z)-(-1)^m r_m(X,Y)\\ +(-1)^m r_m(X,Z)-(-1)^m r_m(Y,Z).$$

We define the following polynomials in the variable t,

$$P(t,X,Y):=\sum_{m\geq 0} r_m(X,Y)\cdot t^m,$$
$$Q(t,X,Y,Z):=\sum_{m\geq 0}\nu_m(X,Y,Z)\cdot t^m.$$

Multiplying the equation $(*)$ by $(-1)^m t^m$ and adding up the equations we obtain the following equation.

Lemma IV.30. *If $X\supset Y\supset Z$ is a triple of manifolds, then*

$$(1+t)Q(t,X,Y,Z)=P(t,X,Y)-P(t,X,Z)+P(t,Y,Z).$$

We are finally ready to complete the proof of Theorem IV.20.

Proof of Theorem IV.20. If $c_1<c_2<c_3<\cdots<c_N$ are the *critical values* of the function $V\colon M\to\mathbb{R}$, then $c_1=\min_{x\in M}V(x)$ and $c_N=\max_{x\in M}V(x)$. We choose real numbers $d_0<\cdots<d_N$, satisfying

$$d_0<c_1<d_1<c_2<d_2<\cdots<c_N<d_N.$$

Since c_1 is the minimum we have $M^{d_0}=\emptyset$ and since c_N is the maximum, $M^{d_N}=M$, and we obtain the following filtration of M:

$$\emptyset=M^{d_0}\subset M^{d_1}\subset\cdots\subset M^{d_N}=M.$$

Applying Lemma IV.30 to the triple $M^{d_0}=\emptyset\subset M^{d_{j-1}}\subset M^{d_j}$ leads to the equation

$$P(t,M^{d_j},M^{d_{j-1}})=P(t,M^{d_j},\emptyset)-P(t,M^{d_{j-1}})+(1+t)Q(t,M^{d_j},M^{d_{j-1}},\emptyset).$$

Adding up these equations we arrive at the identity

$$\sum_{j=1}^{N}P(t,M^{d_j},M^{d_{j-1}})=P(t,M,\emptyset)+(1+t)Q(t)$$
$$=P(t,M)+(1+t)Q(t),$$

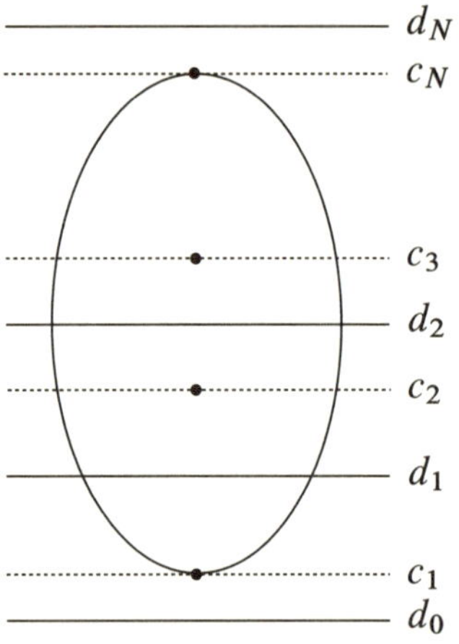

Figure IV.27. Schematically, the levels in the proof of Theorem IV.20.

in which $P(t, M)$ is the Poincaré polynomial of the manifold M. In view of Proposition IV.28 we obtain for the left-hand side the polynomial

$$\sum_{j=1}^{N} P(t, M^{d_j}, M^{d_{j-1}}) = \sum_{k=0}^{\dim M} t^k m_k = m(t, V),$$

which is the Morse polynomial. This completes the sketch of Theorem IV.20. As already pointed out, the details can be found in the book [55] by J. Jost. □

We would like to mention the far-reaching generalization of the above Morse theory to continuous flows on locally compact metric spaces, due to C. Conley. Generalizing the Morse index of a nondegenerate equilibrium point of the gradient flow, C. Conley associates with an isolated invariant set of a continuous flow the so-called Conley index, which is the homotopy type of a pointed topological space. In the case of the nondegenerate equilibrium point above, the Conley index is the pointed sphere having the dimension of the Morse index. Conley's index theory is one of the most powerful tools relating dynamics to invariants of algebraic topology. With a Morse decomposition of a compact invariant set of a continuous flow one can also associate a Morse equation generalizing Theorem IV.20. The main advantage of Conley's theory in the applications are the invariance properties under deformations.

Literature. In the above presentation of Morse theory we followed the corresponding sections in the book [55] by J. Jost, in which one also finds an introduction into the required algebraic topology. The book on Morse theory by J. Milnor [68] is a classic. Presentations of the theory can also be found in the books [21] by L. Conlon, [101] by J. Schwartz, [79] by R. S. Palais and Chun-lian Terng and [9] by

A. Banyaga and D. Hurtubise. The monograph [100] by M. Schwarz is devoted to Morse homology theory. We also recommend the article [12] by R. Bott. Morse theory in infinite dimensions can be found in the books [14] by K. C. Chang, [1] by S. Abbondandolo and in the articles [76], [77] and [78] by R. S. Palais.

As for the Conley index theory, we refer to the original work [19] by C. Conley, to [96] by D. Salamon, to [36] by J. Franks and M. Misiurewicz and to the survey article [69] by K. Mischaikow and M. Mrozek and the references therein. The Conley index theory in infinite dimensions is developed in [94] by K. Rybakowski. The Conley Morse theory for continuous flows on locally compact metric spaces is presented in [20] by C. Conley and E. Zehnder and provides, in particular, an efficient approach to Morse inequalities.

Chapter V
Hamiltonian vector fields and symplectic diffeomorphisms

Chapter V introduces the special class of Hamiltonian vector fields. They are determined by a single function called the Hamiltonian function and play an important role in frictionless systems such as the N-body problem of celestial mechanics. Hamiltonian vector fields are defined on symplectic manifolds. These manifolds are even-dimensional carrying a distinguished 2-form which is closed and nondegenerate and called a symplectic structure. In contrast to Riemannian structures which do exist on every manifold, not every even-dimensional manifold admits a symplectic structure. We shall see that symplectic manifolds of the same dimension are locally indistinguishable (Darboux's theorem), again in sharp contrast to Riemannian manifolds where the curvature is an example of a local invariant. The flow maps generated by the Hamiltonian vector fields leave the Hamilton function (energy conservation) and the symplectic 2-form invariant. They are examples of symplectic mappings. The transformation theory of Hamiltonian vector fields, the Hamiltonian formalism, is treated in detail in the Sections 7 and 8, where we make use of the exterior differential calculus. It is recalled in the Sections 2–6 and is presumably familiar to most of our readers. On energy surfaces the Hamiltonian flow preserves a distinguished volume form (Liouville form), so that in the compact situation the recurrence theorem of Poincaré and the ergodic theorem of Birkhoff are applicable. For practical purposes it is good to know that a symplectic map can be represented by a single function, called a generating function, as is explained in Section 9. Finally, integrable systems are discussed in Section 10. These systems are characterized by the property of possessing sufficiently many integrals such that the task of solving the Hamiltonian equation for all times becomes essentially trivial, as will be illustrated by the proof of the theorem of V. I. Arnold and R. Jost about the existence of action and angle variables. Integrable systems are exceptionally rare, but there are many natural systems close to integrable systems to which the KAM perturbation theory is applicable.

V.1 Symplectic vector spaces

We begin with the symplectic linear algebra.

Definition. A *symplectic vector space* is a pair (V, ω) consisting of a finite dimensional real vector space V and a bilinear form $\omega\colon V \times V \to \mathbb{R}$ which is skew symmetric and nondegenerate. The form ω is called a *symplectic structure* or *symplectic form* on V.

Bilinearity refers to the $\mathbb{R}$-linearity in every argument, skew symmetry means that $\omega(u,v) = -\omega(v,u)$ for all vectors $u, v \in V$ and nondegeneracy requires that from $\omega(u,v) = 0$ for all $v \in V$ it follows that $u = 0$. Equivalently, the nondegeneracy requires that the linear mapping $\omega^\sharp\colon V \to V^*$, which associates with the vector $u \in V$ the linear form $\omega(u,\cdot)$ on V, is an isomorphism.

Examples. (1) Assuming W to be an n-dimensional vector space and W^* to be its dual space, we define the vector space $V := W \times W^*$ and the bilinear form $\omega\colon V \times V \to \mathbb{R}$ by

$$\omega\big((h,h^*),(k,k^*)\big) := h^*(k) - k^*(h).$$

Then the pair (V,ω) is a symplectic vector space of dimension $2n$.

(2) *The symplectic standard space* $(\mathbb{R}^{2n},\omega_0)$. Here the vector space is equal to $V = \mathbb{R}^{2n}$ and the symplectic form is defined by

$$\omega_0(u,v) := \langle Ju, v\rangle, \quad u, v \in \mathbb{R}^{2n},$$

where $\langle\cdot,\cdot\rangle$ denotes the Euclidian scalar product in $\mathbb{R}^{2n}$ and J denotes the block matrix

$$J = \begin{pmatrix} 0 & \mathbb{1} \\ -\mathbb{1} & 0 \end{pmatrix}$$

with respect to $\mathbb{R}^{2n} = \mathbb{R}^n \times \mathbb{R}^n$. This matrix is nondegenerate, because we have $\det J = 1$ and it is skew symmetric, $J^T = -J$. In addition, the matrix J has the properties

$$J^2 = -\mathbb{1} \quad \text{and} \quad J^T = -J = J^{-1}.$$

(3) There is *no* symplectic vector space of dimension $\dim V = 1$. Indeed, choosing a non-vanishing vector $u \neq 0$ in the 1-dimensional space V, it follows that $u = rv$ and hence $\omega(u,v) = 0$ for all vectors v in V in view of the skew symmetry. Therefore the 2-form ω is necessarily degenerate.

As in Euclidian geometry the symplectic orthogonal complement can be defined.

Definition. The *symplectic orthogonal complement* of the linear subspace $E \subset V$ is the linear subspace

$$E^\perp := \{v \in V \mid \omega(e,v) = 0 \text{ for all } e \in E\}.$$

The nondegeneracy of ω is equivalent to $V \cap V^\perp = \{0\}$.

Proposition V.1. *If $E \subset V$ is a linear subspace, then*

$$\dim E + \dim E^\perp = \dim V$$

for the symplectic orthogonal complement $E^\perp$.

Proof. Let $e_1, \dots, e_r$ be a basis of E. It then follows from the bilinearity of ω that the symplectic orthogonal complement is given by

$$\begin{aligned} E^{\perp} &= \{v \in V \mid \omega(e_j, v) = 0, \quad 1 \le j \le r\} \\ &= \bigcap_{1 \le j \le r} \ker \omega(e_j, \cdot). \end{aligned}$$

Therefore $E^{\perp}$ is the kernel of the linear mapping $A \colon V \to \mathbb{R}^r$, defined by $Av = \big(\omega(e_1, v), \dots, \omega(e_r, v)\big)$. The linear functionals $\omega(e_j, \cdot)$ are linearly independent, since $\omega^{\#}$ is an isomorphism. Therefore $\dim \operatorname{Im} A = r$ and the proposition follows from the dimension formula $\dim V = \dim \operatorname{Im} A + \dim \ker A$. □

From the definition of $E^{\perp}$ it follows that $E \subset (E^{\perp})^{\perp}$ and applying Proposition V.1 to the linear subspace $E^{\perp}$ (replacing E) one concludes that $\dim E = \dim (E^{\perp})^{\perp}$ and so,

$$E = (E^{\perp})^{\perp}.$$

We make use of the symbolic notation $u \perp v$ for two vectors satisfying $\omega(u, v) = 0$. In *contrast to* the orthogonality in Euclidian geometry we have in general,

$$E + E^{\perp} \ne V,$$

as the following example shows.

Example. If (V, ω) is a symplectic vector space and $u \ne 0$ a vector in V, we set $E := \langle u \rangle$. Due to the skew symmetry, $\omega(u, u) = 0$ and therefore $E \subset E^{\perp}$. However, $E^{\perp} \ne V$, because ω is nondegenerate.

If $E \subset V$ is a linear subspace, the restriction of ω onto $E \times E$ is still a skew symmetric bilinear form, but in general it is no longer nondegenerate.

Definition. A linear subspace $E \subset (V, \omega)$ is called *symplectic*, if the pair (E, ω) is a symplectic vector space.

Hence, a linear subspace $E \subset V$ is symplectic if and only if $E \cap E^{\perp} = \{0\}$. Due to Proposition V.1, this is equivalent to

$$E \oplus E^{\perp} = V.$$

Setting $E = (E^{\perp})^{\perp}$, we obtain

$$E \text{ symplectic} \iff E^{\perp} \text{ symplectic}.$$

The following proposition shows that every symplectic vector space (V, ω) looks in suitable coordinates like the symplectic standard space $(\mathbb{R}^{2n}, \omega_0)$.

Proposition V.2 (Symplectic basis). *If (V, ω) is a symplectic vector space, then* $\dim V = 2n$ *is an even number and there exists a distinguished basis* $\{e_1, \dots, e_n, f_1, \dots, f_n\}$*, satisfying for all* $i, j = 1, \dots, n$

$$\omega(e_i, e_j) = 0, \quad \omega(f_i, f_j) = 0, \quad \omega(f_i, e_j) = \delta_{ij}.$$

Such a basis is called a ***symplectic basis****. The coordinates with respect to a symplectic basis are called* ***symplectic coordinates****.*

Proof. (1) We choose $e_1 \neq 0$ in V. Since ω is nondegenerate, there exists a vector $u \in V$ satisfying $\omega(u, e_1) \neq 0$. Normalizing we set $f_1 := \omega(u, e_1)^{-1} u$ and obtain $\omega(f_1, e_1) = 1$. Since ω is skew symmetric, the two vectors e_1, f_1 are linearly independent. According to the construction $E := \operatorname{span}\{e_1, f_1\}$ is a symplectic subspace of V of dimension $\dim E = 2$. If $\dim V = 2$, the proposition is proved.

(2) If $\dim V > 2$ the same argument can be applied to the *symplectic* subspace $E^{\perp}$. In view of $E \oplus E^{\perp} = V$ the proof follows in finitely many steps. Due to the nondegeneracy of ω, no 1-dimensional space can remain after the last step, because, as we have seen, 1-dimensional spaces cannot be symplectic. □

With respect to symplectic coordinates, the matrix associated with the symplectic form ω is in a normal form which we shall compute next. We choose the symplectic basis $\{e_1, \dots, e_n, f_1, \dots, f_n\}$ of (V, ω). With respect to this basis, we let

$$u = \sum_{j=1}^{n} (x_j e_j + x_{n+j} f_j) \quad \text{and} \quad v = \sum_{j=1}^{n} (y_j e_j + y_{n+j} f_j)$$

be two vectors belonging to V. Using the definition of a symplectic basis one computes

$$\begin{aligned}
\omega(u, v) &= \sum_{i,j} x_i y_j \omega(e_i, e_j) + \sum_{i,j} x_{n+i} y_j \omega(f_i, e_j) + \sum_{i,j} x_j y_{n+i} \omega(e_j, f_i) \\
&\quad + \sum_{i,j} x_{n+i} y_{n+j} \omega(f_i, f_j) \\
&= \sum_{i,j} x_{n+i} y_j \omega(f_i, e_j) + \sum_{i,j} x_j y_{n+i} \omega(e_j, f_i) \\
&= \sum_{i,j} (x_{n+i} y_j \delta_{ij} - x_j y_{n+i} \delta_{ij}) \\
&= \sum_{j} (x_{n+j} y_j - x_j y_{n+j}) \\
&= \langle Jx, y \rangle
\end{aligned}$$

where $x = (x_1, \dots, x_{2n})$ and $y = (y_1, \dots, y_{2n})$, and where J is the matrix

$$J = \begin{pmatrix} 0 & \mathbb{1} \\ -\mathbb{1} & 0 \end{pmatrix}$$

with respect to the splitting $\mathbb{R}^{2n} = \mathbb{R}^n \times \mathbb{R}^n$. This is the desired normal form we shall often meet later on.

Remark. In the literature, another normal form is sometimes used. In view of the definition of the symplectic basis, the 2-dimensional subspaces $E_j := \operatorname{span}\{e_j, f_j\}$ are symplectic subspaces satisfying $E_i \perp E_j$ for $i \neq j$ and

$$V = E_1 \oplus \cdots \oplus E_n.$$

Representing the vectors u and v in the symplectic coordinates $x_1, \dots, x_{2n}$ with respect to the above decomposition of V, so that

$$u = x_1 e_1 + x_2 f_1 + \cdots + x_{2n-1} e_n + x_{2n} f_n,$$

and similarly for v, one verifies that

$$\omega(u, v) = \langle \hat{J} x, y \rangle$$

where $\hat{J}$ is the block diagonal matrix

$$\hat{J} = \begin{pmatrix} 0 & 1 & & & & & \\ -1 & 0 & & & & & \\ & & 0 & 1 & & & \\ & & -1 & 0 & & & \\ & & & & \ddots & & \\ & & & & & 0 & 1 \\ & & & & & -1 & 0 \end{pmatrix}.$$

Every symplectic structure ω of a symplectic vector space (V, ω) of dimension $2n$ has *the same normal form* with respect to a symplectic basis. This is in contrast to the symmetric bilinear forms of Euclidian geometry where the signature is an invariant. In symplectic coordinates we always have $\omega(u, v) = \langle Jx, y \rangle = \omega_0(x, y)$, so that we are in the symplectic standard space $(\mathbb{R}^{2n}, \omega_0)$.

As in Euclidian geometry we obtain the following rules for linear subspaces $E, F \subset V$ of a symplectic vector space (V, ω). Some of the rules we have already verified and the remaining ones are obvious.

$$\begin{aligned} E \subset F &\Longrightarrow F^{\perp} \subset E^{\perp}, \\ (E^{\perp})^{\perp} &= E, \\ (E \cap F)^{\perp} &= E^{\perp} + F^{\perp}, \\ (E + F)^{\perp} &= E^{\perp} \cap F^{\perp}, \\ \dim E^{\perp} + \dim E &= \dim V. \end{aligned}$$

In a symplectic vector space we distinguish the following subspaces.

Definition. We assume (V, ω) to be a symplectic vector space of dimension $2n$. A linear subspace $E \subset V$ is called

(i) *isotropic*, if $E \subset E^{\perp}$ ($\Longrightarrow \dim E \leq n$),
(ii) *Lagrangian*, if $E = E^{\perp}$ ($\Longrightarrow \dim E = n$),
(iii) *coisotropic*, if $E \supset E^{\perp}$ ($\Longrightarrow \dim E \geq n$),
(iv) *symplectic*, if $E \cap E^{\perp} = \{0\}$ ($\Longrightarrow \dim E = 2k$ even).

The statements about the dimension of the subspaces E are obtained in the following way.

(i) If $E \subset E^{\perp}$, it follows from Proposition V.1 that $\dim E \leq \dim E^{\perp} = \dim V - \dim E$ and hence $2 \dim E \leq \dim V = 2n$.

(ii) Lagrangian means maximal isotropic and is obviously characterized by the two properties $E \subset E^{\perp}$ and $\dim E = \dim E^{\perp} = n$.

(iii) is proved as the statement (i) and

(iv) has already been verified above.

Examples. In the symplectic standard space $(\mathbb{R}^{2n}, \omega_0)$ we set $x = (q, p)$ and $x' = (q', p')$ with respect to $\mathbb{R}^n \times \mathbb{R}^n$, so that

$$\omega_0(x, x') = \langle Jx, x' \rangle = \langle p, q' \rangle - \langle q, p' \rangle.$$

The subspace

$$E = \{x \in \mathbb{R}^{2n} \mid q_1 = \cdots = q_r = 0\}$$

is coisotropic for $r \leq n$ and Lagrangian for $r = n$. The set

$$E = \{x \in \mathbb{R}^{2n} \mid q_1 = p_1 = 0, \ldots, q_r = p_r = 0\}$$

is an example of a symplectic subspace for $r < n$. The subspace

$$E = \{x \in \mathbb{R}^{2n} \mid q_1 = \cdots = q_n = 0,\ p_1 = \cdots = p_r = 0\}$$

is isotropic.

We now consider the *linear* maps $A \colon V \to V$ of (V, ω) which leave the symplectic form ω *invariant*. To be more precise we introduce the notion of a pullback.

We assume V to be a vector space, $b \colon V \times V \to \mathbb{R}$ to be a bilinear form and $A \in \mathcal{L}(V)$ a linear mapping. Then the *pullback* of the bilinear form b under the map A is the bilinear form A^*b on V defined by

$$(A^*b)(u, v) := b(Au, Av) \quad \text{for all } u, v \in V.$$

Definition. A linear map $A \colon V \to V$ of a symplectic vector space (V, ω) is called *symplectic*, if $A^*\omega = \omega$, or explicitly if

$$\omega(Au, Av) = \omega(u, v) \quad \text{for all } u, v \in V.$$

In symplectic coordinates or in the symplectic standard space $(\mathbb{R}^{2n}, \omega_0)$, we obtain a simple criterion for the symplectic character of the map $A \in \mathcal{L}(\mathbb{R}^{2n})$. In view of $\omega_0(u, v) = \langle Ju, v \rangle$, the condition $A^*\omega_0 = \omega_0$ is equivalent to

$$\langle JAu, Av \rangle = \langle Ju, v \rangle, \quad u, v \in \mathbb{R}^{2n}$$

or $\langle A^T JAu, v \rangle = \langle Ju, v \rangle$, for all vectors u, v and hence

$$A^T JA = J.$$

A $(2n \times 2n)$-matrix A is called *symplectic*, if it satisfies $A^T JA = J$. In the special case $n = 1$ the matrix A is symplectic if and only if $\det A = 1$, because for

$$A = \begin{pmatrix} a & b \\ c & d \end{pmatrix}$$

we have

$$A^T JA = \begin{pmatrix} 0 & ad - bc \\ -ad + bc & 0 \end{pmatrix}.$$

In particular, such a matrix A is volume preserving. In the general case it follows from $A^T JA = J$ that $(\det A)^2 = \det J = 1$, so that A is invertible. We shall later show that actually $\det A = 1$.

The set of symplectic matrices constitutes the so-called *symplectic group*

$$\mathrm{Sp}(n) := \{A \in \mathcal{L}(\mathbb{R}^{2n}) \mid A^T JA = J\}$$

which is one of the classical Lie groups.

Proposition V.3. *The set* $\mathrm{Sp}(n)$ *is a group; if* $A, B \in \mathrm{Sp}(n)$ *then also* $AB \in \mathrm{Sp}(n)$, $A^{-1} \in \mathrm{Sp}(n)$ *and* $\mathbb{1}_{2n} \in \mathrm{Sp}(n)$. *In addition,* $A^T \in \mathrm{Sp}(n)$ *and* $J \in \mathrm{Sp}(n)$.

Proof. Let $J = A^T JA$, then $(A^T)^{-1} JA^{-1} = J$ and $A^{-1} \in \mathrm{Sp}(n)$. By inverting this equation we obtain $J^{-1} = AJ^{-1}A^T$ and using $J^{-1} = -J$ we obtain $J = (A^T)^T JA^T$, so that indeed $A^T \in \mathrm{Sp}(n)$. The remaining statements are also easily verified. □

Exercise. (1) We assume A to be a $(2n \times 2n)$-block matrix of the form

$$A = \begin{pmatrix} a & b \\ c & d \end{pmatrix}.$$

Then A is symplectic precisely if the $(n \times n)$-matrices $a^T c$ and $b^T d$ are symmetric and $a^T d - c^T b = \mathbb{1}$.

(2) We now set $b = 0$. Then $A \in \mathrm{Sp}(n)$, precisely if A is a matrix product of the form

$$A = \begin{pmatrix} a & 0 \\ 0 & (a^T)^{-1} \end{pmatrix} \begin{pmatrix} \mathbb{1} & 0 \\ s & \mathbb{1} \end{pmatrix}$$

where a is invertible and where $s = s^T$ is symmetric. Both matrices on the right-hand side are symplectic.

(3) *Polar form.* Let $U \in \mathcal{L}(\mathbb{R}^{2n})$ be a matrix satisfying $\det U \neq 0$. Then U has a unique polar form $U = OP$, where O is orthogonal and P is a positive symmetric matrix. Show that if U is belongs to $\mathrm{Sp}(n)$, then also O and P belong to $\mathrm{Sp}(n)$.

(4) Let O be orthogonal, then $O \in \mathrm{Sp}(n)$ is equivalent to

$$O = \begin{pmatrix} a & b \\ -b & a \end{pmatrix}$$

where $a^T b$ is symmetric and $a^T a + b^T b = \mathbb{1}$. These conditions are also necessary and sufficient for the complex matrix

$$a + ib \in U(n) \subset \mathcal{L}(\mathbb{C}^n)$$

to be a unitary matrix. Here $U(n)$ denotes the group of the unitary $(n \times n)$-matrices. The above construction establishes an *algebraic* isomorphism

$$\mathrm{Sp}(n) \cap O(2n) \cong U(n),$$

where $O(k)$ denotes the group of the orthogonal $(k \times k)$-matrices.

Proposition V.4. *If* $A \in \mathrm{Sp}(n)$, *then* $\det A = 1$, *so that* A *is volume preserving.*

We already know that $(\det A)^2 = 1$ and hence $\det A = \pm 1$. To determine the sign, it is convenient to use the formalism of differential forms.

A brief excursion into differential forms. Take coordinates $z = (z_1, \dots, z_d)$ in $\mathbb{R}^d$. Then the *linear form*

$$dz_j \in \mathcal{L}(\mathbb{R}^d, \mathbb{R}) = (\mathbb{R}^d)^*$$

is defined as

$$dz_j(a) := e_j^*(a) := a_j \quad \text{for all } a = (a_1, \dots, a_d) \in \mathbb{R}^d.$$

This is the derivative of the j-th coordinate function $f_j \colon \mathbb{R}^d \to \mathbb{R}$ defined by $f(z) = z_j$ for $z = (z_1, \dots, z_d)$. By $\{e_1^*, \dots, e_d^*\}$ we denote the *dual basis* of the *canonical* basis $\{e_1, \dots, e_d\}$ of $\mathbb{R}^d$ satisfying $e_j^*(e_i) = \delta_{ij}$.

The 2-form (skew symmetric bilinear form) $dz_i \wedge dz_j$ is defined as

$$(dz_i \wedge dz_j)(a, b) := \det \begin{pmatrix} dz_i(a) & dz_i(b) \\ dz_j(a) & dz_j(b) \end{pmatrix} = a_i b_j - a_j b_i.$$

Obviously, $dz_i \wedge dz_j = -(dz_j \wedge dz_i)$. With this notation, the canonical symplectic structure ω_0 on $\mathbb{R}^{2n}$ can be written as a differential form. In the coordinates $z = (x, y) \in \mathbb{R}^n \times \mathbb{R}^n = \mathbb{R}^{2n}$ it has the representation

$$\omega_0 = \sum_{j=1}^{n} dy_j \wedge dx_j,$$

or explicitly,

$$\omega_0(u, v) = \langle Ju, v\rangle = \sum_{j=1}^{n} (dy_j \wedge dx_j)(u, v) \quad \text{for all } u, v \in \mathbb{R}^{2n}.$$

At this point, we recall the concept of a *k-form* on $\mathbb{R}^d$. We assume $1 \le k \le d$ and $x = (x_1, \dots, x_d) \in \mathbb{R}^d$. The special k-form $dx_{i_1} \wedge \cdots \wedge dx_{i_k}$ is defined by

$$(dx_{i_1} \wedge \cdots \wedge dx_{i_k})(v_1, \dots, v_k) := \det \begin{pmatrix} dx_{i_1}(v_1) & \cdots & dx_{i_1}(v_k) \\ dx_{i_2}(v_1) & \cdots & dx_{i_2}(v_k) \\ \vdots & & \vdots \\ dx_{i_k}(v_1) & \cdots & dx_{i_k}(v_k) \end{pmatrix}$$

for $v_1, \dots, v_k \in \mathbb{R}^d$. The k-forms

$$\{dx_{i_1} \wedge \cdots \wedge dx_{i_k}, \quad 1 \le i_1 < i_2 < \cdots < i_k \le d\}$$

constitute a basis of the vector space $\bigwedge^k(\mathbb{R}^d)$ of all the alternating k-linear forms α on $\mathbb{R}^d$, that is, of the continuous maps $\alpha \colon \mathbb{R}^d \times \cdots \times \mathbb{R}^d \to \mathbb{R}$ (the product taken k times), linear in every argument and satisfying

$$\alpha(\dots, v_i, \dots, v_j, \dots) = -\alpha(\dots, v_j, \dots, v_i, \dots)$$

for all $i \ne j$. The integer k is called the *degree* of the form α and denoted by $\deg \alpha$.

We now assume $1 \le p, q \le d$. For $\alpha \in \bigwedge^p$ and $\beta \in \bigwedge^q$ we can introduce the exterior product $\alpha \wedge \beta \in \bigwedge^{p+q}$. The $(p+q)$-form $\alpha \wedge \beta$ is defined by

$$\begin{aligned}(\alpha \wedge \beta)&(v_1, \dots, v_{p+q}) \\ &= \frac{1}{p!q!} \sum_{\sigma \in \mathcal{S}_{p+q}} \varepsilon(\sigma)\, \alpha(v_{\sigma(1)}, \dots, v_{\sigma(p)})\, \beta(v_{\sigma(p+1)}, \dots, v_{\sigma(p+q)}),\end{aligned}$$

where $\mathcal{S}_n$ is the group of all permutations of n elements and $\varepsilon(\sigma)$ is the sign of the permutation σ. This product is associative and satisfies

$$\beta \wedge \alpha = (-1)^{\deg \alpha \deg \beta} (\alpha \wedge \beta).$$

The *pullback* of a k-form α on $\mathbb{R}^d$ under the linear map $A\colon \mathbb{R}^d \to \mathbb{R}^d$ is the k-form $A^*\alpha$ defined by

$$(A^*\alpha)(v_1, \dots, v_k) := \alpha(Av_1, \dots, Av_k)$$

for vectors $v_1, \dots, v_k \in \mathbb{R}^d$. The map $\alpha \mapsto A^*\alpha$ is linear. Moreover,

$$A^*(\alpha \wedge \beta) = (A^*\alpha) \wedge (A^*\beta).$$

If the form $\alpha \in \bigwedge^d(\mathbb{R}^d)$ is of highest degree, then we claim that its pullback under the map $A \in \mathcal{L}(\mathbb{R}^d)$ is the multiplication by the determinant, so that $A^*\alpha = (\det A)\alpha$. Indeed, due to $\dim \bigwedge^d = 1$, the form α can be written as $\alpha = a(e_1^* \wedge \cdots \wedge e_d^*) =: a\beta$ with a real number $a \in \mathbb{R}$. In view of the definition of the determinants (Leibnitz-formula),

$$A^*\beta(e_1, \dots, e_d) = \beta(Ae_1, \dots, Ae_d) = \det A.$$

Since also $A^*\beta$ has to be of form $A^*\beta = \text{const.} \cdot \beta$, it immediately follows that $A^*\beta = (\det A)\beta$ and in view of the linearity of the pullback we find indeed that

$$A^*\alpha = (\det A)\alpha$$

for every d-form α.

We next turn to the proof of Proposition V.4 which is now very easy.

Proof of Proposition V.4. We introduce the $2n$-form $\Omega := \omega_0 \wedge \cdots \wedge \omega_0$ (the product taken n times). Since this is a $2n$-form on $\mathbb{R}^{2n}$, it follows for every linear map $A \in \mathcal{L}(\mathbb{R}^{2n})$ that on one hand

$$A^*\Omega = (\det A)\,\Omega,$$

and on the other hand, since $A \in \mathrm{Sp}(n)$ is a symplectic map,

$$\begin{aligned} A^*\Omega &= (A^*\omega_0) \wedge \cdots \wedge (A^*\omega_0) \\ &= \omega_0 \wedge \cdots \wedge \omega_0 \\ &= \Omega \end{aligned}$$

so that $\det A = 1$, as claimed in Proposition V.4. □

Having this proposition we investigate the spectrum of symplectic linear maps.

Proposition V.5 (Spectral proposition for $\mathrm{Sp}(n)$). *Let $A \in \mathrm{Sp}(n)$ and let $\sigma(A)$ be the spectrum of A. If λ belongs to the set $\sigma(A)$, then also λ^{-1}, $\bar{\lambda}$ and $\bar{\lambda}^{-1}$ belong to $\sigma(A)$, and all these eigenvalues have the same algebraic multiplicity. If $-1 \in \sigma(A)$ or $1 \in \sigma(A)$, then these eigenvalues have* ***even*** *multiplicity.*

It follows for the dynamics of the map $A \in \mathrm{Sp}(n)$ that there exist *no attractors*! In view of the spectrum of a symplectic map, $A \in \mathrm{Sp}(n)$ is organized in groups of eigenvalues and the following cases can be distinguished.

(a) If $|\lambda| \neq 1$ and $\lambda \notin \mathbb{R}$ then $\lambda, \lambda^{-1}, \bar{\lambda}, \bar{\lambda}^{-1}$ are four different eigenvalues A.

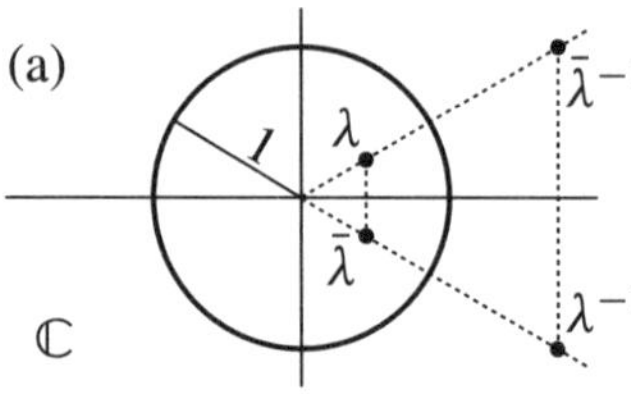

(b) If $|\lambda| \neq 1$ and $\lambda \in \mathbb{R}$, then there are two different eigenvalues λ and λ^{-1}.

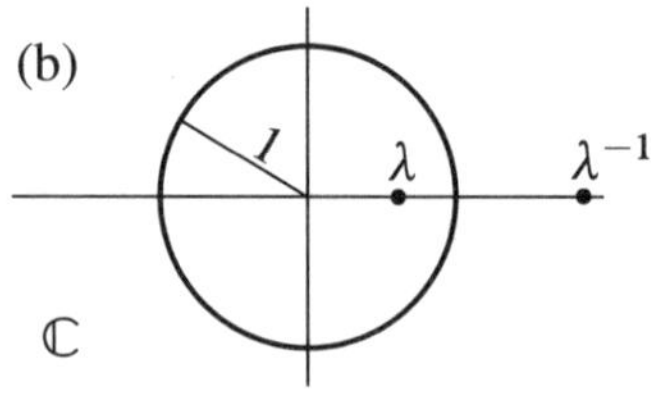

(c) If $|\lambda| = 1$, then $\bar{\lambda} = \lambda^{-1}$ and there are two eigenvalues λ and $\bar{\lambda}$ on the unit circle, respectively one eigenvalue in the cases $\lambda = \pm 1$.

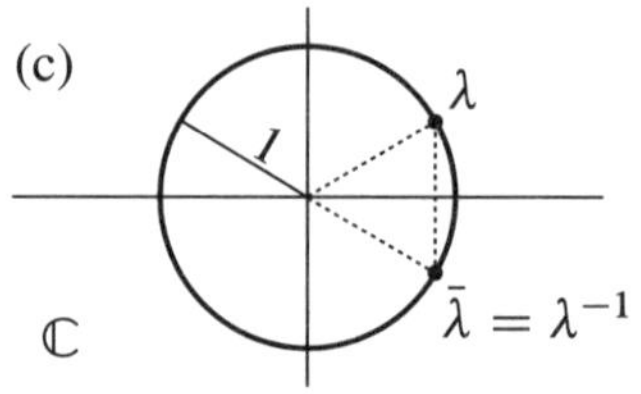

For every eigenspace on which A is contracting, there exists another eigenspace on which A is expanding.

Proof of Proposition V.5. If $A \in \mathrm{Sp}(n)$, then $A^T J A = J$ and hence $JAJ^{-1} = (A^{-1})^T$. Moreover, $\det A^{-1} = \det A = \det J = 1$ and using the rules of determinants we compute, for the characteristic polynomial $p(\lambda)$ of the map A,

$$p(\lambda) = \det(A - \lambda) = \det(JAJ^{-1} - \lambda) = \det((A^{-1})^T - \lambda) = \det(A^{-1} - \lambda)$$

and

$$\det(A^{-1} - \lambda) = \det(A^{-1}(\mathbb{1} - \lambda A)) = \det(\mathbb{1} - \lambda A) = (-1)^{2n} \lambda^{2n} \det(A - \lambda^{-1}).$$

Summing up, the characteristic polynomial satisfies the identity

$$p(\lambda) = \lambda^{2n} p(\lambda^{-1}).$$

If $\lambda \in \sigma(A)$, then $\lambda \neq 0$ and $p(\lambda) = 0$ and from the above formula one concludes $p(\lambda^{-1}) = 0$, so that also λ^{-1} is an eigenvalue of A.

If k is the multiplicity of $\lambda_0 \in \sigma(A)$ then $p(\lambda) = (\lambda - \lambda_0)^k p_1(\lambda)$ for a polynomial p_1 satisfying $p_1(\lambda_0) \neq 0$ and hence

$$\begin{aligned} p(\lambda) &= \lambda^{2n} p(\lambda^{-1}) \\ &= \lambda^{2n}(\lambda^{-1} - \lambda_0)^k p_1(\lambda^{-1}) \\ &= \lambda_0^k \lambda^{2n-k}(\lambda_0^{-1} - \lambda)^k p_1(\lambda^{-1}). \end{aligned}$$

One reads off that also the eigenvalue λ_0^{-1} has multiplicity k.

Let $\lambda_1, \lambda_1^{-1}, \dots, \lambda_k, \lambda_k^{-1}$ be all eigenvalues $\neq \pm 1$ of A listed according to their multiplicities. If l is the multiplicity of $-1 \in \sigma(A)$, then $m := 2n - 2k - l$ is the multiplicity of $1 \in \sigma(A)$. Since det A is the product of all $2n$ eigenvalues of A, we have

$$1 = (\lambda_1 \lambda_1^{-1}) \dots (\lambda_k \lambda_k^{-1})(-1)^l,$$

so that l must be an even integer and consequently also m.

Finally, since the characteristic polynomial has real coefficients, we conclude that with an eigenvalue λ also its complex conjugate $\bar{\lambda}$ is an eigenvalue. The proof of Proposition V.5 is complete. □

Definition. If (V_1, ω_1) and (V_2, ω_2) are two symplectic vector spaces, then a linear map $A \in \mathcal{L}(V_1, V_2)$ is called *symplectic*, if $A^*\omega_2 = \omega_1$, explicitly if

$$\omega_2(Au, Av) = \omega_1(u, v) \quad \text{for all } u, v \in V_1.$$

Remark. Linear symplectic maps are *injective*. Indeed, if u belongs to the kernel of A, then

$$\omega_1(u, v) = A^*\omega_2(u, v) = \omega_2(Au, Av) = \omega_2(0, Av) = 0 \quad \text{for all } v \in V_1,$$

and therefore $u = 0$, since ω_1 is nondegenerate. In particular, dim $V_1 \leq$ dim V_2, if there exists a linear symplectic map $A\colon V_1 \to V_2$.

Proposition V.6. *If (V_1, ω_1) and (V_2, ω_2) are two symplectic vector spaces of the same dimension, then there exists a linear isomorphism $A \in \mathcal{L}(V_1, V_2)$ satisfying $A^*\omega_2 = \omega_1$.*

Such a map A is called a *symplectic isomorphism*. It can be easily verified that the inverse A^{-1} is also symplectic.

Proof of the proposition. If we choose symplectic bases $\{e_i, f_j\}$ in (V_1, ω_1) and $\{\hat{e}_i, \hat{f}_j\}$ in (V_2, ω_2) and define $A\colon V_1 \to V_2$ by $Ae_j = \hat{e}_j$ and $Af_j = \hat{f}_j$ for all j, then it follows that $A^*\omega_2 = \omega_1$. □

Corollary V.7. *All symplectic vector spaces of the same dimension are symplectically isomorphic. The* ***only invariant*** *is the dimension!*

In view of the proof of Proposition V.2 we can choose e_1 and $\hat{e}_1$ freely in the symplectic bases. In the case of $V_1 = V_2$ we therefore find, in view of the proof of Proposition V.6, a symplectic isomorphism $A\colon V_1 \to V_1$ satisfying $Ae_1 = \hat{e}_1$. This proves the following statement.

Proposition V.8. *The symplectic group* $\mathrm{Sp}(n)$ *acts transitively on* $\mathbb{R}^{2n} \setminus \{0\}$. □

In the same way one sees that $\mathrm{Sp}(n)$ acts transitively on the set of all symplectic subspaces of $\mathbb{R}^{2n}$ of fixed dimension $2k$. We choose for a symplectic subspace E a symplectic basis and complete it to a symplectic basis of $\mathbb{R}^{2n}$ by choosing a symplectic basis in the symplectic complement $E^{\perp}$. In the same way one can proceed with a second symplectic subspace F of the same dimension. By mapping the corresponding basis vectors onto each other, we have indeed defined a symplectic map $A \in \mathrm{Sp}(n)$ satisfying $A(E) = F$.

Literature. A presentation of symplectic algebra can be found, for instance, in Appendix A in the book [24] by R. Cushman and L. Bates.

V.2 The exterior derivative *d*

This section introduces Cartan's formalism of differential forms. Proceeding at first locally we consider an open subset $U \subset \mathbb{R}^n$. The spaces $\Omega^k(U)$ of the smooth *differential forms* on U are defined in the following way.

(1) The space $\Omega^0(U)$ of the 0-forms on U is the set $C^\infty(U, \mathbb{R})$ of the smooth functions.

(2) A 1-form on the set U is a map $U \to \bigwedge^1(\mathbb{R}^n) = (\mathbb{R}^n)^*$,

$$x \mapsto \alpha(x) = \sum_{j=1}^{n} \alpha_j(x) dx_j, \quad \alpha_j \in C^\infty(U, \mathbb{R}), \quad dx_j = e_j^*.$$

The space of the 1-forms on U is denoted by $\Omega^1(U)$.

(3) Analogously, a k-form on U is a map $U \to \bigwedge^k(\mathbb{R}^n)$,

$$x \mapsto \alpha(x) = \sum_{1 \le i_1 < i_2 < \cdots < i_k \le n} f_{i_1 \ldots i_k}(x)\, dx_{i_1} \wedge \cdots \wedge dx_{i_k},$$

where $f_{i_1 \ldots i_k} \in C^\infty(U, \mathbb{R})$ are smooth functions. The space of k-forms on U is denoted by $\Omega^k(U)$.

Example. If $f \in C^\infty(U, \mathbb{R})$, then the derivative df is the 1-form

$$x \mapsto df(x) = \sum_{j=1}^{n} \frac{\partial f}{\partial x_j}(x)\, dx_j.$$

By means of the exterior product we can define an algebraic structure on the space of differential forms. If $\alpha \in \Omega^k(U)$ and $\beta \in \Omega^l(U)$, then the product $\alpha \wedge \beta \in \Omega^{k+l}(U)$ is defined pointwise as

$$(\alpha \wedge \beta)(x) := \alpha(x) \wedge \beta(x) \in \textstyle\bigwedge^{k+l}(\mathbb{R}^n).$$

By convention, $f \wedge \alpha = f\alpha$ if $\alpha \in \Omega^k(U)$ and $f \in \Omega^0(U)$. The product of differential forms is *associative* and satisfies

$$\alpha \wedge \beta = (-1)^{\deg\alpha \deg\beta}(\beta \wedge \alpha)$$

where $\deg\alpha = k$ if $\alpha \in \Omega^k(U)$.

Next we define the *pullback of forms*. If $U \subset \mathbb{R}^n$ and $V \subset \mathbb{R}^m$ are open sets and if $\varphi\colon U \to V$ is a C^∞-mapping, the pullback

$$\varphi^*\colon \Omega^k(V) \to \Omega^k(U), \quad k \geq 1$$

is the linear map defined as

$$(\varphi^*\alpha)(x)(v_1, \dots, v_k) := \alpha(\varphi(x))\big(d\varphi(x)v_1, \dots, d\varphi(x)v_k\big)$$

for all forms $\alpha \in \Omega^k(V)$ and vectors $v_j \in \mathbb{R}^k$. If $k = 0$, then $\varphi^*\colon \Omega^0(V) \to \Omega^0(U)$ is by definition the composition

$$(\varphi^* f)(x) = f \circ \varphi(x).$$

The linear map φ^* is an algebra homomorphism, it satisfies

$$\varphi^*(\alpha \wedge \beta) = (\varphi^*\alpha) \wedge (\varphi^*\beta).$$

Next, we turn to the definition of the *exterior derivative*. It is a sequence of linear maps

$$d\colon \Omega^k(U) \to \Omega^{k+1}(U), \quad k \geq 0,$$

which are defined in the following way.

(1) If $f \in \Omega^0(U)$ is a smooth function, then $df \in \Omega^1(U)$ is the ordinary derivative, so that

$$df(x) = \sum_{j=1}^{n} \frac{\partial f}{\partial x_j}(x)\, dx_j \in \mathcal{L}(\mathbb{R}^n, \mathbb{R}) = (\mathbb{R}^n)^*.$$

(2) If $\alpha \in \Omega^k(U)$ for an integer $k \geq 1$ is represented by

$$\alpha(x) = \sum_{1 \leq i_1 < i_2 < \cdots < i_k \leq n} f_{i_1 \ldots i_k}(x)\, dx_{i_1} \wedge \cdots \wedge dx_{i_k}$$

with functions $f_{i_1 \ldots i_k} \in C^\infty(U, \mathbb{R})$, then we define the form $d\alpha \in \Omega^{k+1}(U)$ by

$$\begin{aligned} d\alpha(x) &= \sum_{1 \leq i_1 < i_2 < \cdots < i_k \leq n} df_{i_1 \ldots i_k}(x) \wedge dx_{i_1} \wedge \cdots \wedge dx_{i_k} \\ &= \sum_{1 \leq i_1 < i_2 < \cdots < i_k \leq n} \sum_{j=1}^{n} \frac{\partial f}{\partial x_j}(x) \underbrace{dx_j \wedge dx_{i_1} \wedge \cdots \wedge dx_{i_k}}_{\in \bigwedge^{k+1}(\mathbb{R}^n)}. \end{aligned}$$

Proposition V.9. *$d^2 = 0$, where d^2 denotes the composition $d \circ d \colon \Omega^k(U) \mapsto \Omega^{k+2}(U)$.*

Proof [Proposition of Schwarz]. Since d is a linear map, it suffices to look at the k-form

$$\alpha = f\, dx_{i_1} \wedge \cdots \wedge dx_{i_k} = f\, dx_I$$

in $\Omega^k(U)$, where we have abbreviated $dx_I := dx_{i_1} \wedge \cdots \wedge dx_{i_k}$. By definition of the exterior derivative,

$$d\alpha(x) = \sum_{j=1}^{n} \frac{\partial f}{\partial x_j}(x)\, dx_j \wedge dx_{i_1} \wedge \cdots \wedge dx_{i_k} = \sum_{j=1}^{n} \frac{\partial f}{\partial x_j}(x)\, dx_j \wedge dx_I,$$

and hence

$$d(d\alpha)(x) = \sum_{s=1}^{n} \sum_{j=1}^{n} \frac{\partial^2 f}{\partial x_j \partial x_s}(x)\, dx_s \wedge dx_j \wedge dx_I.$$

In view of $\frac{\partial^2 f}{\partial x_j \partial x_s}(x) = \frac{\partial^2 f}{\partial x_s \partial x_j}(x)$ and the skew symmetry $dx_j \wedge dx_s = -dx_s \wedge dx_j$ the terms cancel each other so that $d(d\alpha)(x) = 0$, as claimed. □

Proposition V.10. *If $\alpha \in \Omega^k(U)$ and $\beta \in \Omega^l(U)$, then*

$$d(\alpha \wedge \beta) = (d\alpha) \wedge \beta + (-1)^k \alpha \wedge (d\beta).$$

Proof [Product rule]. To abbreviate, we set $dx_I = dx_{i_1} \wedge \cdots \wedge dx_{i_k}$ and $dx_J = dx_{j_1} \wedge \cdots \wedge dx_{j_l}$. Due to the linearity of the exterior derivative d, it suffices to consider the two forms $\alpha = f dx_I \in \Omega^k(U)$ and $\beta = g dx_J \in \Omega^l(U)$. In view of the definitions,

$$\begin{aligned} \alpha \wedge \beta &= (f \cdot g) dx_I \wedge dx_J, \\ d(\alpha \wedge \beta) &= d(f \cdot g) dx_I \wedge dx_J, \\ (d\alpha) \wedge \beta &= g df \wedge dx_I \wedge dx_J. \end{aligned}$$

Due to $dg \wedge dx_j = -dx_j \wedge dg$,

$$\begin{aligned} \alpha \wedge (d\beta) &= f dx_I \wedge dg \wedge dx_J \\ &= (-1)^k f dg \wedge dx_I \wedge dx_J \end{aligned}$$

and using the product rule $d(f \cdot g) = f\,dg + g\,df$ the proposition follows. □

Proposition V.11. *If $\varphi\colon U \subset \mathbb{R}^n \to V \subset \mathbb{R}^m$ is a C^∞-map, then the pullback of φ commutes with the exterior derivative d, that is,*

$$\varphi^* \circ d = d \circ \varphi^*.$$

Explicitly, if $\alpha \in \Omega^k(V)$, then $\varphi^*(d\alpha) = d(\varphi^*\alpha)$.

Proof [Chain rule, Proposition V.9, Proposition V.10]. (1) We first consider the special case in which $\alpha = f$ is an element of $\Omega^0(U)$. Then in view of the definition of the pullback of 1-forms, the chain rule and the definition of the pullback of functions, we obtain

$$\begin{aligned} \varphi^*(df)(x) &= df(\varphi(x)) d\varphi(x) \\ &= d(f \circ \varphi)(x) \\ &= d(\varphi^* f)(x). \end{aligned}$$

(2) Next we consider the 1-form $\alpha = df \in \Omega^1(U)$. In view of Proposition V.9, we have $\varphi^*(d\alpha) = \varphi^*(d^2 f) = \varphi^*(0) = 0$. On the other hand, by step (1), we obtain $d(\varphi^* df) = d(d(\varphi^* f)) = d^2(\varphi^* f)$ and hence also $d(\varphi^*\alpha) = 0$, again in view of Proposition V.9.

(3) The claim of the general case now follows by induction using Proposition V.10 and using that the maps d and φ^* are linear and using the formula $\varphi^*(\alpha \wedge \beta) = (\varphi^*\alpha) \wedge (\varphi^*\beta)$. □

The following proposition is a consequence of the definition of the exterior derivative.

Proposition V.12. *The exterior derivative d is a local operator; if $U' \subset U$ are open sets, then $(d\alpha)|_{U'} = d(\alpha|U')$. Therefore, the value $d\alpha(x)$ only depends on an arbitrarily small neighborhood of x.*

Next we introduce the *interior product* $i_X\alpha$ of a vector field X with a k-form α. If $X: U \subset \mathbb{R}^n \to \mathbb{R}^n$ is a C^∞-vector field, then the *linear map*

$$i_X: \Omega^k(U) \to \Omega^{k-1}(U)$$

is defined in the following way. If $f \in \Omega^0(U)$, then we set $i_X f = 0$. For the k-form $\alpha \in \Omega^k(U)$ of degree $k \geq 1$ we define

$$(i_X\alpha)(x)\big(v_1, \dots, v_{k-1}\big) := \alpha(x)\big(X(x), v_1, \dots, v_{k-1}\big), \quad v_j \in \mathbb{R}^n.$$

We collect some of the properties of the interior product in the following proposition whose proof is a simple exercise left to the reader.

Proposition V.13. *The interior product i_X has the following properties:*

(i) $i_{X+Y} = i_X + i_Y$,

(ii) $i_X \circ i_Y = -i_Y \circ i_X$,

(iii) $i_X \circ i_X = 0$,

(iv) $i_X(\alpha \wedge \beta) = (i_X\alpha) \wedge \beta + (-1)^k \alpha \wedge (i_X\beta)$ *for* $\alpha \in \Omega^k(U)$ *and* $\beta \in \Omega^l(U)$.

V.3 The Lie derivative L_X of forms

Let $X: U \subset \mathbb{R}^n \to \mathbb{R}^n$ be a C^∞-vector field on the open set U and let φ^t be the flow of X satisfying

$$\begin{cases} \frac{d}{dt}\varphi^t(x) = X(\varphi^t(x)), & t \in I_x, \\ \varphi^0(x) = x, & x \in U, \end{cases}$$

where I_x is an open interval around 0. The *Lie derivative* L_X of forms is the *linear* map

$$L_X: \Omega^k(U) \to \Omega^k(U), \quad k \geq 0$$

defined by

$$(L_X\alpha)(x) := \frac{d}{dt}\left[(\varphi^t)^*\alpha(x)\right]_{t=0}.$$

From $\varphi^{t+s} = \varphi^t \circ \varphi^s = \varphi^s \circ \varphi^t$ we deduce

$$\frac{d}{dt}(\varphi^t)^*\alpha = (\varphi^t)^* \frac{d}{ds}(\varphi^s)^*\alpha\Big|_{s=0} = (\varphi^t)^*(L_X\alpha).$$

In order to illustrate the Lie derivative in the special case $k = 0$ we take the smooth function $f \in \Omega^0(U)$ and compute using the chain rule

$$\begin{aligned} L_X f(x) &= \frac{d}{dt}(\varphi^t)^* f(x)\Big|_{t=0} \\ &= \frac{d}{dt} f \circ \varphi^t(x)\Big|_{t=0} \\ &= df \circ \varphi^t(x)\, \frac{d}{dt}\varphi^t(x)\Big|_{t=0} \\ &= df(x)\, X(x) \end{aligned}$$

so that

$$L_X f = i_X(df).$$

The following formula for forms of all degrees will prove extremely useful later on.

Proposition V.14 (Cartan's formula). *The Lie derivative satisfies*

$$L_X = i_X \circ d + d \circ i_X.$$

Explicitly, $L_X\alpha = i_X(d\alpha) + d(i_X\alpha)$ *for all differential forms* $\alpha \in \Omega^*(U)$.

Proof [Propositions V.9, V.10, V.11 and V.13]. Since the exterior product $\alpha \wedge \beta$ is bilinear, the Lie derivative satisfies the product rule

$$L_X(\alpha \wedge \beta) = (L_X\alpha) \wedge \beta + \alpha \wedge (L_X\beta).$$

The linear map $D := i_X \circ d + d \circ i_X$ satisfies, due to Proposition V.13 and Proposition V.10, the same product rule

$$D(\alpha \wedge \beta) = (D\alpha) \wedge \beta + \alpha \wedge (D\beta).$$

In view of the linearity of the maps under consideration it therefore suffices to verify the desired formula $L_X = D$ for the forms $\alpha = f \in \Omega^0(U)$ and $\alpha = df \in \Omega^1(U)$.

(1) If $\alpha = f \in \Omega^0(U)$ is a smooth function, then we obtain in view of the previous calculation and using $i_X f = 0$,

$$L_X f = i_X df = i_X df + d(i_X f) = (i_X \circ d + d \circ i_X) f,$$

as desired.

(2) If $\alpha = df \in \Omega^1(U)$ we compute using the formula $d \circ \varphi^* = \varphi^* \circ d$

(Proposition V.11),

$$\begin{aligned} L_X df(x) &= \frac{d}{dt}(\varphi^t)^* df(x)\Big|_{t=0} \\ &= \frac{d}{dt} d[(\varphi^t)^* f](x)\Big|_{t=0} \\ &\overset{(*)}{=} d\Big[\frac{d}{dt}(\varphi^t)^* f\Big|_{t=0}\Big](x) \\ &= d[i_X df](x) \\ &\overset{\text{V.9}}{=} (d \circ i_X + i_X \circ d) df(x); \end{aligned}$$

in step $(*)$ we have used the linearity of d. In the last step we made use of $d^2 = 0$. The proof of the proposition is complete. □

The differential form $\alpha \in \Omega^k(U)$ is called closed if $d\alpha = 0$ and is called exact on U if $\alpha = d\beta$ for a differential form β in $\Omega^{k-1}(U)$. In view of $d^2 = 0$, an exact form is necessarily a closed form. Locally, the converse also holds true, as the next result shows.

Theorem V.15 (Poincaré lemma). *If $V \subset \mathbb{R}^n$ is a starlike open set, then a k-form on V is closed, precisely if it is exact. For all $\alpha \in \Omega^k(V)$ of degree $k \geq 1$ we have*

$$d\alpha = 0 \iff \alpha = d\beta \text{ for some } \beta \in \Omega^{k-1}(V).$$

Proof [Cartan's formula, Proposition V.9]. As already pointed out, an exact form is closed in view of $d^2 = 0$. In order to prove that on V a closed form is exact we make use of the starlike character of the open set. Without loss of generality we may assume (by translation) that V is starlike with respect to the origin 0 in $\mathbb{R}^n$, so that if $x \in V$, then also $\mu x \in V$ for every $\mu \in [0,1]$. We introduce the linear vector field $X(x) = \lambda x$ for some real number $\lambda > 0$. Its flow is equal to $\varphi^t(x) = e^{\lambda t} x$ and therefore leaves the open set V invariant for all $t \leq 0$. The linearized flow is equal to $d\varphi^t(x)v = e^{\lambda t} v$ for all $v \in \mathbb{R}^n$. Hence $\varphi^t(x) \to 0$ and $d\varphi^t(x)v \to 0$ exponentially as $t \to -\infty$. This allows us to define the *linear* map

$$H\colon \Omega^k(V) \to \Omega^{k-1}(V)$$

by

$$H\alpha := i_X\Big(\int_{-\infty}^0 (\varphi^t)^*\alpha \, dt\Big).$$

In view of Cartan's formula $L_X = i_X \circ d + d \circ i_X$ we obtain

$$d(H\alpha) = L_X\Big(\int_{-\infty}^0 (\varphi^t)^*\alpha \, dt\Big) - i_X \circ d\Big(\int_{-\infty}^0 (\varphi^t)^*\alpha \, dt\Big).$$

Because $L_X(\varphi^t)^*\alpha = \frac{d}{dt}\left[(\varphi^t)^*\alpha\right]$ and because the Lie derivative L_X is linear, it follows that

$$d(H\alpha) = (\varphi^t)^*\alpha\Big|_{-\infty}^{0} - i_X\left(\int_{-\infty}^{0}(\varphi^t)^* d\alpha\, dt\right).$$

Due to $(\varphi^0)^* = \mathrm{Id}$ and $(\varphi^t)^*\alpha \to 0$ as $t \to -\infty$ we obtain

$$d(H\alpha) = \alpha - H(d\alpha).$$

Hence we arrive at the formula

$$d \circ H + H \circ d = \mathrm{Id}$$

which holds true on $\Omega^k(V)$. If $d\alpha = 0$, then by the above formula, $d(H\alpha) = \alpha - H(0) = \alpha$ so that $\beta := H\alpha$ is a solution of the equation $d\beta = \alpha$ and the proof is complete. □

V.4 The Lie derivative L_X of vector fields

If X is a vector field on an open set $U \subset \mathbb{R}^n$ and if φ^t is the flow of X, then φ^t is a (local) diffeomorphism and the vector field Y on U can be *transformed* according to the rule

$$(\varphi^t)^*Y(x) = \left[d\varphi^t(x)\right]^{-1} Y(\varphi^t(x)).$$

The *Lie derivative* of Y with respect to X is defined as the *vector field*

$$L_X Y(x) := \frac{d}{dt}\left[(\varphi^t)^*Y(x)\right]_{t=0} =: [X,Y](x).$$

The vector field $[X,Y]$ on the right-hand side is also called the *Lie bracket* of the vector fields X and Y. Due to $\varphi^{t+s} = \varphi^t \circ \varphi^s$,

$$\frac{d}{dt}(\varphi^t)^*Y = \frac{d}{ds}(\varphi^{t+s})^*Y\Big|_{s=0} = (\varphi^t)^*\frac{d}{ds}(\varphi^s)^*Y\Big|_{s=0}$$

so that

$$\frac{d}{dt}(\varphi^t)^*Y = (\varphi^t)^*[X,Y].$$

Proposition V.16. *The Lie bracket $[\cdot,\cdot]$ transforms naturally under a diffeomorphism $u\colon \mathbb{R}^n \to \mathbb{R}^n$, that is,*

$$[u^*X, u^*Y] = u^*[X,Y].$$

Proof. If φ^t is the flow of X, then the transformed vector field u^*X possesses (Proposition IV.5) the flow $\psi^t = u^{-1} \circ \varphi^t \circ u$. We compute

$$\begin{aligned}(\psi^t)^*(u^*Y) &= (d\psi^t)^{-1}(u^*Y) \circ \psi^t \\ &= (du)^{-1}(d\varphi^t)^{-1}du(du)^{-1}Y \circ u \circ \psi^t \\ &= (du)^{-1}[(d\varphi^t)^{-1}Y \circ \varphi^t] \circ u \\ &= u^*(\varphi^t)^*Y.\end{aligned}$$

Therefore,

$$\begin{aligned}\frac{d}{dt}(\psi^t)^*(u^*Y)|_{t=0} &= u^*\Big[\frac{d}{dt}(\varphi^t)^*Y|_{t=0}\Big] \\ &= u^*[X, Y].\end{aligned}$$

The left-hand side is, by definition, equal to $[u^*X, u^*Y]$. □

Proposition V.17. *On $\mathbb{R}^n$ the Lie bracket possesses the representation*

$$[X, Y] = dY(X) - dX(Y).$$

Proof. Let φ^t be the flow of X. The inverse inv: $A \mapsto A^{-1}$ on $\mathrm{GL}(\mathbb{R}^n)$ has the derivative $d\,\mathrm{inv}(A)B = -A^{-1}BA^{-1}$ which follows from the Neumann series. Using the product rule and the chain rule we compute

$$\frac{d}{dt}[(d\varphi^t)^{-1}Y\circ\varphi^t] = -(d\varphi^t)^{-1}\Big[\frac{d}{dt}(d\varphi^t)\Big](d\varphi^t)^{-1}Y\circ\varphi^t + (d\varphi^t)^{-1}dY(\varphi^t)\frac{d}{dt}\varphi^t.$$

Taking the derivative of $\frac{d}{dt}\varphi^t = X(\varphi^t)$ we obtain

$$\frac{d}{dt}(d\varphi^t) = dX(\varphi^t)d\varphi^t$$

which, substituted into the above equation, gives

$$\begin{aligned}\frac{d}{dt}[(d\varphi^t)^{-1}Y \circ \varphi^t] &= -(d\varphi^t)^{-1}dX(\varphi^t)Y(\varphi^t) + (d\varphi^t)^{-1}dY(\varphi^t)X(\varphi^t) \\ &= (d\varphi^t)^{-1}[dY(X) - dX(Y)] \circ \varphi^t.\end{aligned}$$

Evaluating at $t = 0$, the proposition follows in view of $\varphi^0 = \mathrm{Id}$ and $d\varphi^0 = \mathbb{1}$ and the proof is complete. □

Corollary V.18. *The Lie bracket satisfies $[X, Y] = -[Y, X]$.*

V.5 Commuting vector fields

Two vector fields are said to *commute*, if their Lie bracket vanishes.

Proposition V.19. *If X is a vector field having the flow φ^t, then $(\varphi^t)^*Y = Y$ for all t if and only if $[X, Y] = 0$.*

Proof. We recall that

$$\frac{d}{dt}(\varphi^t)^*Y = (\varphi^t)^*[X, Y].$$

If $(\varphi^t)^*Y = Y$ for all t, then the left-hand side vanishes and due to $\varphi^0 = \mathrm{Id}$ we obtain $[X, Y] = 0$. Conversely, if $[X, Y] = 0$, then $\frac{d}{dt}(\varphi^t)^*Y = 0$ so that $(\varphi^t)^*Y$ is constant in t. Hence $(\varphi^t)^*Y = (\varphi^0)^*Y = Y$ for all t. □

Two vector fields commute if and only if their flows commute.

Proposition V.20. *If X and Y are vector fields having the flows φ^t and ψ^t, then $[X, Y] = 0$ is equivalent to $\psi^s \circ \varphi^t = \varphi^t \circ \psi^s$ for all s, t.*

Proof [Uniqueness of the flow, Proposition V.19]. We assume that $\psi^s \circ \varphi^t = \varphi^t \circ \psi^s$ for all s and t. Differentiating both sides in t leads to

$$d\psi^s(\varphi^t)X(\varphi^t) = X(\varphi^t \circ \psi^s),$$

and we deduce for $t = 0$ the equation $d\psi^s X = X \circ \psi^s$, so that

$$X = (d\psi^s)^{-1} X \circ \psi^s = (\psi^s)^* X$$

for all s. In view of Proposition V.19 the claim $[X, Y] = 0$ follows.

Conversely, if $[X, Y] = 0$, then $(\psi^s)^*X = X$ for all s according to Proposition V.19. Since φ^t is the flow of X, the flow of the transformed vector field $(\psi^s)^*X$ is equal to

$$(\psi^s)^{-1} \circ \varphi^t \circ \psi^s$$

for all t. Due to $(\psi^s)^*X = X$ it follows from the uniqueness of the flow that $(\psi^s)^{-1} \circ \varphi^t \circ \psi^s = \varphi^t$ and the proposition is proved. □

V.6 The exterior derivative d on manifolds

Up to now we have dealt with the formalism of differential forms locally on open sets of the Euclidian space. Globally on a manifold we shall glue our local objects together by means of their transformation formulas based on the chart transformations. We consider a smooth manifold M of dimension $\dim M = n$. We recall that M is a Hausdorff topological space equipped with a maximal atlas of *charts* (U, φ). The subsets U are open subsets which constitute a covering of the space

M and the mappings $\varphi\colon U \to \varphi(U) \subset \mathbb{R}^n$ are homeomorphisms. Moreover, if two charts (U, φ) and (V, ψ) overlap, so that $U \cap V \neq \emptyset$, then the associated *chart transformation* is smooth, that is, the bijective mapping

$$\varphi \circ \psi^{-1}\colon \psi(U \cap V) \subset \mathbb{R}^n \to \varphi(U \cap V) \subset \mathbb{R}^n$$

is a C^∞-diffeomorphism between open sets of $\mathbb{R}^n$.

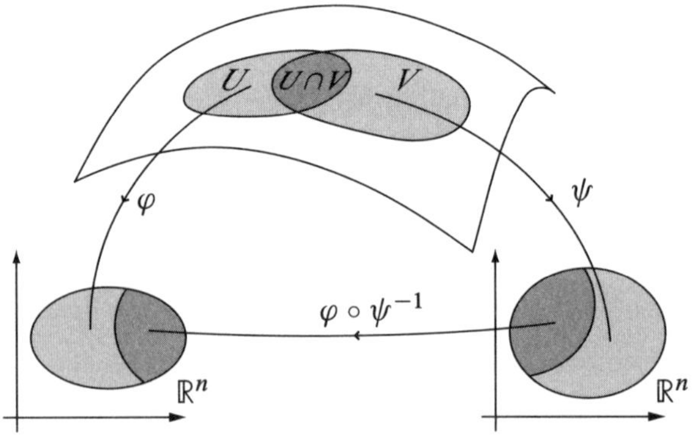

Figure V.1. Manifold with the two charts (U, φ) and (V, ψ).

If T_mM is the *tangent space* at the point $m \in M$, every chart (U, φ) containing $m \in U$ defines a *canonical* linear isomorphism

$$T_m\varphi\colon T_mM \to \mathbb{R}^n.$$

Taking another such chart (V, ψ) around m, the diagram

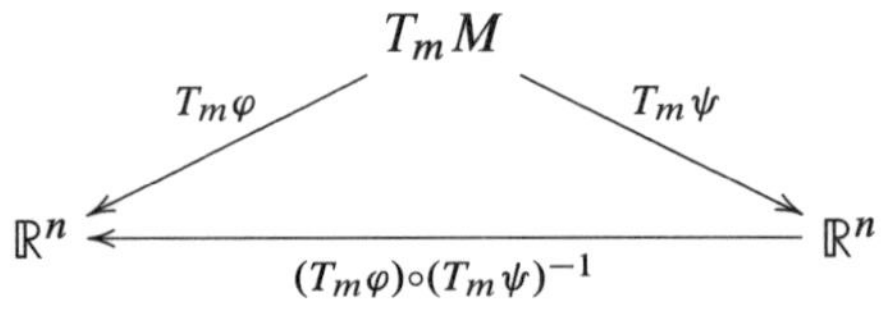

is commutative and

$$T_m\varphi \circ (T_m\psi)^{-1} = d(\varphi \circ \psi^{-1})(\psi(m)) \in \mathcal{L}(\mathbb{R}^n, \mathbb{R}^n)$$

is the linearized chart transformation. The coordinates $x = \varphi(m)$ and $y = \psi(m)$ in $\mathbb{R}^n$ are related by $x = \varphi \circ \psi^{-1}(y)$, and the two vectors $v, w \in \mathbb{R}^n$ satisfying $v = d(\varphi \circ \psi^{-1})(y)w$, represent the same tangent vector of T_mM. In other words v and w are the local coordinates of the same tangential vector in the different charts. Note that the tangent spaces are isomorphic to $\mathbb{R}^n$, but have *no canonical* basis.

A k-form, $0 \leq k \leq n$, on the manifold M is a smooth mapping

$$\omega\colon m \mapsto \omega(m) \in \bigwedge^k (T_m M), \quad m \in M$$

which associates with every point m the multilinear form $\omega(m)(v_1, \dots, v_n) \in \mathbb{R}$ in the tangent space at m. It is linear in every argument $v_j \in T_m M$ and alternating. In the chart (U, φ) the k-form ω is represented by the k-form $\omega_\varphi \in \Omega^k(\varphi(U))$, defined on the open set $\varphi(U) \subset \mathbb{R}^n$ by

$$\omega_\varphi(x)(v_1, \dots, v_k) := \omega(m)\left([T_m\varphi]^{-1} v_1, \dots, [T_m\varphi]^{-1} v_k\right), \quad v_j \in \mathbb{R}^n,$$

where $x = \varphi(m) \in \mathbb{R}^n$. For this we can also write

$$(\varphi^{-1})^* \omega(x)(v_1, \dots, v_k) := \omega_\varphi(x)(v_1, \dots, v_k).$$

The smoothness of the forms on the manifold is defined in the local coordinates as follows. The k-form ω on the manifold M is smooth, if its representatives ω_φ in the charts (U, φ) are smooth k-forms on the open sets $\varphi(U) \subset \mathbb{R}^n$. This requirement is independent of the choice of the charts. Indeed, if ω_ψ is the representative of ω in the chart (V, ψ), then the transformation formula

$$\omega_\varphi = (\psi \circ \varphi^{-1})^* \omega_\psi$$

holds true on $U \cap V$, as is readily verified. Because, by definition, the chart transformation $\psi \circ \varphi^{-1}$ is a smooth diffeomorphism, the smoothness of one representative follows from the smoothness of the other one.

We also see that a k-form ω on the manifold can alternatively be defined by smooth k-forms ω_φ on the open sets $\varphi(U) \subset \mathbb{R}^n$ associated with the charts (U, φ), which are compatible in the sense that on overlapping charts they meet the above transformation formula.

We denote the set of k-forms on the manifold M by $\Omega^k(M)$

Proposition V.21. *There exists a unique* $\mathbb{R}$*-linear map*

$$d\colon \Omega^k(M) \to \Omega^{k+1}(M), \quad k \geq 0$$

having the following properties.

(i) *If* $f \in \Omega^0(M)$ *is a smooth function, then* df *is the ordinary derivative.*

(ii) $d(\omega \wedge \mu) = d\omega \wedge \mu + (-1)^k \omega \wedge d\mu$ *for* $\omega \in \Omega^k(M)$ *and* $\mu \in \Omega^l(M)$.

(iii) $d^2 = 0$.

(iv) d *is a local operation; if* $U \subset M$ *is an open set and* $\omega \in \Omega^k(M)$ *then* $d\omega|_U = d(\omega|_U)$.

(v) $\varphi^* \circ d = d \circ \varphi^*$ *for every smooth map* $\varphi\colon M \to N$ *between manifolds.*

Proof. (1) *Existence.* If $(U_\alpha, \varphi_\alpha)$ and (U_β, φ_β) are two charts satisfying $U_\alpha \cap U_\beta \neq \emptyset$, then we abbreviate the open subsets $V_{\alpha\beta} = \varphi_\alpha(U_\alpha \cap U_\beta)$ and $V_{\beta\alpha} = \varphi_\beta(U_\alpha \cap U_\beta)$ of the $\mathbb{R}^n$ as well as the chart transformation

$$\varphi_{\alpha\beta} = \varphi_\alpha \circ \varphi_\beta^{-1} \colon V_{\beta\alpha} \to V_{\alpha\beta},$$

which is a diffeomorphism of open subsets of $\mathbb{R}^n$. On open subsets of $\mathbb{R}^n$ we have already defined the exterior derivative d above. We recall from Proposition V.11 the crucial fact that the pullback of forms and the exterior derivative d commute, so that the diagram

$$\begin{array}{ccc} \Omega^k(V_{\alpha\beta}) & \xrightarrow{\ \varphi_{\alpha\beta}^*\ } & \Omega^k(V_{\beta\alpha}) \\ {\scriptstyle d}\big\downarrow & & \big\downarrow{\scriptstyle d} \\ \Omega^{k+1}(V_{\alpha\beta}) & \xrightarrow{\ \varphi_{\alpha\beta}^*\ } & \Omega^{k+1}(V_{\beta\alpha}) \end{array}$$

is commutative. In formulas we have $d(\varphi_{\alpha\beta}^*\omega) = \varphi_{\alpha\beta}^*(d\omega)$ for all forms $\omega \in \Omega^k(V_{\alpha\beta})$ and so,

$${\varphi_{\alpha\beta}^*}^{-1} \circ d \circ \varphi_{\alpha\beta}^* = d.$$

Using this formula we shall define the exterior derivative d of the form $\omega \in \Omega^k(M)$ by means of the charts $(U_\alpha, \varphi_\alpha)$ as follows:

$$(d\omega)|_{U_\alpha} := \varphi_\alpha^* \circ d \circ (\varphi_\alpha^{-1})^*(\omega|_{U_\alpha}).$$

In order to verify that we have defined a global object, we have to show that the definition is independent of the choice of the charts. To do so we compute, on $U_\alpha \cap U_\beta$,

$$\begin{aligned} d\omega|_{U_\alpha \cap U_\beta} &= \varphi_\alpha^* \circ d \circ (\varphi_\alpha^{-1})^*(\omega|_{U_\alpha \cap U_\beta}) \\ &= \varphi_\alpha^* \circ {\varphi_{\alpha\beta}^*}^{-1} \circ d \circ \varphi_{\alpha\beta}^* \circ (\varphi_\alpha^{-1})^*(\omega|_{U_\alpha \cap U_\beta}) \\ &= ({\varphi_{\alpha\beta}}^{-1} \circ \varphi_\alpha)^* \circ d \circ ({\varphi_\alpha}^{-1} \circ \varphi_{\alpha\beta})^*(\omega|_{U_\alpha \cap U_\beta}) \\ &= \varphi_\beta^* \circ d \circ {\varphi_\beta^{-1}}^*(\omega|_{U_\alpha \cap U_\beta}). \end{aligned}$$

Hence we have defined a $(k+1)$-form $d\omega$ on M, which possesses the properties (i)–(v), as we already know from the chart representations.

(2) *Uniqueness.* The proof of the uniqueness of d is left as an exercise. It can be found, for example, in [21, Theorem 8.1.6] by L. Conlon. □

V.7 Symplectic manifolds

Definition. A *symplectic manifold* is a pair (M, ω) consisting of a smooth manifold M of dimension $\dim M = 2n$ and a *symplectic structure* ω. A symplectic structure is a 2-form ω on M having the following two properties:

(i) ω is *nondegenerate*,

(ii) ω is *closed*, i.e., $d\omega = 0$.

The differential form ω is called nondegenerate, if in every point $x \in M$ the 2-form $\omega(x)$ on the tangent space $T_x M$ is nondegenerate, i.e., $\omega(x)(u, v) = 0$ for all $v \in T_x M$ implies $u = 0$. Equivalently, the map $u \mapsto \omega(x)(u, \cdot)$ from the tangent space $T_x M$ onto its dual space $(T_x M)^*$ is a linear isomorphism.

By definition, every tangent space $\big(T_x M, \omega(x)\big)$ is a *symplectic* vector space. A first example for a symplectic manifold is the pair (U, ω_0) in which $U \subset \mathbb{R}^{2n}$ is an open set and ω_0 is the standard symplectic structure on $\mathbb{R}^{2n}$. Actually this is what every symplectic manifold looks like locally, according to the following theorem of Darboux.

Theorem V.22 (Darboux). *If ω is a nondegenerate 2-form on the manifold M of dimension* $\dim M = 2n$, *then $d\omega = 0$ if and only if there exists around every point $p \in M$ a chart (U, φ) satisfying $\varphi(p) = 0 \in \mathbb{R}^{2n}$ and*

$$\omega|_U = \varphi^* \omega_0,$$

where $\omega_0 = \sum_{j=1}^n dy_j \wedge dx_j$ is the constant symplectic standard structure on $\mathbb{R}^{2n}$.

Definition. These special coordinates are called *symplectic coordinates* or *Darboux coordinates*. A chart (U, φ) on which $\omega|_U = \varphi^* \omega_0$ is called a *Darboux chart*.

In view of Darboux's theorem we do not have to look for *local* invariants!

Corollary V.23. *Two symplectic manifolds of the same dimension have* **locally** *the same normal form, namely $(\mathbb{R}^{2n}, \omega_0)$. They are therefore locally indistinguishable: Symplectic manifolds do not possess local symplectic invariants except the dimension!*

The situation is completely different from Riemannian manifolds. Two Riemannian metrics are in general *not* locally isometric. The *Gauss curvature* is an example of a local invariant.

Proof of Theorem V.22 [Deformation trick by R. Thom and J. Moser, Cartan's formula, Poincaré lemma]. We first assume that the nondegenerate 2-form is closed so that $d\omega = 0$. In local coordinates $\omega = \omega(z)$ is a nondegenerate 2-form on $\mathbb{R}^{2n}$, which depends on the point $z \in \mathbb{R}^{2n}$. By a translation we may assume that the point p in M corresponds to $z = 0$, and by a *linear* transformation of $\mathbb{R}^{2n}$ we can achieve, in view of Proposition V.2, that in the point $z = 0$ the 2-form is already in normal form,

$$\omega(0) = \omega_0 = \sum_{j=1}^n dy_j \wedge dx_j .$$

We are looking for a local diffeomorphism φ around 0 satisfying

$$\varphi(0) = 0 \text{ and } \varphi^*\omega = \omega_0.$$

Using the deformation trick we are looking for more. We interpolate between ω and ω_0 by the family of 2-forms

$$\omega_t = \omega_0 + t(\omega - \omega_0), \quad 0 \le t \le 1,$$

so that $\omega_t = \omega_0$ if $t = 0$ and $\omega_t = \omega$ if $t = 1$. By assumption, $d\omega_t = 0$ for all t. We look for a family φ^t, $0 \le t \le 1$ of local diffeomorphisms leaving the origin fixed and satisfying

$$(*) \quad \begin{cases} (\varphi^t)^*\omega_t = \omega_0, & 0 \le t \le 1, \\ \varphi^0 = \mathrm{Id}\,. \end{cases}$$

Then the time 1 map φ^1 of the family is the desired local diffeomorphism.

To determine the family φ^t we look for a *time dependent vector field* X_t, which generates the flow φ^t such that

$$\frac{d}{dt}\varphi^t = X_t(\varphi^t), \quad \varphi^0 = \mathrm{Id}\,.$$

Differentiating the identity $(*)$ in the variable t and using Cartan's formula we find

$$\begin{aligned} 0 &= \frac{d}{dt}[(\varphi^t)^*\omega_t] \\ &= \Big(\frac{d}{dt}(\varphi^t)^*\Big)\omega_t + (\varphi^t)^*\Big(\frac{d}{dt}\omega_t\Big) \\ &= (\varphi^t)^*\,[L_{X_t}\omega_t + \omega - \omega_0] \\ &= (\varphi^t)^*\,[d(i_{X_t}\omega_t) + i_{X_t}(d\omega_t) + \omega - \omega_0] \\ &= (\varphi^t)^*\,[d(i_{X_t}\omega_t) + \omega - \omega_0]\,. \end{aligned}$$

We have used that $d\omega_t = 0$. Therefore, the vector field X_t has to satisfy the *linear* equation

$$d(i_{X_t}\omega_t) + (\omega - \omega_0) = 0.$$

By assumption, $d(\omega - \omega_0) = d\omega - d\omega_0 = 0$ and hence in view of the Poincaré lemma there exists locally around 0 a 1-form λ satisfying

$$\omega - \omega_0 = d\lambda \quad \text{and} \quad \lambda(0) = 0.$$

Since $\omega_t(0) = \omega_0$ for all t, the 2-forms $\omega_t(z)$ are nondegenerate for all t in a neighborhood of $z = 0$. Therefore, there exists a unique time dependent vector field X_t, solving the equation

$$i_{X_t}\omega_t = \omega_t(X_t, \cdot) = -\lambda.$$

Consequently,

$$0 = d(i_{X_t}\omega_t) + d\lambda = d(i_{X_t}\omega_t) + \omega - \omega_0$$

and we have found the desired vector field. From $\lambda(0) = 0$ we conclude that $X_t(0) = 0$, since ω_t is nondegenerate. We therefore find an open neighborhood of $z = 0$, on which the flow φ^t of the vector field X_t exists for all $0 \le t \le 1$. It satisfies $\varphi^0 = \mathrm{Id}$ and $\varphi^t(0) = 0$. Reversing now all our steps, this flow satisfies

$$\frac{d}{dt}(\varphi^t)^*\omega_t = 0, \quad 0 \le t \le 1$$

and therefore

$$(\varphi^t)^*\omega_t = (\varphi^0)^*\omega_0 = \omega_0, \quad 0 \le t \le 1.$$

Due to $\omega = \omega_1$ the time 1 map of the flow is the desired diffeomorphism solving

$$(\varphi^1)^*\omega = \omega_0$$

and $\varphi^1(0) = 0$.

Conversely, assuming that $\varphi^*\omega = \omega_0$, we conclude $d\omega = d(\varphi^*(\omega_0)) = \varphi^*(d\omega_0) = \varphi^*(0) = 0$ so that the 2-form ω is closed. This completes the proof of Darboux's theorem. □

V.8 Symplectic maps

Definition. A smooth map $u\colon M_1 \to M_2$ between two symplectic manifolds (M_1, ω_1) and (M_2, ω_2) is called *symplectic*, if

$$u^*\omega_2 = \omega_1.$$

Explicitly,

$$\omega_2(u(x))\big(du(x)v_1, du(x)v_2\big) = \omega_1(x)(u, v), \quad x \in M_1,\ u, v \in T_x M_1.$$

Consequently, the derivative $du(x)\colon T_x M_1 \to T_{u(x)} M_2$ is a symplectic linear map from the symplectic vector space $\big(T_x M_1, \omega_1(x)\big)$ into the symplectic vector space $\big(T_{u(x)} M_2, \omega_2(u(x))\big)$ for every $x \in M_1$.

Recalling the linear symplectic mappings in Section VI.1, we conclude from the nondegeneracy of ω_1 that a symplectic map is an *immersion* and

$$\dim M_1 \le \dim M_2$$

if there exists a symplectic map $u\colon M_1 \to M_2$.

A differentiable map $u\colon M \to M$ on a symplectic manifold (M,ω) is called symplectic, if it leaves the symplectic form invariant, that is, if

$$u^*\omega = \omega.$$

In the symplectic coordinates (U,φ) this is equivalent to the requirement that the linearized map $du(x) \in \mathrm{Sp}(n)$ is a symplectic matrix for every $x \in \varphi(U) \subset \mathbb{R}^{2n}$, so that

$$du(x)^T J\, du(x) = J,$$

where J is the matrix

$$J = \begin{pmatrix} 0 & \mathbb{1} \\ -\mathbb{1} & 0 \end{pmatrix}.$$

In particular, u is a local diffeomorphism.

We see that a symplectic map satisfies a special *nonlinear partial differential equation* of first order. It allows us to give another equivalent definition of a symplectic manifold.

Definition. We assume M to be a manifold of dimension $\dim M = 2n$. A *symplectic structure* on M is a covering of M consisting of charts $(U_\alpha, \varphi_\alpha)_{\alpha\in A}$,

$$\varphi_\alpha\colon U_\alpha \to \mathbb{R}^{2n}, \quad \alpha \in A,$$

whose chart transformations

$$\varphi_{\alpha\beta} = \varphi_\alpha \circ \varphi_\beta^{-1}\colon \varphi_\beta(U_\alpha \cap U_\beta) \to \varphi_\alpha(U_\alpha \cap U_\beta)$$

are all symplectic local diffeomorphisms in $(\mathbb{R}^{2n}, \omega_0)$ and hence satisfy $d\varphi_{\alpha\beta}(x) \in \mathrm{Sp}(n)$, or equivalently

$$\varphi_{\alpha\beta}^*\omega_0 = \omega_0, \quad \alpha, \beta \in A.$$

Proof of the equivalence of the definitions. For a symplectic structure $(U_\alpha, \varphi_\alpha)_{\alpha\in A}$ on M we define the 2-form ω_α on U_α by

$$\omega_\alpha := \varphi_\alpha^*\omega_0.$$

In this way we define a global 2-form ω on M, since on $U_\alpha \cap U_\beta$ we have

$$\omega_\alpha = \varphi_\alpha^*\omega_0 = (\varphi_\alpha \circ \varphi_\beta^{-1} \circ \varphi_\beta)^*\omega_0 = \varphi_\beta^* \circ (\varphi_\alpha \circ \varphi_\beta^{-1})^*\omega_0 = \varphi_\beta^*\omega_0 = \omega_\beta.$$

Because ω_0 is nondegenerate and because $d\varphi_\alpha(x)$ is a bijective linear map, the 2-form ω is nondegenerate. Moreover, ω is closed, since $d\omega_\alpha = d(\varphi_\alpha^*\omega_0) = \varphi_\alpha^*(d\omega_0) = 0$ so that ω is a symplectic form on M and (M,ω) is a symplectic manifold. The other direction of the equivalence follows from Darboux's theorem. □

Every manifold can be equipped with a Riemannian metric. In sharp contrast, not every manifold of even dimension admits a symplectic structure, as the following example of the sphere S^4 shows.

Example. The 4-dimensional sphere S^4 does not admit a symplectic structure.

Proof. Arguing by contradiction we assume that ω is a symplectic structure on S^4. Since ω is nondegenerate, the 4-form $\Omega := \omega \wedge \omega \neq 0$ is a volume form on S^4 and hence

$$\int_{S^4} \omega \wedge \omega = \int_{S^4} \Omega \neq 0.$$

On the other hand ω is closed and since the second cohomology group of the sphere S^4 vanishes, $H^2(S^4) = 0$, there exists according to the de Rham theorem a 1-form α on S^4 satisfying $\omega = d\alpha$. Using $d\omega = 0$ we compute

$$d(\omega \wedge \alpha) = d\omega \wedge \alpha + \omega \wedge d\alpha = \omega \wedge d\alpha = \omega \wedge \omega$$

and due to $\partial S^4 = \emptyset$ we arrive, in view of Stoke's theorem, at the contradiction

$$\int_{S^4} \omega \wedge \omega = \int_{S^4} d(\omega \wedge \alpha) = 0. \qquad \square$$

Analogously one sees that every compact manifold M whose second cohomology group $H^2(M) = 0$ vanishes, does not admit a symplectic structure. Examples are all spheres S^{2n} for $n \geq 2$, while in dimension $2n = 2$ the volume forms on S^2 and on the oriented Riemann surfaces are symplectic structures.

Important examples of symplectic diffeomorphisms are the flow maps of Hamiltonian vector fields introduced next.

Definition. A smooth function $H\colon M \to \mathbb{R}$ on a symplectic manifold (M, ω) is called a *Hamiltonian function*. The unique C^∞-vector field X_H on M, defined by

$$i_{X_H}\omega = -dH,$$

or explicitly by

$$\omega(x)\big(X_H(x), \cdot\big) = -dH(x) \quad \text{for all } x \in M,$$

is called the associated *Hamiltonian vector field*.

The vector field X_H is uniquely defined, since, in view of the nondegeneracy of the symplectic form ω, the map $X(x) \mapsto \omega(x)(X(x), \cdot)$ is a linear isomorphism from the tangent space T_xM onto its dual space $(T_xM)^*$.

In a Darboux chart, i.e., locally in $(\mathbb{R}^{2n}, \omega_0)$, the equation $i_{X_H}\omega_0 = -dH$ becomes

$$\begin{aligned}\langle JX_H(x), v\rangle &= -dH(x)v \\ &= -\langle \nabla H(x), v\rangle\end{aligned}$$

for all $v \in \mathbb{R}^{2n}$ where the gradient ∇H of the function H is defined with respect to the Euclidian scalar product $\langle \cdot, \cdot \rangle$ in $\mathbb{R}^{2n}$. We conclude that $JX_H(x) = -\nabla H(x)$ and multiplying by J we obtain, due to $J^2 = -\mathbb{1}$, the *local representation*

$$X_H(x) = J\nabla H(x)$$

of a Hamiltonian vector field *in symplectic coordinates.*

We see that X_H is a very special vector field on M. Since J is antisymmetric, the Hamiltonian vector field X_H has properties that are completely different from the properties of the gradient vector field ∇H studied in Chapter IV.

Proposition V.24. *The flow maps φ^t of a Hamiltonian vector field X_H are symplectic maps and hence satisfy $(\varphi^t)^*\omega = \omega$.*

Proof [Cartan's formula]. Abbreviating $X = X_H$ we compute

$$\begin{aligned}\frac{d}{dt}(\varphi^t)^*\omega &= (\varphi^t)^* L_X\omega \\ &= (\varphi^t)^*[i_X(d\omega) + d(i_X\omega)] \\ &= 0,\end{aligned}$$

since $d\omega = 0$ and $d(i_X\omega) = d(-dH) = 0$. Using $\varphi^0 = \mathrm{Id}$ we therefore conclude $(\varphi^t)^*\omega = (\varphi^0)^*\omega = \omega$ for all t, as claimed. □

Exercise. We assume X to be a vector field on (M, ω). Its flow φ^t is symplectic if and only if $L_X\omega = 0$. In this case X can be written *locally* as a Hamiltonian vector field X_F. (Hint: Make use of Cartan's formula and Poincaré's lemma.)

Proposition V.25. *The interior product $i_X\alpha$ of vector fields with forms transforms naturally, i.e., for a diffeomorphism $u\colon M \to N$ we have*

$$u^*(i_X\alpha) = i_{u^*X}(u^*\alpha)$$

for all vector fields X on N and all forms $\alpha \in \Omega^*(N)$.

Proof. In view of the definitions, we have on one hand

$$u^*(i_X\alpha)(v_1, \dots, v_{k-1}) = (\alpha \circ u)(X \circ u, du v_1, \dots, du v_{k-1}),$$

on the other hand,

$$\begin{aligned} i_{u^*X}(u^*\alpha)(v_1,\dots,v_{k-1}) &= (u^*\alpha)(u^*X, v_1,\dots,v_{k-1}) \\ &= (\alpha\circ u)(du(u^*X), duv_1,\dots,duv_{k-1}) \\ &= (\alpha\circ u)(X\circ u, duv_1,\dots,duv_{k-1}), \end{aligned}$$

where we have used that $du(u^*X) = du(du^{-1}X\circ u) = X\circ u$. □

We next show that under the symplectic maps the Hamiltonian vector fields are transformed into Hamiltonian vector fields. All one needs to do is to transform the Hamiltonian function! We recall that by definition of the pullback, $u^*H = H\circ u$ is the composition.

Proposition V.26 (Transformation theory). *If $u\colon M\to M$ is a symplectic map, then*

$$u^*X_H = X_{u^*H}$$

for every smooth function $H\colon M\to\mathbb{R}$. Conversely, a local diffeomorphism u is symplectic if the above formula holds true for all functions $H\colon M\to\mathbb{R}$.

Proof [Proposition V.25, Proposition V.11]. We assume that $u^*\omega=\omega$. By means of the definition of the Hamiltonian vector field and using the Propositions V.25 and V.11 we compute

$$\begin{aligned} i_{X_{u^*H}}\omega &= -d(u^*H) \\ &= -u^*(dH) \\ &= u^*(i_{X_H}\omega) \\ &= i_{u^*X_H}(u^*\omega) \\ &= i_{u^*X_H}\omega. \end{aligned}$$

Since ω is nondegenerate, we conclude that $u^*X_H = X_{u^*H}$, as claimed.

In order to verify the converse we assume that for a given local diffeomorphism u the formula $u^*X_H = X_{u^*H}$ holds true for all functions H. By the previous computation, $i_{u^*X_H}(u^*\omega) = i_{u^*X_H}\omega$. And hence, $\omega(u^*X_H,\cdot) = u^*\omega(u^*X_H,\cdot)$ for all functions H. By choosing in symplectic coordinates the functions $H(x) = x_j$ for $1\le j\le 2n = \dim M$, one sees that $\omega(v_j,\cdot) = u^*\omega(v_j,\cdot)$ for a basis $\{v_1,\dots,v_{2n}\}$ of the tangent space. Therefore $\omega - u^*\omega = 0$, as we wanted to prove. □

Definition. If (M,ω) is a symplectic manifold and $F,G\colon M\to\mathbb{R}$ are smooth functions, then the *Poisson bracket* of F and G is the smooth function $\{F,G\}\colon M\to\mathbb{R}$, defined by

$$\{F,G\}(x) := \omega(x)\big(X_F(x), X_G(x)\big),\quad x\in M.$$

The Poisson bracket can also be written as

$$\{F, G\} = \omega(X_F, X_G) = -dF(X_G) = dG(X_F).$$

Proposition V.27. *If $F, G \colon M \to \mathbb{R}$ are smooth functions, then*

$$[X_F, X_G] = X_{\{F,G\}}.$$

Explicitly, $[X_F, X_G] = X_H$ where the Hamiltonian function H is defined by $H(x) = \omega(x)\big(X_F(x), X_G(x)\big) \in \mathbb{R}$.

Proof [Proposition V.24, Proposition V.26]. If φ_F^t is the flow of X_F, it follows from Propositions V.24 and V.26 that

$$(\varphi_F^t)^* X_G = X_{G \circ \varphi_F^t}.$$

In view of the definition of the Lie bracket we therefore obtain

$$[X_F, X_G] = \frac{d}{dt}\Big|_{t=0} (\varphi_F^t)^* X_G = X_{\frac{d}{dt}|_{t=0} G \circ \varphi_F^t}.$$

For the last term we compute

$$\begin{aligned} \frac{d}{dt}\Big|_{t=0} G \circ \varphi_F^t &= dG X_F \\ &= -\omega(X_G, X_F) \\ &= \omega(X_F, X_G), \end{aligned}$$

where we have used that $\omega(X_G, \cdot) = -dG$ by the definition of the Hamiltonian vector field X_G. □

Proposition V.27 states that the linear map $H \mapsto X_H$ is a homomorphism of Lie-algebras from the space of smooth functions $C^\infty(M, \mathbb{R})$ equipped with the Poisson bracket $\{\cdot, \cdot\}$ into the linear space of Hamiltonian vector fields on M equipped with the Lie bracket $[\cdot, \cdot]$. This leads to the Jacobi identity for the Poisson bracket. The following proposition sums up some properties of the Poisson bracket. The Jacobi identity (ii) below is equivalent to the closedness of the 2-form ω, the proof of which is left as an exercise to the reader. A proof can be found, for example in [56] by R. Jost.

Proposition V.28. *We consider the smooth functions $F, G, H \colon M \to \mathbb{R}$ on the symplectic manifold (M, ω). The Poisson bracket $\{\cdot, \cdot\}$ has the following properties:*

(i) $\{F, G\} = -\{G, F\}$,

(ii) $\{\{F, G\}, H\} + \{\{G, H\}, F\} + \{\{H, F\}, G\} = d\omega(X_F, X_G, X_H) = 0$,

(iii) $[X_F, X_G] = X_{\{F,G\}}$.

Under a symplectic mapping the Poisson bracket is transformed in a natural way.

Proposition V.29. *A local diffeomorphism* $u\colon M \to M$ *is symplectic if and only if*

$$\{u^*F, u^*G\} = u^*\{F, G\}$$

for all smooth functions $F, G\colon M \to \mathbb{R}$.

Proof [Proposition V.26]. We assume that u is symplectic. Recalling the transformation formula of Hamiltonian vector fields under symplectic transformations we compute

$$\begin{aligned}\{F, G\} \circ u &= dG \circ u(X_F \circ u) = d(G \circ u)(du)^{-1} X_F \circ u \\ &= d(G \circ u) X_{F \circ u} = \{G \circ u, F \circ u\}\end{aligned}$$

as claimed. In order to prove the converse, we assume that the diffeomorphism u satisfies $\{F, G\} \circ u = \{G \circ u, F \circ u\}$ for all functions F and G and deduce, using the definition of the Poisson bracket,

$$d(G \circ u)X_{F \circ u} = dG \circ u(X_F \circ u) = d(G \circ u)(du)^{-1} X_F \circ u = d(G \circ u)u^* X_F$$

for all functions G. Consequently, $u^* X_F = X_{u^*F}$ for all functions F, so that $u^*\omega = \omega$ in view of Proposition V.26. The proof of Proposition V.29 is complete. □

Proposition V.30 (Energy conservation). *The function* $H\colon M \to \mathbb{R}$ *is an* ***integral*** *of the Hamiltonian vector field* X_H, *i.e.,*

$$dH(x)X_H(x) = 0, \quad x \in M.$$

Equivalently, $H(\varphi^t(x)) = H(x)$ *for all* x *and* t, *where* φ^t *is the flow of* X_H. *Hence the Hamiltonian function is constant along the orbits of the Hamiltonian vector field.*

Proof [Skew symmetry of ω]. In view of the definition of the Hamiltonian vector field, $dH(x)\cdot = -\omega(X_H(x), \cdot)$. Hence, due to the skew symmetry of ω,

$$dH(x)X_H(x) = -\omega\big(X_H(x), X_H(x)\big) = 0.$$

If φ^t is the flow of X_H we compute

$$\begin{aligned}\frac{d}{dt} H(\varphi^t(x)) &= dH(\varphi^t(x)) \frac{d}{dt} \varphi^t(x) \\ &= dH(\varphi^t(x)) X_H(\varphi^t(x)) \\ &= -\omega(X_H, X_H) \circ \varphi^t(x) \\ &= 0,\end{aligned}$$

so that $H(\varphi^t(x)) = H(\varphi^0(x)) = H(x)$. □

The *geometrical significance* of the energy conservation is the following. Given a Hamiltonian function $H: M \to \mathbb{R}$, then the level set

$$S = S_E = \{x \in M \mid H(x) = E\} \subset M$$

belonging to the prescribed energy value $E \in \mathbb{R}$ is called an *energy surface*. From the energy conservation we deduce the invariance of the energy surface under the Hamiltonian flow,

$$\varphi^t(S) = S$$

for all t if S is compact.

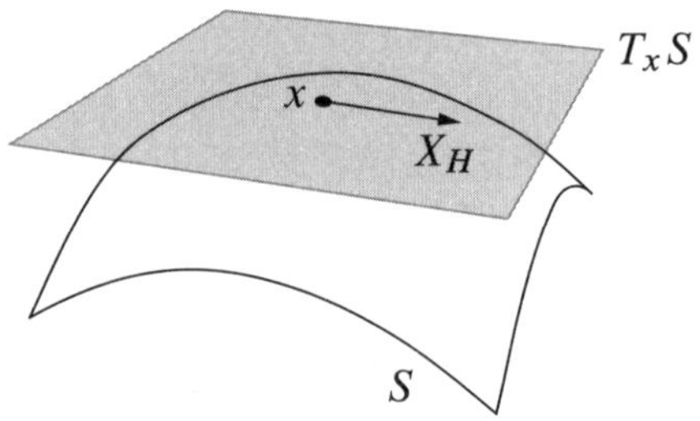

Figure V.2. The Hamiltonian vector field X_H belongs to the tangent space of a regular energy surface.

If $dH(x) \neq 0$ for all $x \in S$, the energy surface is called *regular*. It is a submanifold of codimension 1 of M and has at the point $x \in S$ the tangent space

$$T_xS = \ker dH(x) = \{v \in T_xM \mid dH(x)v = 0\}.$$

In view of $dH(x)X_H(x) = 0$ the Hamiltonian vector field is tangent to S,

$$X_H(x) \in T_xS, \quad x \in S,$$

so that its restriction $X_H|_S$ is a vector field on S. We next show that a compact regular energy surface S possesses a canonical volume form which is invariant under the Hamiltonian flow. As a consequence, the flow on S enjoys many recurrence properties in view of Poincaré's Theorem I.14. This encourages us to search for periodic orbits on S later on.

Proposition V.31 (Liouville). *If $S \subset (M, \omega)$ is a compact regular energy surface of a Hamilton function H, then there exists a finite regular Borel measure μ on S, which is invariant under the flow φ^t of the Hamiltonian vector field X_H.*

Proof [Proposition V.24, Proposition V.25, Proposition V.30]. By hypothesis, the energy surface

$$S = \{x \in M \mid H(x) = \text{const.}\}$$

is compact and regular, hence in particular $dH(x) \neq 0$ for $x \in S$. We choose a smooth vector field X transversal to S and satisfying

$$dH(X) = 1 \quad \text{on } S,$$

and define the $(2n-1)$-form σ on S by

$$\sigma := j^*(i_X \Omega),$$

where $\Omega = \omega \wedge \cdots \wedge \omega$ is the volume form on M associated with the symplectic structure ω, and where $j : S \to M$ is the injection map. The $(2n-1)$-form σ is a volume form on S and, in view of the following lemma, independent of the choice of the vector field X.

Lemma V.32. *If Y_1 and Y_2 are vector fields on M satisfying $dH(Y_1) = dH(Y_2) = 1$ on S, then*

$$j^*(i_{Y_1 - Y_2}\Omega) = 0.$$

Proof. The vector field $Y = Y_1 - Y_2$ satisfies $dH(Y) = 0$ on S and is therefore tangent to S. Consequently, it follows for the $2n-1$ linear independent tangent vectors $y_2, \dots, y_{2n}$ of S that $j^*(i_Y\Omega)(y_2, \dots, y_{2n}) = \Omega(Y, y_2, \dots, y_{2n}) = 0$, because $\dim S < 2n$. □

Lemma V.33. *The flow φ^t of the Hamiltonian vector field X_H satisfies $\varphi^t : S \to S$, as we already know. We claim that*

$$(\varphi^t)^*\sigma = \sigma$$

for all t. In other words the volume form σ on S is invariant under the Hamiltonian flow.

Proof. From $\varphi^t \circ j = j \circ \varphi^t$ on S we conclude $j^* \circ (\varphi^t)^* = (\varphi^t)^* \circ j^*$. Therefore in view of $(\varphi^t)^*\Omega = \Omega$ and using Proposition V.25 we compute

$$\begin{aligned}
(\varphi^t)^*\sigma &= j^* \circ (\varphi^t)^*(i_X\Omega) \\
&= j^* i_{(\varphi^t)^*X}((\varphi^t)^*\Omega) \\
&= j^*(i_{(\varphi^t)^*X}\Omega) \\
&= \sigma + j^*(i_{(\varphi^t)^*X - X}\Omega).
\end{aligned}$$

The second term vanishes due to Lemma V.32, because on S we have $dH((\varphi^t)^*X) = 1$. Indeed, it follows from the energy conservation $H \circ \varphi^t = H$ by differentiation that $dH(\varphi^t(x))\, d\varphi^t(x) = dH(x)$ and hence for all $x \in S$,

$$\begin{aligned}
dH(x)\big((\varphi^t)^*X(x)\big) &= dH(x)\big(d\varphi^t(x)^{-1}X \circ \varphi^t(x)\big) \\
&= dH(\varphi^t(x))\big(d\varphi^t(x)d\varphi^t(x)^{-1}X \circ \varphi^t(x)\big) \\
&= dH(X) \circ \varphi^t(x) = 1,
\end{aligned}$$

in view of the assumption about the vector field X. This proves the lemma. □

The volume form σ induces a regular Borel measure μ on S, defined by $\mu(A) = \int_A |\sigma|$. This measure is invariant under φ^t, since

$$\mu(\varphi^t(A)) = \int_{\varphi^t(A)} |\sigma| = \int_A (\varphi^t)^*|\sigma| = \int_A |\sigma| = \mu(A), \quad t \in \mathbb{R}.$$

In addition $\mu(S) < \infty$, since S is compact and the proof of Proposition V.31 is complete. □

V.9 Generating functions of symplectic maps in $\mathbb{R}^{2n}$

For practical purposes it is often useful to know that a symplectic diffeomorphism can be represented locally by a single function. To see this we consider the symplectic map $u\colon (\xi, \eta) \mapsto \big(a(\xi,\eta), b(\xi,\eta)\big) = (x, y)$ of the symplectic standard space $(\mathbb{R}^{2n}, \omega_0)$, that is,

$$\begin{aligned} x &= a(\xi,\eta), \\ y &= b(\xi,\eta). \end{aligned}$$

If

$$\det a_\xi \neq 0$$

at some point where a_ξ denotes the partial derivative in the variable ξ, then we can solve, in view of the implicit function theorem, the first equation locally in ξ, to obtain a function $\xi = \alpha(x,\eta)$ satisfying in a neighborhood of the reference point the identities $x = a\big(\alpha(x,\eta),\eta\big)$ and $\xi = \alpha\big(a(\xi,\eta),\eta\big)$, and moreover $a_\xi \circ \alpha_x = \mathrm{Id}$. Inserting the function $\xi = \alpha(x,\eta)$ into the second equation we arrive at the following implicit representation of the map u:

$$\begin{aligned} \xi &= \alpha(x,\eta), \\ y &= \beta(x,\eta) := b\big(\alpha(x,\eta),\eta\big). \end{aligned}$$

The advantage of scrambling the variables in this manner is that the condition for the map to be symplectic is much easier to express in term of the functions α and β than in terms of a and b. Indeed, as we shall show next the map is symplectic if and only if there exists a function $W = W(x,\eta)$ satisfying $\xi = \alpha(x,\eta) = W_\eta(x,\eta)$ and $y = \beta(x,\eta) = W_x(x,\eta)$. We should emphasize that all our statements are of local nature.

Proposition V.34. *If $u = (a,b)\colon (\xi,\eta) \mapsto (x,y)$ is a symplectic map on $\mathbb{R}^{2n}$ satisfying $\det a_\xi \neq 0$ at some point, then there exists locally a function $W = W(x,\eta)$ satisfying*

$$\det W_{x\eta} \neq 0,$$

which represents u in the neighborhood of this point in the implicit form

$$\xi = W_\eta(x,\eta),$$
$$y = W_x(x,\eta).$$

Conversely, a function $W(x,\eta)$ satisfying $\det W_{x\eta} \neq 0$ *defines in this manner implicitly a local symplectic diffeomorphism $u = (a,b)$ satisfying* $\det a_\xi \neq 0$.

Proof. Let the symplectic map $u(\xi,\eta)$ satisfy the assumptions of the proposition. We define the 1-form σ on $\mathbb{R}^{4n}$ in the coordinates $(\xi,\eta,x,y) \in (\mathbb{R}^n)^4$ by

$$\sigma = \sum_{j=1}^{n} y_j\, dx_j + \xi_j\, d\eta_j.$$

Then,

$$d\sigma = \sum_{j=1}^{n} dy_j \wedge dx_j - \sum_{j=1}^{n} d\eta_j \wedge d\xi_j,$$

which is the difference of the canonical symplectic forms ω_0 on the subspaces $\{(x,y)\}$ and $\{(\xi,\eta)\}$ of dimension $2n$. We introduce the injections $i, j: \mathbb{R}^{2n} \to \mathbb{R}^{4n}$ by

$$i(x,\eta) = \big(\alpha(x,\eta), \eta, x, \beta(x,\eta)\big),$$
$$j(\xi,\eta) = \big(\xi, \eta, a(\xi,\eta), b(\xi,\eta)\big),$$

and the local diffeomorphism φ of $\mathbb{R}^{2n}$ by

$$\varphi(x,\eta) = (\xi,\eta) = \big(\alpha(x,\eta),\eta\big).$$

Then $i = j \circ \varphi = (\mathrm{Id}, u) \circ \varphi$ and consequently

$$\begin{aligned} d(i^*\sigma) &= i^* d\sigma \\ &= \varphi^*(j^* d\sigma) \\ &= \varphi^*(u^*\omega_0 - \omega_0). \end{aligned}$$

Since u is symplectic, $u^*\omega_0 - \omega_0 = 0$ and hence $d(i^*\sigma) = 0$ and we find by the Poincaré lemma a function $W = W(x,\eta)$ satisfying $i^*\sigma = dW(x,\eta)$. Explicitly,

$$\sum_{j=1}^{n} \beta_j(x,\eta)\, dx_j + \alpha_j(x,\eta)\, d\eta_j = \sum_{j=1}^{n} W_{x_j}(x,\eta)\, dx_j + W_{\eta_j}(x,\eta)\, d\eta_j.$$

Consequently, $\alpha = W_\eta$ and $\beta = W_x$ as claimed. Moreover, $W_{\eta x} = \alpha_x = a_\xi^{-1}$ is invertible.

The converse follows by reversing all the steps. If the given function W satisfies $\det W_{x\eta} \neq 0$ and if $\alpha = W_\eta$ and $\beta = W_x$ then $i^*\sigma = dW$ and hence $d(i^*\sigma) = d^2W = 0$ and therefore $u^*\omega_0 - \omega_0 = 0$, so that the map u is symplectic. □

The function $W(x, \eta)$ is called a *generating function* of the symplectic map u.

Example. The identity mapping $u = \mathrm{Id}\colon (\xi, \eta) \mapsto (x, y) = (\xi, \eta)$ satisfies $a_\xi = \mathrm{Id}$ and is generated by the function $W(x, \eta) = \langle x, \eta \rangle$.

Thus, any function $W(x, \eta = \langle x, \eta \rangle + w(x, \eta)$ with a function w which is small along with its first and second derivatives defines a symplectic local diffeomorphism near the identity map. We can use the function w to parametrize symplectic diffeomorphisms near the identity. The symplectic rotation

$$\begin{aligned} x &= \eta, \\ y &= -\xi, \end{aligned}$$

satisfies $a_\xi = 0$, so that Proposition V.34 is not applicable. However, using $a_\eta = \mathrm{Id}$, we can construct another generating function $V = V(\xi, x)$, which represents the rotation implicitly by

$$\begin{aligned} \eta &= -V_\xi(\xi, x), \\ y &= V_x(\xi, x), \end{aligned}$$

and satisfies $\det V_{x\xi} \neq 0$. This is proved analogously to Proposition V.34 by using instead of σ the 1-form $\sigma' = \sum_{j=1}^n y_j\, dx_j - \eta_j\, d\xi_j$. For every symplectic diffeomorphism there is some generating function representing it implicitly as we shall show next.

Definition. The *group of elementary symplectic matrices* is the subgroup of $\mathrm{Sp}(n)$ generated by the following symplectic linear maps $(\xi, \eta) \mapsto (x, y)$.

(i) The rotation

$$\begin{aligned} x_1 &= \eta_1, \\ x_j &= \xi_k, \quad 2 \le j \le n, \\ y_1 &= -\xi_1, \\ y_j &= \eta_k, \quad 2 \le j \le n, \end{aligned}$$

of the symplectic (e_1, f_1) plane.

(ii) The permutations of the coordinates that leave the splitting $\mathbb{R}^n \oplus \mathbb{R}^n$ invariant,

$$\begin{aligned} x_j &= \xi_{\pi(j)}, \\ y_j &= \eta_{\pi(j)}, \end{aligned}$$

where $\pi \in \mathbb{S}_n$ is a permutation.

All these mappings are symplectic and can be represented by symplectic matrices which we abbreviate by E.

Proposition V.35. *If $U \in \mathrm{Sp}(n)$ is a linear symplectic map of $(\mathbb{R}^{2n}, \omega_0)$, there exists an elementary symplectic transformation E such that*

$$UE = \begin{pmatrix} A & B \\ C & D \end{pmatrix} \quad \textit{with } A \in \mathbb{R}^{n\times n} \textit{ invertible.}$$

Proof. We write the matrix U in block form as $U = \left(\begin{smallmatrix} A & B \\ C & D \end{smallmatrix}\right)$ with respect to $\mathbb{R}^n \oplus \mathbb{R}^n$. Since U is regular, the $(2n \times n)$-matrix (A, B), which is generated by writing the two matrices side by side, has rank n so that the matrix has n linearly independent column vectors $\{v_1, \dots, v_n\} =: \mathcal{F}$. We would like to move them to the left by means of elementary transformations E, so that

$$UE = \begin{pmatrix} A' & B' \\ C' & D' \end{pmatrix}$$

with A' consisting of linearly independent column vectors, so that it is invertible. We permute the columns of U by multiplying from the right with matrices E.

We denote by a_j and b_j the j-th column vector of A respectively of B. In view of the definition of the group of elementary transformations we can reach our goal if $\mathcal{F}$ can be chosen in such a way that

$$a_j \in \mathcal{F} \Longrightarrow b_j \notin \mathcal{F} \quad \text{and} \quad b_j \in \mathcal{F} \Longrightarrow a_j \notin \mathcal{F}$$

for all $1 \le j \le n$. Because with U also U^T is symplectic, it follows that AB^T is a symmetric matrix. Therefore the following lemma due to C. de Boor implies the proposition. □

Lemma V.36. *Let $A, B \in \mathbb{R}^{n\times n}$ be matrices having the column vectors a_j respectively b_j for $1 < j < n$. If $AB^T = BA^T$ and if r is the rank of the $(2n \times n)$-matrix (A, B), then there exists a set $\mathcal{F} = \{v_1, \dots, v_r\}$ of r linearly independent column vectors of (A, B) satisfying*

$$a_j \in \mathcal{F} \Longrightarrow b_j \notin \mathcal{F}$$

and

$$b_j \in \mathcal{F} \Longrightarrow a_j \notin \mathcal{F}$$

for all $1 \le j \le n$, so that every index j appears at the most once in $\mathcal{F}$.

Proof. If r_A is the rank of A we choose r_A linearly independent column vectors a_j of A which span the image of A. Let $J \subset \{1, \dots, n\}$ be the set of the corresponding indices j, so that $\operatorname{Im} A = \operatorname{span}\{a_j \mid j \in J\}$. We shall show that $\operatorname{Im}(A, B) \subset \operatorname{span}\{a_j, b_s \mid j \in J, s \notin J\}$. Then

$$\dim \operatorname{span}\{a_j, b_s \mid j \in J, s \notin J\} \ge \dim \operatorname{Im}(A, B) = r$$

and the lemma follows. Due to

$$\operatorname{Im}(A, B) = \operatorname{Im} A + \operatorname{Im} B = \operatorname{span}\{a_j \mid j \in J\} + \operatorname{Im} B$$

we only have to show that

$$\operatorname{Im} B \subset \operatorname{span}\{a_j, b_s \mid j \in J, s \notin J\}.$$

In order to do this we choose a basis $c_1, \dots, c_n$ of $\mathbb{R}^n$ satisfying $\langle a_j, c_k \rangle = \delta_{jk}$ for all $j, k \in J$. If $j \in J$, the vector $A^T c_j$ has the components

$$(A^T c_j)_s = \begin{cases} 1, & s = j \in J, \\ 0, & s \in J,\ s \neq j, \end{cases}$$

and hence

$$B(A^T c_j) = b_j + \sum_{s \notin J} (A^T c_j)_s\, b_s.$$

We see that $b_j \in \operatorname{span}\{b_s \mid s \notin J\} + \operatorname{Im}(BA^T)$ for $j \in J$. From the assumption $AB^T = BA^T$ we conclude $\operatorname{Im}(BA^T) \subset \operatorname{Im} A$ and hence,

$$b_j \in \operatorname{span}\{b_s \mid s \notin J\} + \operatorname{Im} A, \quad j \in J.$$

Consequently $\operatorname{Im} B = \operatorname{span}\{b_j \mid 1 \leq j \leq n\}$ is contained in $\operatorname{span}\{a_j, b_s \mid j \in J, s \notin J\}$ and the lemma is proved. □

By applying the proposition to a map $u = (a, b)$ whose linearization is equal to $du = U$ at some point, we achieve for $u \circ E = (a', b')$ the desired condition $\det a'_\xi \neq 0$ and have proved the following statement.

Corollary V.37. *Let u be a symplectic map on $\mathbb{R}^{2n}$. Using a suitable elementary symplectic transformation E, the composition $u \circ E$ can locally be represented by a generating function as in Proposition* V.34.

As an application we shall determine the most general symplectic map $u = (a, b)$ for which the first part

$$x = a(\xi),$$

is prescribed, satisfies $\det a_\xi \neq 0$ and a is independent of η. We represent the map u by means of the generating function $V(\xi, y)$ by

$$\begin{aligned} x &= V_y(\xi, y), \\ \eta &= V_\xi(\xi, y). \end{aligned}$$

From the assumption $a(\xi) = x = V_y(\xi, y)$ we deduce that V has the form

$$V(\xi, y) = \langle a(\xi), y \rangle - v(\xi),$$

with an arbitrary function v depending only on ξ. From the second equation we therefore deduce

$$\eta = V_\xi = a_\xi^T y - v_\xi,$$

so that we obtain for the symplectic map u the explicit representation

$$\begin{aligned} x &= a(\xi), \\ y &= (a_\xi^T)^{-1}\big(\eta + v_\xi(\xi)\big). \end{aligned}$$

This is the most general symplectic map compatible with the map $x = a(\xi)$ above.

Example. The most general symplectic transformation of $\mathbb{R}^{2n}$, which is the identity on the configuration space, namely $a = \mathrm{Id}$, has the form $x = \xi$ and

$$y = \eta + v_\xi(\xi),$$

with a function $v\colon \mathbb{R}^n \to \mathbb{R}$.

We assume next that the map $a = (a_1, \dots, a_n)$ from $\mathbb{R}^{2n}$ into $\mathbb{R}^n$ depends on ξ and on η and look for conditions under which it can be extended to a local symplectic diffeomorphism $u = (a, b)$ of $\mathbb{R}^{2n}$. That such an extension does exist, if the functions a_j are independent and in involution, is the statement of the following theorem which goes back to Liouville.

Theorem V.38 (Liouville). *We consider n functions $a_j = a_j(\xi, \eta)$ on $(\mathbb{R}^{2n}, \omega_0)$ whose Poisson brackets vanish, so that*

$$\{a_i, a_j\} = 0, \quad 1 \le i, j \le n.$$

We define the map $a = (a_1, \dots, a_n)$ and assume that the matrix $da = (a_\xi, a_\eta)$ has rank n. Then there exists a symplectic local diffeomorphism $u = (a, b)$ of the form

$$\begin{aligned} x &= a(\xi, \eta), \\ y &= b(\xi, \eta), \end{aligned}$$

which extends the map $x = a(\xi, \eta)$.

Proof. Since (a_ξ, a_η) has full rank, we can, after applying an elementary symplectic transformation, assume that $\det a_\xi \neq 0$. Consequently, the equation $x = a(\xi, \eta)$ can be solved for $\xi = \alpha(x, \eta)$ and we shall show next that the matrix α_η is symmetric. If F, G are real-valued functions, then the Poisson bracket in the symplectic standard space $(\mathbb{R}^{2n}, \omega_0)$ looks as follows:

$$\{F, G\} = \langle JX_F, X_G\rangle = \langle J\nabla F, \nabla G\rangle = G_\xi F_\eta^T - G_\eta F_\xi^T.$$

Thus the condition $\{a_i, a_j\} = 0$ for all i, j becomes

$$a_\xi a_\eta^T - a_\eta a_\xi^T = 0.$$

Multiplication with a_ξ^{-1} from the left-hand side and with $(a_\xi^T)^{-1}$ from the right-hand side leads to

$$a_\eta^T(a_\xi^T)^{-1} - a_\xi^{-1}a_\eta = 0,$$

so that the matrix $a_\xi^{-1}a_\eta$ is symmetric. Differentiating the identity $\xi = \alpha\big(a(\xi,\eta),\eta\big)$ we obtain $\mathbb{1} = \alpha_x a_\xi$ and $0 = \alpha_x a_\eta + \alpha_\eta$ and therefore the matrix $\alpha_\eta = -\alpha_x a_\eta = -a_\xi^{-1}a_\eta$ is indeed symmetric.

Explicitly, $\frac{\partial \alpha_i}{\partial \eta_j} = \frac{\partial \alpha_j}{\partial \eta_i}$, and consequently the vector field $\eta \mapsto \alpha(x,\eta)$ is a gradient vector field and there exists a function $W(x,\eta)$ such that

$$\alpha(x,\eta) = W_\eta(x,\eta).$$

This function satisfies $W_{x\eta} = \alpha_x = a_\xi^{-1}$ and hence $\det W_{x\eta} \neq 0$. In view of Proposition V.34 the function W generates a local symplectic map u which extends the map $x = a(\xi,\eta)$ and the proof is complete. □

Corollary V.39. *Assume that the $n+1$ functions $a_0, a_1, \dots, a_n$ on $(\mathbb{R}^{2n}, \omega_0)$ satisfy*

$$\{a_i, a_j\} = 0, \quad 0 \le i, j \le n,$$

and assume that the matrix (a_ξ, a_η) has rank n where the map a is defined by $a = (a_1, \dots, a_n)$. Then the function a_0 can be written locally as a function of $a_1, \dots, a_n$, in other words, there exists a function f such that

$$a_0(\xi,\eta) = f\big(a_1(\xi,\eta), \dots, a_n(\xi,\eta)\big).$$

In particular, there exist at the most n linearly independent functions whose Poisson brackets all vanish.

Proof. In view of Liouville's theorem the map $x = a(\xi,\eta)$ can be extended to a symplectic map $u = (a,b)\colon (\xi,\eta) \mapsto (x,y)$. Because its Jacobi-determinant is equal to 1, we can invert u locally. The inverse map u^{-1} is again symplectic and satisfies $a \circ u^{-1}(x,y) = x$. Next we define the function f by $f(x,y) := a_0 \circ u^{-1}(x,y)$ and show that $f = f(x)$ is independent of the variable y. It then follows that

$$a_0(\xi,\eta) = a_0 \circ u^{-1}(x,y) = f(x) = f(a(\xi,\eta)),$$

as we want to prove. In order to show $\frac{\partial}{\partial y_j} f = 0$ for all $1 \le j \le n$, we take the coordinate functions $x \mapsto x_j$ on $\mathbb{R}^{2n}$ and obtain for the Poisson brackets $\{f, x_j\} = \frac{\partial}{\partial y_j} f$. The Poisson brackets vanish, since in view of Proposition V.29,

$$\{f, x_j\} = \{a_0 \circ u^{-1}, a_j \circ u^{-1}\} = \{a_0, a_j\} \circ u^{-1} = 0,$$

because u^{-1} is symplectic and $\{a_0, a_j\} = 0$, by assumption. We have shown that f only depends on the variable x and so the proof is complete. □

V.10 Integrable systems, action–angle variables

Integrable systems are a very special class of Hamiltonian systems. They are characterized by the property of possessing sufficiently many integrals, so that the problem of solving the differential equation becomes essentially trivial. For general vector fields on $\mathbb{R}^m$ this would require $m-1$ integrals, for a Hamiltonian vector field on $\mathbb{R}^{2n}$ however, n integrals are sufficient. We begin by recalling the definition of an integral.

Definition. The smooth function $F\colon M \to \mathbb{R}$ is an *integral* of the vector field X on the manifold M, if

$$dF(x)\,X(x) = 0, \quad x \in M$$

and $dF(x) \neq 0$ for all $x \in M$.

The integral F is constant along every orbit of the vector field X, because for the flow φ^t of X it follows using the chain rule that

$$\frac{d}{dt}F(\varphi^t(x)) = dF(\varphi^t(x))\,\frac{d}{dt}\varphi^t(x) = dF(\varphi^t(x))\,X(\varphi^t(x)) = 0.$$

In the case of a Hamiltonian vector field X_H on a symplectic manifold, the function F is an integral, if and only if

$$\{H, F\}(x) = 0$$

and $dF(x) \neq 0$ for all $x \in M$.

Definition. A Hamiltonian vector field X_H is called *integrable* on the subset $D \subset M$ of a symplectic manifold (M, ω) of dimension $2n$, if it possesses n integrals $F_j \in C^\infty(D)$ for $1 \leq j \leq n$ having the following three properties on D.

(i) $dF_1, \dots, dF_n$ are linearly independent,

(ii) $\{H, F_j\} = 0$,

(iii) $\{F_j, F_k\} = 0$,

for all $j = 1, \dots, n$ and $k = 1, \dots, n$.

It is important to observe that we require these properties in the large, since locally, in the neighborhood of a point $p \in M$ at which $X_H(p) \neq 0$ every Hamiltonian vector field X_H is integrable. As a rule one therefore requires of the flow φ^t of X_H that $\varphi^t(D) = D$ holds true for all t.

Example. We consider for an open set $U \subset \mathbb{R}^n$ the symplectic manifold

$$M = T^n \times U$$

where T^n is the n-torus $\mathbb{R}^n/\mathbb{Z}^n$. We make use of the coordinates $(x, y) \in \mathbb{R}^n \times \mathbb{R}^n$ in the covering space. The symplectic form is induced from the standard form ω_0

of $\mathbb{R}^{2n}$. If H is a function on M which is *independent* of the x-coordinates of the torus so that

$$H = H(y_1, \dots, y_n), \quad y \in U,$$

then the n functions F_j on M, defined by

$$F_j(x, y) = y_j, \quad 1 \le j \le n,$$

satisfy the properties (i) to (iii). The Hamiltonian equations of X_H can be solved right away, explicitly, and for all times. Indeed, the system

$$\begin{cases} \dot{x} = \frac{\partial}{\partial y} H(y), \\ \dot{y} = -\frac{\partial}{\partial x} H(y) = 0 \end{cases}$$

has the solutions

$$\big(x(t), y(t)\big) = \big(x + t\omega(y), y\big)$$

for all $t \in \mathbb{R}$ and all initial conditions $(x, y) \in T^n \times U$, where

$$\omega(y) := \frac{\partial}{\partial y} H(y) \in \mathbb{R}^n$$

are the so-called frequencies of the torus $T^n \times \{y\}$. We see that every torus $T^n \times \{y\}$ is invariant under the flow of the vector field. Moreover, on every torus the flow is a Kronecker flow.

The example is typical and we shall see that many integrable systems can be brought into this form by choosing suitable coordinates. In physics the variables $(x, y) \in T^n \times U$ are called *action–angle variables*. The last term is related to the action integral

$$\int_\gamma \sum_{s=1}^n y_s \, dx_s$$

over suitable closed curves $\gamma \subset T^n \times U$. If we choose as a closed curve on the torus $T^n \times \{y\}$ the special curve $\gamma(t) = x(t)$ for $0 \le t \le 1$ having the coordinates $x_j(t) = t$ and $x_s(t) = 0$ for $s \ne j$, then

$$\int_\gamma \sum_{s=1}^n y_s \, dx_s = \int_0^1 y_j \, dt = y_j.$$

Hence, the variables y_j are simply obtained as action integrals over suitable closed curves on the manifolds $\{F_j(x, y) = \text{const.}\}$.

If X_H is an integrable system on M possessing the n integrals $F_j\colon M \to \mathbb{R}$ for $j = 1, \dots, n$, we define the map $F\colon M \to \mathbb{R}^n$ by

$$F(x) = \big(F_1(x), \dots, F_n(x)\big) \in \mathbb{R}^n.$$

From property (i) it follows that the level sets

$$N_c := \{x \in M \mid F(x) = c\} \subset M$$

for given constants $c \in \mathbb{R}^n$ are submanifolds M of dimension n. The Hamiltonian vector fields X_{F_j} are linearly independent and, due to property (iii), that is, due to $dF_j(X_{F_k}) = 0$ for all i and j, they span the tangent space $T_x N_c$ at the point $x \in N_c$,

$$T_x N_c = \operatorname{span}\{X_{F_j}(x) \mid 1 \leq j \leq n\}.$$

From $\omega(X_{F_j}, X_{F_k}) = 0$ it follows in addition that $\omega(v, w) = 0$ for all pairs of vectors $v, w \in T_x N_c$. Therefore $T_x N_c$ is a Lagrangian subspace of the symplectic vector space $T_x M$. The manifold N_c is consequently called a Lagrangian manifold. The assumption (ii) requires that $0 = \omega(X_H, X_{F_j}) = dF_j(X_H)$ for all $j = 1, \dots, n$, so that X_H is tangential to all submanifolds N_c and the flow of X_H leaves every submanifold N_c invariant.

Under an additional topological assumption on N_c the flow of X_H can be determined explicitly and for all times, as the following theorem of Arnold–Jost shows. This theorem goes back to V. I. Arnold in the special case $M = \mathbb{R}^{2n}$. Arnold required an additional assumption which, however, is superfluous, as R. Jost [57] has proved in the general case of a symplectic manifold (M, ω).

Theorem V.40 (Arnold–Jost). *Let (M, ω) be a symplectic manifold of dimension $2n$. We assume, moreover, that there exist n functions $F_j : M \to \mathbb{R}$ satisfying the properties* (i) *and* (iii) *in the definition of an integrable system in every point of M and that one of the level sets, as for instance $N_0 = F^{-1}(0) \subset M$, is compact and connected. Then the following holds true.*

(a) *N_0 is an embedded torus T^n.*

(b) *There exists an open neighborhood U of N_0 in M which can be described by action and angle variables (x, y) in the following way. There exist two open neighborhoods D_1 and D_2 of 0 in $\mathbb{R}^n$ having the following properties. There exists a diffeomorphism $\mu \colon D_2 \to D_1$ satisfying $\mu(0) = 0$. If $x = (x_1, \dots, x_n) \in \mathbb{R}^n$ are the coordinates of the torus $T^n = \mathbb{R}^n/\mathbb{Z}^n$ and if $y = (y_1, \dots, y_n)$ are the coordinates on $D_1 \subset \mathbb{R}^n$, then there exists a diffeomorphism*

$$\psi \colon T^n \times D_1 \to U = \bigcup_{c \in D_2} N_c \cap U \subset M,$$

which is symplectic,

$$\psi^* \omega = \sum_{j=1}^{n} dy_j \wedge dx_j = \omega_0,$$

and which satisfies

$$(*) \quad \mu \circ F \circ \psi(x, y) = y.$$

In particular, ψ induces diffeomorphisms $T^n \times \{0\} \to N_0$ and $T^n \times \{y\} \to N_c \cap U$, where $y = \mu(c)$.

Corollary V.41. *An integrable Hamiltonian vector field X_H possessing the integrals F_j for $1 \le j \le n$, as in Theorem V.40, is, by the symplectic diffeomorphism ψ, transformed into the following Hamiltonian system on $(T^n \times D_1, \omega_0)$,*

$$H \circ \psi(x, y) = h(y),$$

where the Hamiltonian function h only depends on the action variables y and not on the angle variables x.

Proof. Since ψ is symplectic, it follows from Proposition V.29 for the functions $f_j = F_j \circ \psi \colon T^n \times D_1 \to \mathbb{R}$ that

$$\{h, f_j\} = \{H \circ \psi, F_j \circ \psi\} = \{H, F_j\} \circ \psi = 0,$$

so that the functions f_j are integrals of the Hamiltonian h. Therefore, due to $(*)$ also the coordinate functions y_j are integrals of h and consequently,

$$0 = \{h, y_j\} = \frac{\partial}{\partial x_j} h,$$

so that the function h is independent of the variables x_j for all $1 \le j \le n$, as claimed. □

On $\psi^{-1}(U) = T^n \times D_1$ the Hamiltonian system looks extremely simple,

$$X_h : \quad \begin{cases} \dot{x} = \frac{\partial h}{\partial y}(y), \\ \dot{y} = 0 \end{cases}$$

as in the above example. The solutions are given by

$$\big(x(t), y(t)\big) = \big(x(0) + t\omega(y(0)), y(0)\big).$$

Geometrically every torus $T^n \times \{y\} \subset T^n \times D_2$ is invariant under the flow φ^t of the Hamiltonian vector field X_h and the flow on the torus $T^n \times \{y\}$ is the Kronecker flow

$$\varphi^t|_{T^n \times \{y\}} \colon x \mapsto x + t\omega,$$

having the frequencies $\omega = \frac{\partial h}{\partial y}(y) \in \mathbb{R}^n$ depending on the torus $T^n \times \{y\}$.

Proof of Theorem V.40. The proof proceeds in four steps. We first show that N_0 is an n-dimensional torus. In a second step we introduce convenient symplectic coordinates (x, y) in the neighborhood of a chosen point $p \in N_0$, using Liouville's Theorem V.38. In a third step, these coordinates are extended to an open neighborhood of N_0 by means of the flows of the Hamiltonian vector fields X_{F_j} of the integrals. Finally, the normalization of the periods requires a further symplectic transformation, and this is the fourth and last step of the proof.

(1) We denote by $\varphi_j^{s_j}$ the flow of the Hamiltonian vector field X_{F_j}, for $1 \leq j \leq n$. Due to $[X_{F_j}, X_{F_k}] = X_{\{F_j,F_k\}}$ and $\{F_j, F_k\} = 0$ the vector fields X_{F_j} commute, hence in view of Proposition V.20 their flows also commute. For $s = (s_1, \dots, s_n) \in \mathbb{R}^n$ we introduce the family of symplectic mappings $\varphi^s : M \to M$ as the composition

$$\varphi^s = \varphi_1^{s_1} \circ \varphi_2^{s_2} \circ \cdots \circ \varphi_n^{s_n},$$

wherever it is defined. It satisfies $\varphi^0 = \mathrm{Id}$ and

$$\varphi^{s+t} = \varphi^s \circ \varphi^t$$

for all $s, t \in \mathbb{R}^n$ in the domain of definition. Moreover, since the functions F_j are commuting integrals,

$$F(\varphi^s(x)) = F(x), \quad x \in M$$

for all $s \in \mathbb{R}^n$ in the domain of definition, so that the maps φ^s leave the level sets N_c invariant. We now choose a point $p \in N_0$. Because $N_0 \subset M$ is compact and invariant, the symplectic maps $\varphi^s(p)$ do exist for all parameters $s \in \mathbb{R}^n$ and we can define the action A of $\mathbb{R}^n$ on N_0 by

$$A : \mathbb{R}^n \to N_0, \quad s \mapsto \varphi^s(p).$$

The tangent map $dA(s) : \mathbb{R}^n \to T_{\varphi^s(p)}N_0$ maps the vectors $e_j \in \mathbb{R}^n$ onto the tangent vectors $X_{F_j}(\varphi^s(p)) \in T_{\varphi^s(p)}N_0$. These vectors are linearly independent, because the differentials dF_j are linearly independent and therefore span the whole tangent space. Hence the map A is a local diffeomorphism whose image in N_0 is closed as well as open. By assumption, the manifold N_0 is connected, hence it follows that $A(\mathbb{R}^n) = N_0$, so that the map A is surjective. However, the map A is not injective, otherwise $\mathbb{R}^n$ would be compact. Consequently, there exists a non-trivial isotropy group

$$\Gamma = \{s \in \mathbb{R}^n \mid \varphi^s(p) = p\}$$

which is necessarily a discrete subgroup of $\mathbb{R}^n$, hence a lattice. (More details can be found, for example, in [21] by L. Conlon.) The lattice is generated by the linearly independent vectors $\gamma_1, \dots, \gamma_d \in \mathbb{R}^n$, so that

$$\Gamma = \{\gamma = \textstyle\sum_{j=1}^d n_j \gamma_j \mid n_j \in \mathbb{Z}\}.$$

From $\varphi^{s+\gamma}(p) = \varphi^s(p)$ for all $s \in \mathbb{R}^n$ and $\gamma \in \Gamma$ it follows that the map A induces a diffeomorphism between $\mathbb{R}^n/\Gamma$ and N_0. Since due to the assumption, N_0 is compact, we conclude that $d = n$ so that $N_0 \cong \mathbb{R}^n/\Gamma$ is an n-dimensional torus, as claimed in the theorem.

(2) In the second step we shall introduce in the neighborhood of the chosen point $p \in N_0$ suitable symplectic coordinates. By adding constants to the integrals we may assume that $F(p) = 0$. □

Lemma V.42 (Liouville coordinates). *There exists an open neighborhood $W \subset (M, \omega)$ of the distinguished point p and a diffeomorphism $\psi\colon W \to V$ onto an open neighborhood of $0 \in \mathbb{R}^{2n}$ such that*

(i) $\psi(p) = 0$,

(ii) $\psi^*\omega_0 = \omega$,

(iii) $F(\psi^{-1}(x, y)) = y$,

where (x, y) are the symplectic coordinates in $(\mathbb{R}^{2n}, \omega_0)$.

Proof [Theorem of Darboux, theorem of Liouville]. If $\psi_0\colon W' \subset M \to \mathbb{R}^{2n}$ is a symplectic chart satisfying $\psi_0(p) = 0$, as in Darboux's theorem and if $f(x, y) = F(\psi_0^{-1}(x, y))$ are the integrals in the chart, then $f(0) = 0$ and in view of Liouville's Theorem V.38, there exists a symplectic extension $\varphi\colon \mathbb{R}^{2n} \to \mathbb{R}^{2n}$ of the form $\varphi(x, y) = \big(g(x, y), f(x, y)\big)$. By translation we may assume that $\varphi(0) = 0$. Then $f(x, y) = \pi_2 \circ \varphi(x, y)$ where π_2 is the projection $\pi_2(x, y) = y$. For the inverse φ^{-1} it follows that

$$f(\varphi^{-1}(x, y)) = \pi_2 \circ \varphi \circ \varphi^{-1}(x, y) = \pi_2(x, y) = y.$$

Therefore $\psi := \varphi \circ \psi_0$ is the desired diffeomorphism. □

Corollary V.43. *In the Liouville coordinates the flow φ^s looks as follows:*

$$\varphi^s \circ \psi^{-1}(x, y) = \psi^{-1}(x + s, y)$$

for $s \in \mathbb{R}^n$, if (x, y) and $(x + s, y)$ in V.

Proof. In the Liouville coordinates the Hamiltonian function F_j is represented by the function

$$f_j(x, y) = F_j(\psi^{-1}(x, y)) = y_j.$$

Therefore, the flow of the transformed Hamiltonian vector field X_{f_j} at time s_j is given by

$$(x, y) \mapsto (x + s_j e_j, y),$$

as claimed. □

(3) We return to our chosen point $p \in N_0$ and to the Liouville chart (W, ψ) of Lemma V.42, where ψ maps the open neighborhood W of p onto the open neighborhood V of $0 \in \mathbb{R}^{2n}$ and satisfies $\psi(p) = 0$. To extend the inverse map $\psi^{-1}: V \to W$, we introduce the map

$$\vartheta: \mathbb{R}^n \times D_2 \to M, \quad \vartheta(x, y) := \varphi^x \circ \psi^{-1}(0, y),$$

where D_2 is a small neighborhood of $0 \in \mathbb{R}^n$ and φ^x for $x \in \mathbb{R}^n$ is, as in the first step, the flow of the Hamiltonian vector fields X_{F_j}. If $y = 0$, we obtain $\vartheta(x, 0) = \varphi^x(p)$ and the flow does exist for all $x \in \mathbb{R}^n$, as we have already seen. We shall show that also for small $y \in D_2$ the map $\vartheta(x, y)$ is defined for all $x \in \mathbb{R}^n$. In view of Corollary V.43,

$$\vartheta(x, y) = \psi^{-1}(x, y),$$

for small $x \in \mathbb{R}^n$. Geometrically we apply the flow maps φ^x to the section $S = \{\psi^{-1}(0, y) \in M \mid y \in D_2\}$ which, in view of Corollary V.43 is transversal to the vector fields X_{F_j}.

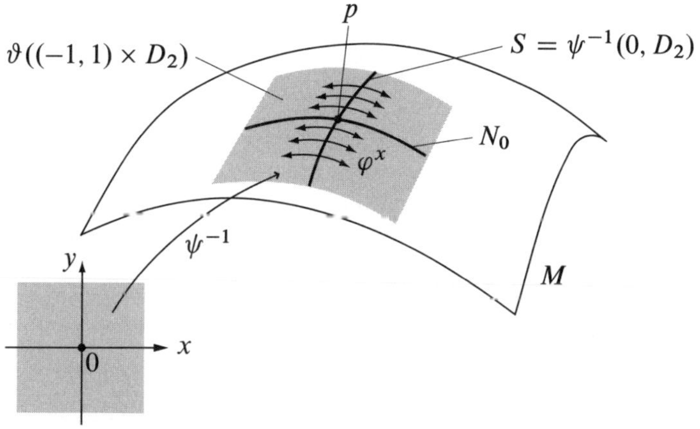

Figure V.3. The geometrical meaning of the map ϑ, flowing out from the transversal section S of N_0.

From the invariance $F \circ \varphi^x = F$ we conclude by means of Lemma V.42 that

$$F(\vartheta(x, y)) = F(\varphi^x \circ \psi^{-1}(0, y)) = F(\psi^{-1}(0, y)) = y,$$

so that on its domain of definition the map ϑ satisfies

$$\vartheta: \mathbb{R}^n \times \{y\} \to N_y = F^{-1}(y)$$

for $y \in D_2$. In the special case $y = 0$ the map $\vartheta(x, 0) = \varphi^x(p)$ is, due to step (1), defined for all $x \in \mathbb{R}^n$ and maps $\mathbb{R}^n \times \{0\}$ onto the submanifolds $N_0 = F^{-1}(0)$. From this it follows for D_2 sufficiently small that the map ϑ is defined on $K \times D_2$, where $K \subset \mathbb{R}^n$ is a compact set containing the generators $\gamma_j \in \Gamma$ of the lattice.

We show that ϑ is defined on $\mathbb{R}^n \times D_2$. If $y = 0$, then $\vartheta(x + \gamma, 0) = \vartheta(x, 0)$ for all $x \in \mathbb{R}^n$ and all $\gamma \in \Gamma$. We construct for y in a small neighborhood of 0 an n-dimensional lattice $\Gamma(y) \subset \mathbb{R}^n$ which is close to $\Gamma(0) = \Gamma$ and satisfies $\vartheta\big(x + \gamma(y), y\big) = \vartheta(x, y)$ for all $x \in \mathbb{R}^n$ and for all $\gamma(y) \in \Gamma(y)$.

If $\gamma = \gamma_j \in \Gamma$ is a generator of the lattice, then $\varphi^\gamma(p) = p$ and $\psi(p) = 0$. In the Liouville chart $V \subset \mathbb{R}^{2n}$ the map

$$(*) \quad \psi \circ \varphi^\gamma \circ \psi^{-1} \colon (x, y) \mapsto (\xi, \eta)$$

is a local symplectic diffeomorphism near the fixed point 0 in $\mathbb{R}^{2n}$.

Lemma V.44. *The symplectic map* $(*)$ *has the form*

$$\xi = x - v(y), \quad v(y) = \frac{\partial}{\partial y} Q(y), \quad \eta = y$$

for a smooth real-valued function Q *which satisfies* $Q(0) = 0$. *Moreover,* $v(0) = 0$.

Proof. By definition $\varphi^\gamma \circ \psi^{-1}(x, y) = \psi^{-1}(\xi, \eta)$ and from $F \circ \varphi^x = F$ it follows using Lemma V.42 that

$$y = F \circ \psi^{-1}(x, y) = F \circ \varphi^\gamma \circ \psi^{-1}(x, y) = F \circ \psi^{-1}(\xi, \eta) = \eta.$$

In Section V.9. we have seen that every symplectic map which extends the relation $\eta = y$ is necessarily of the desired form. The statement $v(0) = 0$ follows from $\psi \circ \varphi^\gamma \circ \psi^{-1}(0) = 0$, and the lemma is proved. □

By definition, $\varphi^\gamma \circ \psi^{-1}(x, y) = \psi^{-1}(\xi, \eta)$, and hence in view of Lemma V.44 and Corollary V.43,

$$(**) \quad \psi^{-1}(x, y) = \varphi^{\gamma + v(y)} \circ \psi^{-1}(x, y)$$

for $(x, y) \in V' \subset V$. Carrying out this construction for every generator γ_j of Γ we obtain the maps $v_j(y) = \frac{\partial}{\partial y} Q_j(y)$, which satisfy the identity $(**)$ with $\gamma_j + v_j(y)$ replacing $\gamma + v(y)$. Defining $\gamma_j(y) = \gamma_j + v_j(y)$, we therefore introduce the lattice $\Gamma(y) \subset \mathbb{R}^n$ as

$$\Gamma(y) = \big\{\gamma(y) = \textstyle\sum_{j=1}^n n_j \gamma_j(y) \mid n_j \in \mathbb{Z}\big\}.$$

In view of Lemma V.44,

$$\gamma_j(y) = \gamma_j + v_j(y) = \frac{\partial}{\partial y} W_j(y),$$

with the real-valued functions $W_j(y) = Q_j(y) + \langle y, \gamma_j \rangle$ satisfying $W_j(0) = 0$ for all $1 \le j \le n$. From $v_j(0) = 0$ we deduce $\gamma_j(0) = \gamma_j$ and so

$$\Gamma(0) = \Gamma$$

if $y = 0$. We conclude for small y that the lattice vectors $\gamma_j(y)$ are linearly independent and satisfy, due to $(**)$, the equations

$$(\sharp) \quad \psi^{-1}(x, y) = \varphi^{\gamma_j(y)} \circ \psi^{-1}(x, y)$$

for all $(x, y) \in V' \subset V$ and for all $1 \le j \le n$.

Lemma V.45. *The map $\vartheta(x, y) = \varphi^x \circ \psi^{-1}(0, y)$ is defined on $\mathbb{R}^n \times D_2$ and satisfies*

(i) $\vartheta^* \omega = \omega_0$,

(ii) $\vartheta\big(x + \gamma(y), y\big) = \vartheta(x, y)$ *for all $\gamma(y) \in \Gamma(y)$ and all $(x, y) \in \mathbb{R}^n \times D_2$.*

Proof. The map ϑ is defined for $(x, y) \in K \times D_2$ where $K \subset \mathbb{R}^n$ is a compact set containing the generators of $\Gamma = \Gamma(0)$ in its interior, if $D_2 \subset \mathbb{R}^n$ is a sufficiently small neighborhood of $0 \in \mathbb{R}^n$. For small x we find in view of the identity $(\sharp)$,

$$\begin{aligned}
\vartheta(x + \gamma_j(y), y) &= \varphi^{x+\gamma_j(y)} \circ \psi^{-1}(0, y) \\
&= \varphi^x \circ \varphi^{\gamma_j(y)} \circ \psi^{-1}(0, y) \\
&= \varphi^x \circ \psi^{-1}(0, y) \\
&= \vartheta(x, y)
\end{aligned}$$

for every generator $\gamma_j(y)$ of Γ. Therefore, we can indeed continue the flow $x \mapsto \vartheta(x, y) = \varphi^x \circ \psi^{-1}(0, y)$ to all $x \in \mathbb{R}^n$ and it remains to prove that $\vartheta^* \omega = \omega_0$. If $(x, y) \in \mathbb{R}^n \times D_2$, we choose $x_0 \in \mathbb{R}^n$ near x such that $(x - x_0, y)$ lies in V. This way we can represent ϑ locally as a composition of symplectic maps, since in view of Corollary V.43 we obtain

$$\begin{aligned}
\vartheta(x, y) &= \varphi^{x_0 + (x - x_0)} \circ \psi^{-1}(0, y) \\
&= \varphi^{x_0} \circ \psi^{-1}(x - x_0, y) \\
&= \varphi^{x_0} \circ \psi^{-1} \circ \tau_{x_0}(x, y),
\end{aligned}$$

with the translation $\tau_{x_0}(x, y) = (x - x_0, y)$ in $\mathbb{R}^{2n}$. As a composition of symplectic maps, the map ϑ is symplectic and the lemma is proved. □

We have already seen that $\vartheta(\mathbb{R}^n \times \{y\}) \subset F^{-1}(y)$. Next we show that ϑ induces a diffeomorphism

$$\vartheta \colon \mathbb{R}^n / \Gamma(y) \to \vartheta(\mathbb{R}^n \times \{y\}) \subset F^{-1}(y)$$

for every small $y \in D_2$. It then follows in particular that the image is diffeomorphic to the torus T^n. Since $\vartheta(\cdot, y)$ is a local diffeomorphism, it only remains to show that ϑ is injective, so that $\Gamma(y)$ is indeed the isotropy group of $\vartheta(\cdot, y)$. By construction the lattice $\Gamma(y)$ is n-dimensional and we have to show that the periods $\gamma_j(y)$ are in fact minimal. Arguing by contradiction we assume that there exist bounded sequences (x_j, y_j) and (x_j', y_j) satisfying $y_j \to 0$ and having the properties

$$\vartheta(x_j, y_j) = \vartheta(x_j', y_j) \quad \text{and} \quad x_j - x_j' \notin \Gamma(y_j).$$

Going over to subsequences we may assume that $x_j \to x^*$ and $x_j' \to y^*$ so that $\vartheta(x^*, 0) = \vartheta(y^*, 0)$ and hence $x^* - y^* \in \Gamma(0) = \Gamma$. Therefore $x_j - x_j'$ is near a lattice point in $\Gamma(y_j)$. Since the lattice is discrete, $x_j - x_j' \in \Gamma(y_j)$ must hold true for large j, in contradiction to the assumption. Consequently, we do, indeed, have a diffeomorphism $\mathbb{R}^n / \Gamma(y) \to \vartheta(\mathbb{R}^n \times \{y\})$.

(4) Finally, we normalize all lattices $\Gamma(y)$ to $\mathbb{Z}^n$ by a symplectic diffeomorphism

$$\sigma\colon \mathbb{R}^n \times D_1 \to \mathbb{R}^n \times D_2, \quad (\xi, \eta) \mapsto (x, y).$$

The lattice $\Gamma(y)$ is generated by the lattice vectors

$$\gamma_j(y) = \gamma_j + v_j(y) = \frac{\partial}{\partial y} W_j(y).$$

The diffeomorphism σ should satisfy

$$\sigma(\mathbb{Z}^n \times \{\eta\}) = \Gamma(y) \times \{y\}, \quad \eta = \mu(y)$$

where we still have to define the diffeomorphism μ. We define the map σ implicitly by means of the generating function

$$S(\xi, y) = \sum_{j=1}^{n} \xi_j W_j(y),$$

so that

$$\eta_j = \frac{\partial}{\partial \xi_j} S(\xi, y) = W_j(y) \quad \text{and} \quad x_j = \frac{\partial}{\partial y_j} S(\xi, y) = \sum_{s=1}^{n} \xi_s \frac{\partial}{\partial y_j} W_s(y).$$

The first line defines the map $\eta = \mu(y) = (W_1(y), \dots, W_n(y))$. It is a local diffeomorphism in the neighborhood of $0 = \mu(0)$, in the statement of the theorem denoted by $\mu\colon D_2 \to D_1$. This follows because the Jacobi matrix of μ is equal to

$$d\mu(y) = \big(\gamma_1(y), \dots, \gamma_n(y)\big)$$

and the column vectors $\gamma_j(y)$ are linearly independent. Since the matrix $S_{\xi y}(\xi, y)$ is equal to $d\mu(y)$ and therefore is invertible for all ξ if y is small, the function S generates a symplectic map $\sigma\colon \mathbb{R}^n \times D_1 \to \mathbb{R}^n \times D_2$. It satisfies

$$\sigma(e_j, \eta) = (\gamma_j(y), y), \quad \eta = \mu(y).$$

Hence $\sigma(\mathbb{Z}^n \times \{\eta\}) = \Gamma(y) \times \{y\}$ for $\eta = \mu(y)$ as desired. The composition

$$\Phi = \vartheta \circ \sigma\colon (\mathbb{R}^n \times D_1, \omega_0) \to (M, \omega)$$

is symplectic, satisfies $\Phi(\xi + j, \eta) = \Phi(\xi, \eta)$ for all $j \in \mathbb{Z}^n$, and therefore induces a symplectic diffeomorphism

$$\Phi\colon \mathbb{R}^n/\mathbb{Z}^n \times D_1 \to U \subset M$$

onto an open neighborhood of N_0. The map Φ defines the desired action- angle variables $(\xi, \eta) \in T^n \times D_1$. Indeed, $\Phi(T^n \times \{\eta\}) = F^{-1}(y) \cap U$ where $\eta = \mu(y)$ and

$$\begin{aligned}
\mu \circ F \circ \Phi(\xi, \eta) &= \mu \circ F \circ \vartheta \circ \sigma(\xi, \eta)\\
&= \mu \circ F \circ \vartheta(x, y)\\
&= \mu \circ F \circ \varphi^x \circ \psi^{-1}(0, y)\\
&= \mu \circ F \circ \psi^{-1}(0, y)\\
&= \mu(y)\\
&= \eta.
\end{aligned}$$

This completes the proof of the theorem of Arnold and Jost.

Outlook. Integrable systems are extremely rare and therefore attract special attention and we refer to J. Moser's lecture [72] given at the ICM 1998 in Berlin. On the other hand, there are many systems in nature, which are not integrable, but which are near an integrable system. The most prominent is our planetary system. We therefore wonder how such an integrable system behaves under perturbation. We recall that an integrable system X_h on the phase space $T^n \times D$ of the angle and action variables (x, y) can be solved immediately. The Hamiltonian function $h(y)$ depends only on the action variable $y \in D$, so that the flow leaves every single torus $T^n \times \{y\}$ invariant. On the torus the flow is the Kronecker flow $x \mapsto x + t\omega(y)$, where $\omega(y) = \frac{\partial}{\partial y} h(y) \in \mathbb{R}^n$ are the frequencies associated with the torus $T^n \times \{y\}$.

If the frequencies $\omega(y) = (\omega_1, \dots, \omega_n)$ are rationally independent or, more explicitly, if $\langle \omega, j \rangle \notin \mathbb{Z}$ for all $j \in \mathbb{Z}^n \setminus \{0\}$, then every orbit on this torus is dense, in view of the proposition by Kronecker (Corollary I.7). Otherwise, there exist non-trivial relations of the form $\langle \omega, j \rangle \in \mathbb{Z}$ for integer vectors $j \in \mathbb{Z} \setminus \{0\}$, so-called resonance conditions. In this case, the closure of every orbit lies densely in a torus of lower dimension. If there exist $n-1$ independent resonance conditions, then every orbit is periodic on the corresponding torus. In the nondegenerate case defined by the condition

$$\det \frac{\partial}{\partial y}\omega(y) = \det \frac{\partial^2}{\partial y^2}h(y) \neq 0, \quad y \in D$$

the frequencies $\omega(y)$ actually do depend on the action variables y and all the different types of tori are mixed together. It turns out that these tori $T^n \times \{y\}$, depending on their frequency vectors $\omega(y)$, react completely differently to a Hamiltonian perturbation H of h of the form

$$H(x, y) = h(y) + h_1(x, y),$$

where the perturbation term h_1 is small and periodic in the angle variable, so that $h_1(x + j, y) = h_1(x, y)$ for all integer vectors $j \in \mathbb{Z}$. As an example we consider an irrational torus $T^n \times \{y\}$, whose frequency vector is strongly irrational in the sense that it satisfies the Diophantine inequalities

$$|\langle\omega(y), j\rangle| \geq \gamma |j|^{-\tau}$$

for all $j \in \mathbb{Z}^n \setminus \{0\}$, with two constants $\gamma > 0$ and $\tau > n - 1$. One can prove that such a torus *survives* under the perturbation h_1 as an invariant parametrized torus, on which the flow is conjugated to the unperturbed flow, if only the perturbation h_1 is sufficiently small and sufficiently smooth. Therefore, the torus is only slightly deformed. For this purpose, one constructs an embedding

$$\varphi\colon T^n \times \{y\} \to T^n \times D$$

of the unperturbed torus into the phase space, which is in the neighborhood of the inclusion map of the torus and solves the partial differential equation

$$d\varphi(x)\omega = J\nabla H(\varphi(x)) \quad \text{for all } x \in T^n$$

where $\omega = \omega(y)$ are the frequencies of the unperturbed system,

Such an embedding φ maps the solution $\xi(t) = \xi + t\omega$ on the unperturbed torus $T^n \times \{y\}$ onto the solution $x(t) := \varphi(\xi(t))$ on the perturbed torus $\varphi(T^n \times \{y\})$, since using the above partial differential equation,

$$\begin{aligned} \dot{x}(t) &= d\varphi(\xi(t))\,\dot{\xi}(t) \\ &= d\varphi(\xi(t))\,\omega \\ &= J\nabla H(\varphi(\xi(t))) \\ &= J\nabla H(x(t)). \end{aligned}$$

The partial differential equation is an analytically very subtle boundary value problem for periodic boundary condition. Solutions can be constructed by means of powerful iteration methods, called KAM theory. The acronym KAM stands for Kolmogorov–Arnold–Moser. The KAM theory guarantees not only one, but an abundance of such invariant tori for the vector field X_H, under suitable conditions on the smoothness of H and the smallness of the perturbation h_1. The set of all these invariant tori form a set of positive Lebesgue measure in $T^n \times D$ which, however, is nowhere dense in $T^n \times D$, so that the *complement* of this set of stable orbits for $n \geq 3$ is connected. It immediately follows that perturbed integrable systems *cannot be ergodic*, as the famous physicist E. Fermi has claimed. So far, the system still looks like an integrable system, at least as far as the measure is concerned.

However, in the complement of the set of invariant KAM tori something completely different happens under the perturbation. Here the invariants and stability properties break down and unstable phenomena show up generically. They can give rise to a slow so-called Arnold diffusion through the phase space and even to escaping solutions.

That in the Hamiltonian systems near integrable systems stable and unstable orbits coexist and cannot be separated from each other is best illustrated by the celebrated sketch due to V. I. Arnold and A. Avez in their book [8]. It demonstrates vividly the complexity of the orbit structure of a smooth nonlinear area preserving diffeomorphism φ near an elliptic fixed point in the plane $\mathbb{R}^2$. As explained in Section VIII.5 such a map arises, for example, as a transversal section map of a stable periodic orbit on the 3-dimensional energy surface of a Hamiltonian system. Instead of studying nearby solutions for all times one can, just as well, study all the iterates of the section map.

The elliptic fixed point 0 at the center of Figure V.4 is surrounded by smooth closed curves which are close to circles and which are invariant under the map φ. Their existence was established by J. Moser, 1962, in [70]. For a recent proof in the analytic case we point out [62] by M. Levi and J. Moser. On every invariant curve the map is conjugated to the rigid rotation of a circle by an irrational rotation which depends on the curve. These invariant curves represent the stable part of the map φ and fill out a Cantor set of relatively large measure and reflect the fact that in a neighborhood of the elliptic fixed point the map φ is close to an integrable map. One concludes, in particular, that the fixed point is topologically stable. However, between these invariant curves one sees generically orbits of elliptic and hyperbolic periodic points. Taking a closer look, one discovers that the stable and unstable invariant manifolds issuing from the hyperbolic periodic points intersect transversally in homoclinic points. Recalling the discussion in Chapter III, the homoclinic points give rise to invariant hyperbolic sets for higher iterates of the map φ, near which the unstable and unpredictable orbit structure is described by means of topological Bernoulli systems. The elliptic fixed point is a cluster point of homoclinic points. The picture is repeated near the elliptic periodic orbits for

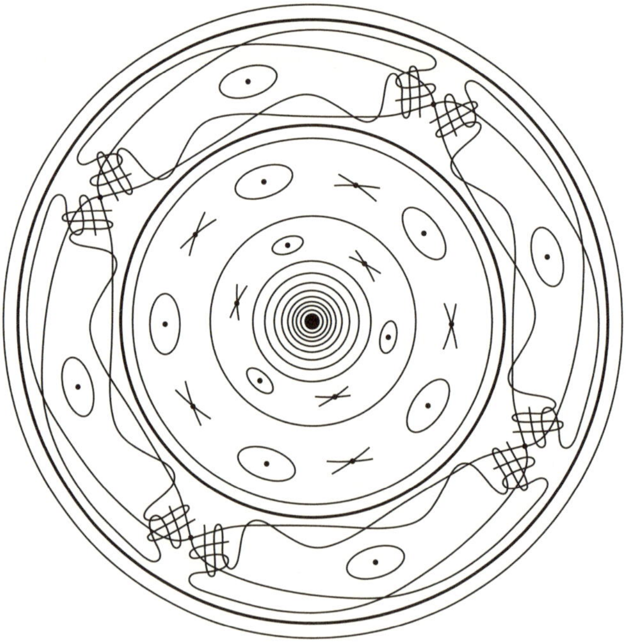

Figure V.4. The sketch of Arnold and Avez, after [8].

higher iterates of the map φ, and so on. The existence of the homoclinic orbits in the picture was established in 1973 by E. Zehnder in[120]. A modern version is due to C. Genecand [39]; further references are A. Chenciner [17] and J. Moser [73].

Literature. The monographs [2] by R. Abraham and J. Marsden, [45] by C. Godbillon, [66] by K. Meyer and G. Hall, [107] by C. L. Siegel and J. Moser and [73] by Moser are classics in Hamiltonian mechanics and present also some of the interesting history of the topics. Examples of integrable systems and applications to celestial mechanics are contained in the lecture notes [74] by J. Moser and E. Zehnder. An efficient introduction into the Hamiltonian formalism on symplectic manifolds is in Chapter 3 of [30] by J. J. Duistermaat. The monograph [65] by D. McDuff and D. Salamon goes beyond the formalism and introduces symplectic topology. Introductions into the KAM theory containing the references to the old masters are the lectures [61] by R. de la Llave and [86] by J. Pöschel and the paper [97] by D. Salamon. The KAM theory in the configuration space is treated in [98] by D. Salamon and E. Zehnder.

Chapter VI

Questions, phenomena, results

So far we have introduced the concepts of symplectic manifolds and symplectic diffeomorphisms and have presented the Hamiltonian formalism which was used and taught in physics and mathematics over centuries. We now turn to more recent developments which are at the beginning of the new field of symplectic topology. In the present, very short, Chapter VI we shall raise some questions and present some phenomena and results, in order to motivate the detailed construction of symplectic invariants later on in Chapter VII.

VI.1 Geometric questions

We consider two compact domains K_1 and K_2 in $\mathbb{R}^m$ which possess smooth boundaries and which are diffeomorphic by an orientation preserving diffeomorphism $u\colon K_1 \to K_2$. We look for conditions on these two domains which allow us to conclude that there exists a diffeomorphism $\psi\colon K_1 \to K_2$ that is volume preserving and hence satisfies $\det d\psi(x) = 1$ for all $x \in K_1$, or equivalently,

$$\psi^*\Omega = \Omega \quad \text{where } \Omega = dx_1 \wedge \cdots \wedge dx_m$$

is the volume form on $\mathbb{R}^m$. In order to derive *necessary conditions* we assume that $\psi\colon K_1 \to K_2$ is a volume preserving diffeomorphism. Then it follows from the variable transformation formula for every integrable function $f\colon K_2 \to \mathbb{R}$ that

$$\int_{K_1} f(\psi(x))\, dx = \int_{K_1} f(\psi(x))\, |\det d\psi(x)|\, dx = \int_{K_2=\psi(K_1)} f(x)\, dx.$$

If $f = \chi_{K_2}$ is the characteristic function of the domain K_2, then $f \circ \psi = \chi_{K_1}$ and we obtain

$$\operatorname{vol}(K_1) = \operatorname{vol}(K_2)$$

so that the domains must have the same volume. We see that the *total volume* is an *invariant* of volume preserving diffeomorphisms. It turns out that this condition is also *sufficient*, as the following statement shows.

Theorem VI.1 (Dacorogna–Moser, 1990). *If K_1 and K_2 are two compact domains in $\mathbb{R}^m$ which possess smooth boundaries, are diffeomorphic and have the same volume,* $\operatorname{vol}(K_1) = \operatorname{vol}(K_2)$, *then there exists a volume preserving diffeomorphism $\psi\colon K_1 \to K_2$, hence satisfying $\psi^*\Omega = \Omega$.*

For the proof of a more general theorem with precise regularity conditions we refer to Dacorogna and Moser [25]. That the total volume is the only invariant of volume preserving diffeomorphisms is also demonstrated by the following result for compact manifolds without boundaries.

Theorem VI.2 (Moser). *If M is an oriented, compact and connected manifold (without boundary) and if Ω_0 and Ω_1 are two volume forms on M having the same total volume such that*

$$\int_M \Omega_0 = \int_M \Omega_1,$$

then there exists a volume preserving diffeomorphism u of M such that

$$\Omega_0 = u^* \Omega_1.$$

In other words, the only invariant of volume forms is their total volume.

Proof. We proceed as in the proof of Darboux's theorem and deform the volume form Ω_0 into Ω_1 defining

$$\Omega_t = (1-t)\Omega_0 + t\Omega_1, \quad 0 \le t \le 1.$$

These forms are volume forms since locally Ω_0 and Ω_1 are represented by $\Omega_0(x) = a(x)dx_1 \wedge \cdots \wedge dx_m$ and $\Omega_1(x) = b(x)dx_1 \wedge \cdots \wedge dx_m$ with non-vanishing smooth functions a and b which, by the assumption on the total volume, must have the same sign. We shall construct a family φ^t of diffeomorphisms satisfying

$$(*) \quad (\varphi^t)^* \Omega_t = \Omega_0, \quad \varphi^0 = \mathrm{Id}$$

for $0 \le t \le 1$, so that the diffeomorphism $u = \varphi^1$ will solve our problem. Since M is compact, connected and oriented we conclude from $\int_M (\Omega_1 - \Omega_0) = 0$ that

$$\Omega_1 - \Omega_0 = d\gamma$$

for some $(m-1)$-form γ on M. This is a special case of the de Rham theorem. Since Ω_t are volume forms, we find a unique time dependent vector field X_t on M solving the linear equation

$$i_{X_t} \Omega_t = -\gamma$$

for $0 \le t \le 1$. Denote by φ^t the flow of the vector field X_t satisfying $\varphi^0 = \mathrm{Id}$. Since $d\alpha_t = 0$ for volume forms we find, again using Cartan's formula,

$$\begin{aligned} \frac{d}{dt}[(\varphi^t)^* \alpha_t] &= (\varphi^t)^* \Big[L_{X_t} \alpha_t + \frac{d}{dt} \alpha_t \Big] \\ &= (\varphi^t)^* [d(i_{X_t} \alpha_t) + \Omega_1 - \Omega_0] \end{aligned}$$

which vanishes, because $d(i_{X_t}\alpha_t) + \Omega_1 - \Omega_0 = d(i_{X_t}\alpha_t + \gamma) = 0$ by our choice of the vector field X_t. Therefore, $(*)$ holds true and the proof is finished. □

We can ask the same question for symplectic instead of volume preserving diffeomorphisms in the symplectic standard space $(\mathbb{R}^{2n}, \omega_0)$. If $n = 1$ we do not obtain anything new since here the concepts *symplectic and area preserving* are the same. Therefore, we assume $n \geq 2$ and consider two compact domains K_1 and K_2 of $\mathbb{R}^{2n}$ which possess smooth boundaries and are diffeomorphic. Under which conditions on K_1 and K_2 can we obtain a diffeomorphism $\psi : K_1 \to K_2$ satisfying

$$\psi^* \omega_0 = \omega_0,$$

or explicitly, solving the nonlinear partial differential equation

$$d\psi(x)^T J d\psi(x) = J$$

for $x \in K_1$? From $\det d\psi(x) = 1$ we conclude that $\mathrm{vol}(K_1) = \mathrm{vol}(K_2)$ is again a necessary condition. Is it also sufficient?

Turning to a seemingly simpler problem we take two open sets $U, V \subset \mathbb{R}^{2n}$ and look for conditions under which there exists a *symplectic embedding* $\varphi : U \to V$ that is a diffeomorphism $\varphi : U \to \varphi(U) \subset V$ satisfying $\varphi^* \omega_0 = \omega_0$ on U. Clearly,

$$\mathrm{vol}(U) = \mathrm{vol}(\varphi(U)) \leq \mathrm{vol}(V)$$

is a necessary condition. Are there other obstructions than the volume?

Let us first look at some examples and take the Euclidian open ball

$$B(R) = \{(x, y) \in \mathbb{R}^n \times \mathbb{R}^n \mid |x|^2 + |y|^2 < R^2\} \subset \mathbb{R}^{2n}$$

of radius R centered at the origin. Is it possible to embed this open ball symplectically into the *isotropic cylinder*

$$Z_{\mathrm{iso}}(r) = \{(x, y) \in \mathbb{R}^n \times \mathbb{R}^n \mid x_1^2 + x_2^2 < r^2\} \subset \mathbb{R}^{2n}$$

of a small radius $r > 0$?

We point out that the coordinates refer to the symplectic basis

$$\{e_1, \ldots, e_n, f_1 \ldots, f_n\}$$

which satisfies $\omega_0(e_i, f_j) = \delta_{ij}$, $\omega_0(e_i, e_j) = 0$ and $\omega_0(f_i, f_j) = 0$. The plane $E = \mathrm{span}\{e_1, e_2\}$ on which the cylinder stands satisfies $\omega_0|_E = 0$, so that the symplectic form ω_0 *does not induce a symplectic structure* on E.

For this isotropic cylinder the answer to our question is *positive*. Indeed, we simply take the linear map

$$\varphi(x, y) = (\varepsilon x, \varepsilon^{-1} y).$$

Then $\varphi(B(R)) \subset Z_{\mathrm{iso}}(r)$ for every radius r, if only the parameter $\varepsilon > 0$ is chosen sufficiently small. The map φ is symplectic, since it is of the matrix form

$$\begin{pmatrix} A & \\ & (A^T)^{-1} \end{pmatrix}$$

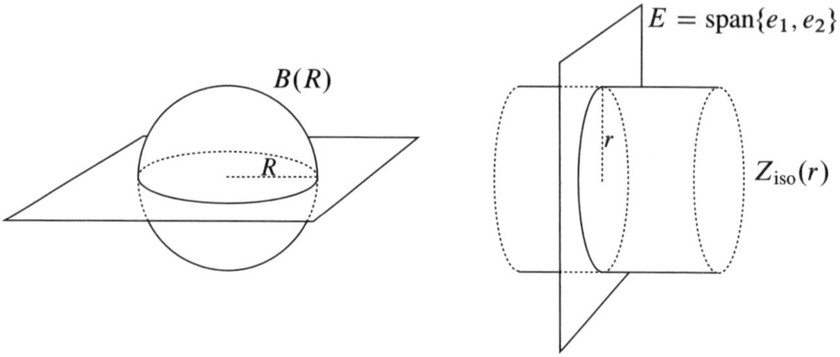

Figure VI.1. The ball $B(R)$ and the isotropic cylinder $Z_{\text{iso}}(r)$.

with respect to the splitting $\mathbb{R}^n \times \mathbb{R}^n$. We see that we can embed the open ball $B(R)$ symplectically into the open isotropic cylinder $Z_{\text{iso}}(r)$, regardless of the smallness of r.

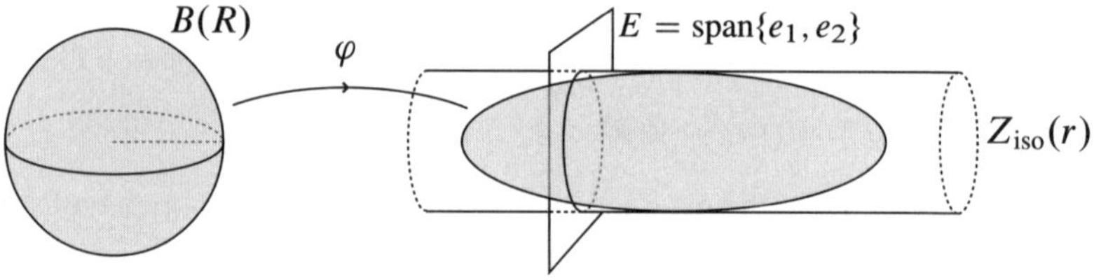

Figure VI.2. Symplectic embedding of a large ball $B(R)$ into a narrow isotropic cylinder $Z_{\text{iso}}(r)$.

We next take a completely different cylinder, namely the *symplectic cylinder*

$$Z(r) = \{(x, y) \in \mathbb{R}^n \times \mathbb{R}^n \mid x_1^2 + y_1^2 < r^2\}$$

of radius $r > 0$. This time the cylinder stands on the *symplectic* plane $E = \text{span}\{e_1, f_1\}$, where $\omega_0(e_1, f_1) = 1$, so that $\omega_0|_E$ is a symplectic form on E.

Analogous to the above example we are tempted to try with a map of the form

$$\psi(x, y) = (\varepsilon x_1, \varepsilon^{-1} x_2, x_3, \dots, x_n, \varepsilon y_1, \varepsilon^{-1} y_2, y_3, \dots, y_n).$$

Choosing ε sufficiently small, one achieves indeed that $\psi(B(R)) \subset Z(r)$. The map ψ is volume preserving, since $\det d\psi(x, y) = \varepsilon\varepsilon^{-1}\varepsilon\varepsilon^{-1} = 1$. However, ψ is *symplectic only if* $\varepsilon = 1$, in which case $\psi = \text{Id}$ and hence $\psi(B(R)) \subset Z(r)$ symplectically only if $r \geq R$.

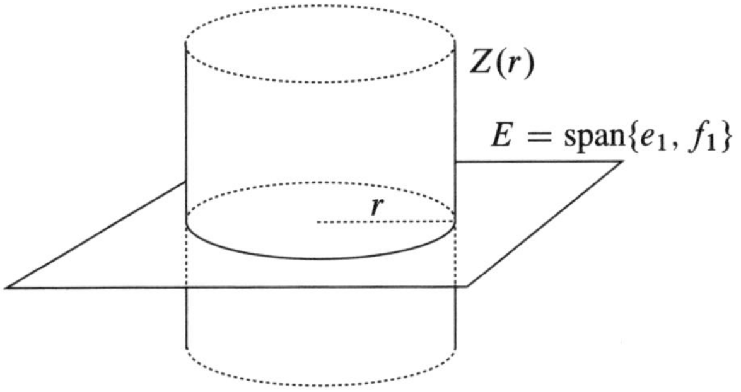

Figure VI.3. The symplectic cylinder $Z(r)$.

Surprisingly, it is not possible to find a symplectic diffeomorphism that does the job in a more clever way, because M. Gromov discovered in 1985 his seminal *non-squeezing theorem* which we shall prove in the next chapter. It says the following:

- *If* $\varphi\colon B(R) \to \mathbb{R}^{2n}$ *is a symplectic embedding satisfying* $\varphi(B(R)) \subset Z(r)$, *then* $r \geq R$.

Even though $\mathrm{vol}(B(R))$ is finite, while $\mathrm{vol}(Z(r)) = \infty$, the open ball can only be embedded symplectically into the symplectic cylinder $Z(r)$, if the radius of the cylinder is bigger than or equal to the radius of the ball. This holds true in particular for every *flow map* φ^t of a Hamiltonian differential equation, independent of the choice of the Hamiltonian function. We see, in particular, that symplectic maps constitute a much smaller class than volume preserving maps.

From the theorem we draw the *conclusion* that there must be other symplectic invariants apart from the volume! Some of them will be constructed in the next chapter.

VI.2 Approximation of measure preserving diffeomorphisms

Is it possible to approximate measure preserving diffeomorphisms of $\mathbb{R}^{2n}$ by symplectic maps of $(\mathbb{R}^{2n}, \omega_0)$ locally uniformly, that is, in the topology of C^0_{loc}?

We look at the sequence $\varphi_j : \mathbb{R}^{2n} \to \mathbb{R}^{2n}$ of symplectic maps for $j \geq 1$, so that

$$d\varphi_j(x)^T J d\varphi_j(x) = J, \quad x \in \mathbb{R}^{2n}.$$

If the sequence converges to the map φ in the C^1_{loc}-sense, then clearly

$$d\varphi(x)^T J d\varphi(x) = J, \quad x \in \mathbb{R}^{2n}$$

so that the limit map is again symplectic.

If the sequence converges merely in the C^0_{loc}-sense to the map φ, then the limit φ is a continuous map which is still measure preserving. Indeed, from $\det d\varphi_j(x) = 1$ for all j we deduce for every smooth function $f \in C_c^\infty(\mathbb{R}^{2n})$ having compact support that

$$\int_{\mathbb{R}^{2n}} f(\varphi_j(x))\, dx = \int_{\mathbb{R}^{2n}} f(x)\, dx.$$

Since $f(\varphi_j(x)) \to f(\varphi(x))$ for every $x \in \mathbb{R}^{2n}$, we obtain in view of the Lebesgue convergence theorem that

$$\int_{\mathbb{R}^{2n}} f(\varphi(x))\, dx = \int_{\mathbb{R}^{2n}} f(x)\, dx.$$

This holds true for all functions $f \in C_c^\infty(\mathbb{R}^{2n})$. Because $C_c^\infty \subset L^1$ is dense, the equation holds true for all integrable functions $f \in L^1$. Therefore, the limit map φ is *measure preserving*, as claimed.

However, it is a striking phenomenon that the limit map is even symplectic if it is assumed to be differentiable. Hence the symplectic nature survives under *topological limits* in view of the following theorem.

Theorem VI.3 (Eliashberg–Gromov–Ekeland–Hofer). *We consider a sequence φ_j of symplectic mappings in the symplectic standard space, so that $\varphi_j^*\omega_0 = \omega_0$, which converges locally uniformly to the map φ. If the limit map φ is differentiable in the point x_0, then*

$$d\varphi(x_0) \in \mathrm{Sp}(n).$$

Explicitly, $d\varphi(x_0)$ is a linear symplectic map. If, in particular, the limit map is differentiable, it is necessarily a symplectic map and hence satisfies $\varphi^*\omega_0 = \omega_0$.

It follows that the group of symplectic diffeomorphisms of a compact symplectic manifold is C^0-closed in the group of all diffeomorphisms of the manifold. The elegant proof by I. Ekeland and H. Hofer is based on the symplectic invariants which will be introduced in the next chapter and we refer to [52, S. 59].

VI.3 A dynamical question

We recall that the smooth function $H\colon \mathbb{R}^{2n} \to \mathbb{R}$ determines the Hamiltonian vector field X_H by the formula $\omega_0(X_H, \cdot) = -dH$. If the energy surface

$$S := \{x \in \mathbb{R}^{2n} \mid H(x) = 0\} \subset \mathbb{R}^{2n}$$

is *compact* and *regular*, so that $dH(x) \neq 0$ on S, then $X_H(x) \neq 0$ for all $x \in S$. In view of the energy conservation, the Hamiltonian vector field is tangent to the energy surface

$$X_H(x) \in T_xS, \quad x \in S$$

and its flow φ^t leaves the surface invariant,

$$\varphi^t(S) = S, \quad t \in \mathbb{R}.$$

In view of Liouville's theorem (Proposition V.31) there exists a finite measure μ on S, which is invariant under the flow maps φ^t. Therefore, in view of Poincaré's recurrence theorem (Theorem I.14) *almost every point is recurrent*. Consequently, there exists for almost every point $x \in S$ a sequence of times $t_j \to \infty$ satisfying

$$\lim_{j \to \infty} \varphi^{t_j}(x) \to x.$$

The question arises whether there exists a point $x \in S$ and a time $T > 0$ satisfying

$$\varphi^T(x) = x.$$

From the uniqueness of the solutions it then follows that

$$\varphi^{t+T}(x) = \varphi^t(x), \quad t \in \mathbb{R},$$

so that $\varphi^t(x)$ is a *periodic solution of the vector field X_H on the energy surface S.*

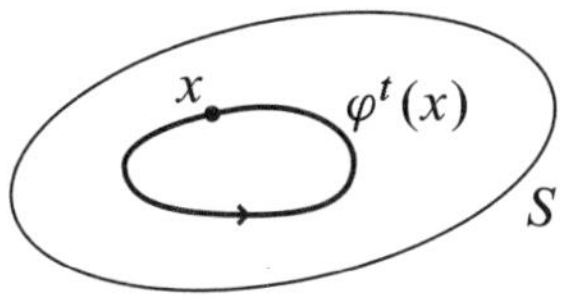

Figure VI.4. A periodic solution on the energy surface S.

Thickening the regular and compact hypersurface S by considering a whole family of hypersurfaces

$$S_\lambda = \{x \in \mathbb{R}^{2n} \mid H(x) = \lambda\}, \quad \lambda \in I,$$

parametrized by an open interval $I = (-\varepsilon, \varepsilon)$ around 0 for some $\varepsilon > 0$, we shall prove in Chapter VIII the following global existence theorem of periodic solutions on energy surfaces (Theorem VIII.4).

- *For almost every $\lambda \in I$ the energy surface S_λ carries a periodic solution of the Hamiltonian vector field X_H.*

We see that a compact regular energy surface S gives rise to an abundance of periodic solutions of X_H in the neighborhood of S! However, we point out that the given energy surface S does not necessarily carry a periodic solution, as examples show. The above existence statement will be deduced from a distinguished symplectic invariant which also explains the non-squeezing phenomenon of Gromov.

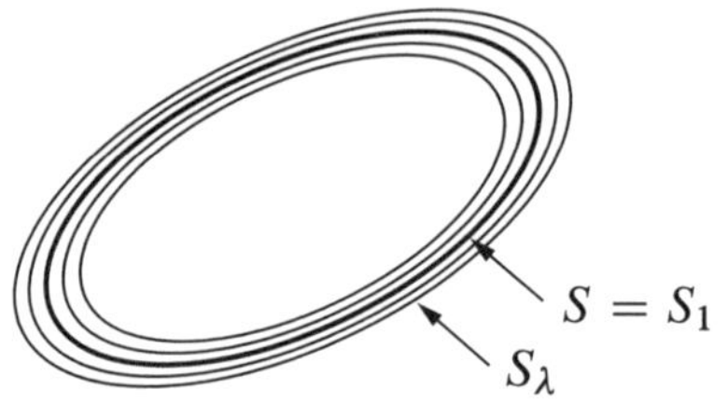

Figure VI.5. Thickening of the energy surface S.

It will be useful for the analysis later on to formulate the existence problem of periodic orbits on a given hypersurface in more geometrical terms. In order to do so we shall introduce the canonical line bundle of a hypersurface $S \subset \mathbb{R}^{2n}$. The hypersurface S is a submanifold of codimension 1, so that the dimension of the tangent space at $x \in S$ is equal to $\dim T_x S = 2n - 1$. Therefore, the restriction of the symplectic structure ω_0 onto $T_x S$ is necessarily degenerate. Because ω_0 is nondegenerate on the tangent space $T_x \mathbb{R}^{2n}$ we conclude

$$\dim\big(\ker(\omega_0|_{T_x S})\big) = 1, \quad x \in S$$

for the kernel of the restriction of the symplectic form ω_0 onto $T_x S$, defined by

$$\ker(\omega_0|_{T_x S}) = \{u \in T_x S \mid \omega_0(u, v) = 0 \text{ for all } v \in T_x S\} \subset T_x S.$$

This line in the tangent space $T_x S$ is denoted by $\mathcal{L}_S(x)$ and the *canonical line bundle* of the hypersurface S is the subbundle of the tangent bundle TS defined by

$$\mathcal{L}_S := \ker(\omega_0|_{TS}) := \{(x, u) \in TS \mid \omega_0(u, v) = 0 \text{ for all } v \in T_x S\}.$$

The *canonical line bundle* of the hypersurface S gives at every point of S the direction of *every* Hamiltonian vector field X_H which has S as its regular energy

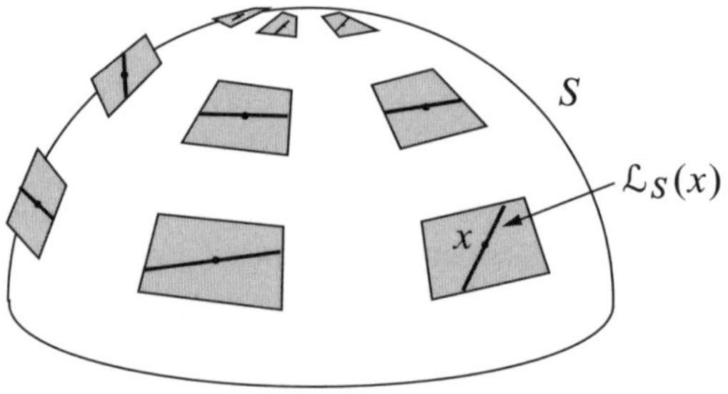

Figure VI.6. Illustration of the line bundle $\mathcal{L}_S$.

surface. Indeed, if

$$S = \{x \in \mathbb{R}^{2n} \mid H(x) = c\}, \quad dH(x) \neq 0, \quad x \in S,$$

it follows for $x \in S$ that

$$\omega_0(X_H(x), v) = -dH(x)v = 0 \quad \text{for every } v \in T_xS \equiv \ker dH(x)$$

and therefore $X_H(x) \in \ker(\omega_0|_{T_xS}) = \mathcal{L}_S(x)$.

As a consequence, the *orbits* of a Hamiltonian vector field X_H on a regular energy surface $S \subset \mathbb{R}^{2n}$ depend only on the hypersurface S and on the symplectic structure ω_0 (except for the parametrization of the individual solutions) and do *not* depend on the choice of the Hamilton function H which represents the hypersurface S as

$$S = \{x \in \mathbb{R}^{2n} \mid H(x) = c\}, \quad dH(x) \neq 0 \quad \text{if } x \in S$$

for a constant c.

This observation will allow us later on to *choose* a Hamiltonian function which is well suited for the analysis. The orbits of the vector field X_H on S, that is the images of the solution curves, lie on the characteristics of the line bundle $\mathcal{L}_S$. A *characteristic* of $\mathcal{L}_S$ is a differentiable curve $\gamma\colon I \subset \mathbb{R} \to S$ satisfying

$$0 \neq \dot{\gamma}(t) \in \mathcal{L}_S(\gamma(t)), \quad t \in I.$$

The existence problem of periodic solutions can now be formulated purely geometrically as follows. Does a compact smooth hypersurface $S \subset (\mathbb{R}^{2n}, \omega_0)$ possess a *closed characteristic* of the canonical line bundle $\mathcal{L}_S$?

VI.4 A connection between geometry and Hamiltonian dynamics

If two compact domains with smooth boundaries in the symplectic standard space are diffeomorphic by a symplectic diffeomorphism of $\mathbb{R}^{2n}$, then the canonical line bundles of their boundaries are isomorphic. Hence the actions of possible closed characteristics on the boundaries are numerical invariants, as we shall briefly explain next. In order to define the action of a closed curve in $\mathbb{R}^{2n}$ we introduce the 1-form

$$\lambda = \frac{1}{2}\sum_{j=1}^{n}(y_j\,dx_j - x_j\,dy_j)$$

on $\mathbb{R}^{2n}$. Then

$$d\lambda = \sum_{j=1}^{n} dy_j \wedge dx_j = \omega_0$$

is our canonical symplectic form on $\mathbb{R}^{2n}$. If $u\colon \mathbb{R}^{2n} \to \mathbb{R}^{2n}$ is a symplectic map of $\mathbb{R}^{2n}$, then the 1-form $\lambda - u^*\lambda$ is *closed*, because

$$d(\lambda - u^*\lambda) = d\lambda - u^* d\lambda = \omega_0 - u^*\omega_0 = 0.$$

Thus, in view of the Poincaré lemma there exists a smooth function $F\colon \mathbb{R}^{2n} \to \mathbb{R}$ satisfying

$$\lambda - u^*\lambda = dF.$$

Consequently, if γ is a closed curve in $\mathbb{R}^{2n}$, then

$$\int_\gamma \lambda - \int_\gamma u^*\lambda = \int_\gamma dF = 0$$

and therefore

$$\int_\gamma \lambda = \int_\gamma u^*\lambda = \int_{u(\gamma)} \lambda.$$

The *action of a closed curve* γ in $(\mathbb{R}^{2n}, \omega_0)$ is defined as the real number

$$A(\gamma) := \int_\gamma \lambda.$$

We claim that a differentiable map $u\colon \mathbb{R}^{2n} \to \mathbb{R}^{2n}$ is symplectic if and only if it leaves all actions invariant, that is,

$$A(u(\gamma)) = A(\gamma)$$

for all closed curves γ of $\mathbb{R}^{2n}$. Indeed, if u is symplectic, then

$$A(\gamma) - A(u(\gamma)) = \int_\gamma \lambda - \int_{u(\gamma)} \lambda = 0,$$

as we have already seen. Conversely, if $A(\gamma) = A(u(\gamma))$, then $\int_\gamma (\lambda - u^*\lambda) = 0$ for all closed curves γ. Therefore, the 1-form $\lambda - u^*\lambda$ is exact and hence closed and we conclude that $\omega_0 - u^*\omega_0 = d\lambda - d(u^*\lambda) = 0$, as claimed. If $\gamma\colon [0,1] \to \mathbb{R}^{2n}$ is a closed curve so that $\gamma(0) = \gamma(1)$, its action is represented as

$$\begin{aligned} A(\gamma) &= \frac{1}{2}\int_0^1 \langle -J\dot\gamma, \gamma\rangle\, dt \\ &= \frac{1}{2}\int_0^1 \omega_0(\gamma, \dot\gamma)\, dt. \end{aligned}$$

We return to the existence question of symplectic diffeomorphisms $u\colon K_1 \to K_2$ between two compact domains $K_1, K_2 \subset \mathbb{R}^{2n}$ having smooth boundaries ∂K_1

and ∂K_2. These boundaries are *compact hypersurfaces*. If u is a symplectic diffeomorphism of $\mathbb{R}^{2n}$, then its linearized

$$du\colon \mathcal{L}_{\partial K_1} \to \mathcal{L}_{\partial K_2}$$

defines an isomorphism of the line bundles. Therefore u maps the characteristics of ∂K_1 bijectively onto the characteristics of ∂K_2. If γ_1 is a closed characteristic in $\mathcal{L}_{\partial K_1}$, then

$$A(u(\gamma_1)) = A(\gamma_1).$$

Consequently, there exist *symplectic invariants* on the boundaries. We have numerical conditions the domains K_1 and K_2 must satisfy, in order to be symplectically diffeomorphic. The Hamiltonian flows induced on the boundaries ∂K_1 and ∂K_2 must be equivalent up to reparametrization of the orbits. Therefore, the complete Hamiltonian system on the boundary ∂K_1 is a symplectic invariant! This is why we started with embeddings of *open* domains.

Finally, we present some examples of characteristics which will be useful later on.

Examples. (1) *Surface of a sphere.* Let $\partial B(R) = \{z \in \mathbb{R}^{2n} \mid |x|^2 + |y|^2 = R^2\}$ be the sphere of radius $R > 0$ where $z = (x, y) \in \mathbb{R}^n \times \mathbb{R}^n$. The Hamiltonian systems defined by the function

$$H(z) = \frac{1}{2}(|x|^2 + |y|^2),$$

has the boundary $\partial B(R) = \{z \mid H(z) = R^2/2\}$ as a regular and compact energy surface. The Hamiltonian equation is *linear* and given by

$$\dot{z} = J\nabla H(z) = Jz,$$

and the Hamiltonian flow can be written explicitly as

$$\varphi^t(z) = e^{tJ}z = (\cos t)z + (\sin t)Jz.$$

All the solutions are periodic of the same period 2π. The characteristics γ on the sphere $\partial B(R)$ are all closed and have the same actions equal to

$$A(\gamma) = \frac{1}{2}\int_0^{2\pi} \langle -J\dot{z}(t), z(t)\rangle\, dt = \int_0^{2\pi} H(z(t))\, dt = \pi R^2.$$

(2) *Symplectic cylinder.* If $Z(r) = \{z \in \mathbb{R}^{2n} \mid x_1^2 + y_1^2 < r^2\}$ is the symplectic cylinder we take the Hamiltonian function

$$H(z) = \frac{1}{2}(x_1^2 + y_1^2),$$

so that the boundary

$$\partial Z(r) = \{z \mid x_1^2 + y_1^2 = r^2\} = \{z \mid H(z) = r^2/2\}$$

is a regular energy surface of the Hamiltonian system. The Hamiltonian equation

$$\dot{z} = J\nabla H(z) = J(x_1, 0, \dots, 0, y_1, 0, \dots, 0)$$

is linear and looks explicitly as follows:

$$\begin{aligned} \dot{x}_1 &= \ \ y_1, \\ \dot{y}_1 &= -x_1, \\ \dot{x}_i &= \ \ 0, \quad 2 \le i \le n, \\ \dot{y}_i &= \ \ 0, \quad 2 \le i \le n. \end{aligned}$$

We see that all the solutions on $\partial Z(r)$ are again periodic, lie on circles in planes parallel to the plane $E = \operatorname{span}\{e_1, f_1\}$ and have the same action equal to

$$A(\gamma) = \pi r^2, \quad \gamma \subset \partial Z(r).$$

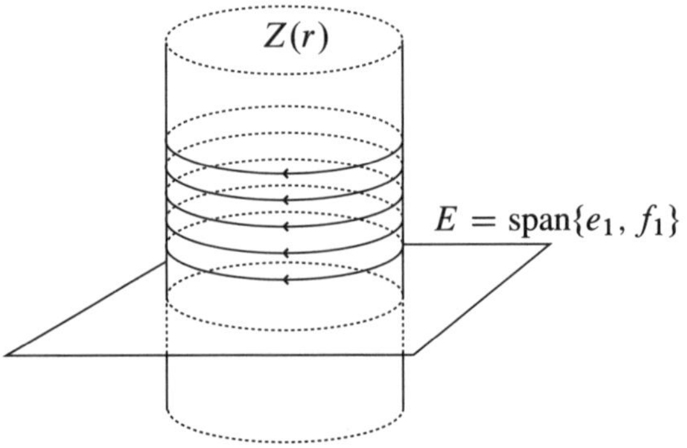

Figure VI.7. Some solutions on $\partial Z(r)$.

(3) *Isotropic cylinder.* If $Z_{\text{iso}}(r) = \{z \in \mathbb{R}^{2n} \mid x_1^2 + x_2^2 < r^2\}$ is the isotropic cylinder, we take the Hamiltonian function

$$H(z) = \frac{1}{2}(x_1^2 + x_2^2).$$

Then the boundary $\partial Z_{\text{iso}}(r) = \{z \mid x_1^2 + x_2^2 = r^2\} = \{z \mid H(z) = r^2/2\}$ of the cylinder is a regular energy surface. The Hamiltonian equation is again linear,

$$\dot{z} = J\nabla H(z) = J(x_1, x_2, 0, \dots, 0) = (0, \dots, 0, -x_1, -x_2, 0, \dots, 0),$$

and explicitly given by

$$\begin{aligned} \dot{y}_1 &= -x_1, \\ \dot{y}_2 &= -x_2, \\ \dot{x}_i &= 0, \qquad 1 \le i \le n, \\ \dot{y}_i &= 0, \qquad 3 \le i \le n. \end{aligned}$$

This time the solutions are straight *lines* orthogonal to the plane $E = \text{span}\{e_1, e_2\}$. There are no closed characteristics and no finite actions!

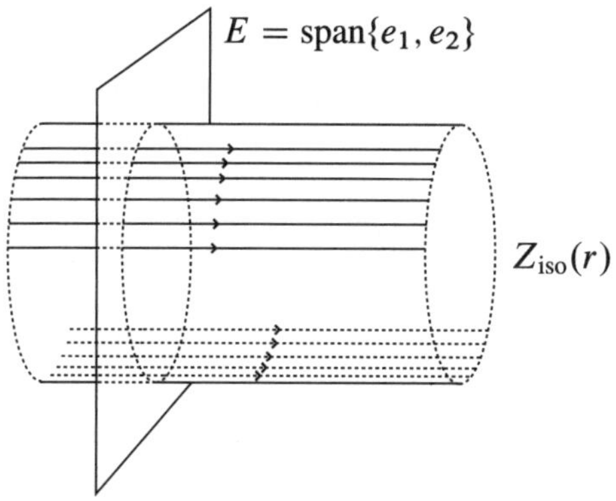

Figure VI.8. Some solutions on $\partial Z_{\text{iso}}(r)$.

In view of the examples above it is tempting to explain the non-squeezing phenomenon of Gromov from the dynamical point of view as follows. On the boundary $\partial B(R)$ of the open ball the characteristics are all closed and have actions equal to πR^2. Similarly, on the boundary $\partial Z_{\text{iso}}(r)$ of the symplectic cylinder the characteristics are closed and their actions are equal to πr^2. If the map φ is a symplectic embedding of the closed ball of radius R, then on the boundary of its image the characteristics are also closed and have actions equal to πR^2. Imagine now that there is an optimal embedding of the open ball $B(R)$ into the open symplectic cylinder, in which case one is inclined to think that the boundary of the embedded ball touches the boundary of the cylinder and that this is happening along a closed characteristic. We then conclude in view of the invariance of the actions that $r = R$. A proof of the non-squeezing theorem of Gromov along these lines does not exist and we shall prove it in the next chapter by means of a symplectic invariant called symplectic capacity.

Chapter VII
Symplectic invariants

This chapter is devoted to a special class of invariants of symplectic manifolds, introduced in the framework of subsets of the symplectic standard space $(\mathbb{R}^{2n}, \omega_0)$ by I. Ekeland and H. Hofer [33] 1989 and [34] 1990. They are called symplectic capacities. A symplectic capacity associates with every symplectic manifold a non-negative real number or infinity, so that three axioms are satisfied. In view of its monotonicity axiom a symplectic capacity represents, in particular, an obstruction to symplectic embeddings. The Gromov non-squeezing phenomenon is an immediate consequence, as we shall first demonstrate among other simple consequences. Then we shall construct a special symplectic capacity c_0, the so-called Hofer–Zehnder capacity which is of dynamical nature. It measures the minimal oscillation of Hamiltonian functions needed to conclude the existence of a fast periodic solution of the associated Hamiltonian vector field. In dimension 2, the capacity c_0 agrees with the total area. The construction of c_0 is based on a variational principle for the action functional of classical mechanics which is bounded neither below nor above. Critical points can be guaranteed by mini-max arguments. The tools from the calculus of variations will be developed in detail and from scratch. Applications of the dynamical capacity c_0 to Hamiltonian vector fields are postponed to Chapter VIII where it will be used to establish global periodic orbits on and near compact hypersurfaces of symplectic manifolds.

VII.1 Symplectic capacities and first applications

In the following we denote by $\mathrm{SM}(2n)$ the class of *all* symplectic manifolds (M, ω) of dimension $2n$. This includes compact and non-compact manifolds, manifolds with boundaries and manifolds without boundaries. Examples are the symplectic standard space $(\mathbb{R}^{2n}, \omega_0)$, the manifold (U, ω_0) where $U \subset \mathbb{R}^{2n}$ is an open subset, the manifold (D, ω_0) where $D \subset \mathbb{R}^{2n}$ is a domain having smooth boundaries, or the same subsets of $\mathbb{R}^{2n}$ equipped with a different symplectic structure defined by a closed 2-form

$$\omega(x) = \sum_{i,j=1}^{2n} a_{ij}(x) dx_i \wedge dx_j,$$

where a_{ij} are smooth functions and where $\omega(x)$ is for every x a nondegenerate skew symmetric bilinear form. In matrix notation,

$$\omega(x)(u, v) = \langle J(x)u, v\rangle,$$

with a skew symmetric matrix $J(x) = -J(x)^T$ satisfying $\det J(x) \neq 0$.

Definition. A *symplectic capacity* is a map

$$c\colon \mathrm{SM}(2n) \to [0, \infty], \quad (M, \omega) \mapsto c(M, \omega),$$

associating with a symplectic manifold of dimension $2n$ a non-negative real number or infinity, satisfying the following three axioms.

(A1) *Monotonicity.* If there exists a symplectic embedding $(M, \omega) \to (N, \tau)$ into a second symplectic manifold of dimension $2n$, then

$$c(M, \omega) \leq c(N, \tau).$$

(A2) *Conformality.* For every symplectic manifold (M, ω) and for every real number $\alpha \neq 0$ we have

$$c(M, \alpha\omega) = |\alpha|\, c(M, \omega).$$

(A3) *Normalization.* If $B(1) \subset \mathbb{R}^{2n}$ is the open unit ball centered at the origin and if $Z(1) \subset \mathbb{R}^{2n}$ is the symplectic open unit cylinder, then

$$c(B(1), \omega_0) = \pi = c(Z(1), \omega_0).$$

We recall that $B(R) = \{(x, y) \in \mathbb{R}^n \times \mathbb{R}^n \mid |x|^2 + |y|^2 < R^2\}$ and $Z(r) = \{(x, y) \in \mathbb{R}^n \times \mathbb{R}^n \mid x_1^2 + y_1^2 < r^2\}$ with respect to the symplectic coordinates $(x, y) = (x_1, \dots, x_n, y_1, \dots, y_n)$ of the symplectic standard space $(\mathbb{R}^{2n}, \omega_0)$. The map $\varphi\colon (M, \omega) \to (N, \tau)$ is a symplectic embedding, if $\varphi\colon M \to N$ is an injective immersion satisfying $\varphi^*\tau = \omega$, and if $\varphi\colon M \to \varphi(M) \subset N$ is a homeomorphism onto its image $\varphi(M)$.

The three axioms (A1)–(A3) do not determine a unique capacity function. The existence proof of a capacity function is neither trivial nor intuitive. We postpone it for the moment, and assuming that the existence of a capacity is guaranteed, we first illustrate the concept and deduce some simple consequences.

Proposition VII.1. *A symplectic capacity c is a symplectic invariant which differs from the volume if $n > 1$.*

Proof [Monotonicity]. If $\varphi\colon (M, \omega) \to (N, \tau)$ is a symplectic diffeomorphism, then $c(M, \omega) \leq c(N, \tau)$ in view of the monotonicity axiom (A1). Since the inverse diffeomorphism $\varphi^{-1}\colon N \to M$ is also symplectic, the converse inequality holds also true, so that $c(M, \omega) = c(N, \tau)$. From (A3) it follows that the volume is not a symplectic capacity if $n > 1$, because $\mathrm{vol}(Z(1)) = \infty$, while the capacity of $Z(1)$ is finite. □

In the special case $n = 1$, that is, $\dim M = 2$, the total volume is an example of a symplectic capacity c,

$$c(M, \omega) = \left| \int_M \omega \right|, \quad \dim M = 2.$$

If, for instance, $U \subset \mathbb{R}^2$ is an open set or a domain having smooth boundaries, then $c(U, \omega_0) = c(U, dx \wedge dy) = |\operatorname{area}(U)|$ is its symplectic capacity.

Lemma VII.2. *If c is a symplectic capacity, $U \subset (\mathbb{R}^{2n}, \omega_0)$ an open subset and $\lambda \neq 0$ a real number, then*

$$c(\lambda U, \omega_0) = \lambda^2 c(U, \omega_0).$$

Proof [Conformality]. This is a consequence of the conformality axiom. The diffeomorphism

$$\varphi \colon \lambda U \to U, \quad x \mapsto \lambda^{-1} x$$

satisfies $\varphi^*(\lambda^2 \omega_0) = \lambda^2 \varphi^* \omega_0 = \lambda^2 \lambda^{-2} \omega_0 = \omega_0$. Therefore

$$\varphi \colon (\lambda U, \omega_0) \to (U, \lambda^2 \omega_0)$$

is a *symplectic* diffeomorphism, so that

$$c(\lambda U, \omega_0) = c(U, \lambda^2 \omega_0) = \lambda^2 c(U, \omega_0),$$

by Proposition VII.1 and axiom (A2). □

Proposition VII.3. *Let c be symplectic capacity. In the symplectic standard space $(\mathbb{R}^{2n}, \omega_0)$ we have, for every $r > 0$,*

$$c(B(r)) = \pi r^2,$$
$$c(\overline{B(r)}) = \pi r^2,$$
$$c(Z(r)) = \pi r^2.$$

Proof. From Lemma VII.2 and (A3) we conclude

$$c(B(r)) = c(rB(1)) = r^2 c(B(1)) = \pi r^2.$$

Analogously, one verifies that $c(Z(r)) = \pi r^2$. If $\varepsilon > 0$, then $B(r) \subset \overline{B(r)} \subset B(r + \varepsilon)$ and, in view of the monotonicity axiom (A1),

$$\pi r^2 = c(B(r)) \leq c(\overline{B(r)}) \leq c(B(r + \varepsilon)) = \pi (r + \varepsilon)^2.$$

This holds true for every $\varepsilon > 0$, so that $c(\overline{B(r)}) = \pi r^2$, as claimed. □

Remark. *In* $\mathbb{R}^2$ *it therefore follows for every symplectic capacity* c *that*

$$c(B(r)) = c(\overline{B(r)}) = \text{area}(B(r)), \quad n = 1.$$

One can conclude that for every bounded connected domain $D \subset (\mathbb{R}^2, \omega_0)$ *with smooth boundary, the symplectic capacity agrees with the area,*

$$c(D) = \text{area}(D).$$

For a proof we refer to [52].

Proposition VII.4. *If* $\emptyset \neq U \subset (\mathbb{R}^{2n}, \omega_0)$ *is an open and bounded subset, then*

$$0 < c(U) < \infty$$

for every symplectic capacity c.

Proof. The set U contains a small open ball $B_{x_0}(r)$ around a point $x_0 \in U$ and is contained in a large open ball $B(R)$ centered at 0. Since the translation in $\mathbb{R}^{2n}$ is a symplectic diffeomorphism, we have $c(B_{x_0}(r)) = c(B(r))$ and conclude the estimates

$$0 < \pi r^2 \leq c(U) \leq \pi R^2 < \infty$$

from the inclusions $B_{x_0}(r) \subset U \subset B(R)$, using the monotonicity axiom. □

Examples. (1) If $U \subset (\mathbb{R}^{2n}, \omega_0)$ is an open set satisfying

$$B(r) \subset U \subset Z(r),$$

one concludes by the monotonicity and Proposition VII.3 for every symplectic capacity c, that $c(U) = \pi r^2$ – regardless of the volume, the geometrical properties and the topological properties of U!

(2) *Ellipsoid.* We consider the *ellipsoid* E in $(\mathbb{R}^{2n}, \omega_0)$ which is the set

$$E = \{z \mid q(z) < 1\} \subset \mathbb{R}^{2n}$$

with the quadratic function $q \colon \mathbb{R}^{2n} \to \mathbb{R}$ defined by

$$q(z) = \sum_{j=1}^{n} \frac{x_j^2 + y_j^2}{r_j^2}, \qquad z = (x_1, \dots, x_n, y_1, \dots, y_n) \in \mathbb{R}^{2n}$$

where the constants $r_j = r_j(E) > 0$ are ordered according to $0 < r_1(E) \leq \dots \leq r_n(E)$. Then $B(r_1) \subset E \subset Z(r_1)$ and using the monotonicity one concludes that the capacity of the ellipsoid is equal to $c(E, \omega_0) = \pi r_1(E)^2$.

In view of their monotonicity properties the symplectic invariants $c(M, \omega)$ represent in particular *obstructions to symplectic embeddings*. For example, an immediate consequence of the axioms is M. Gromov's non-squeezing theorem.

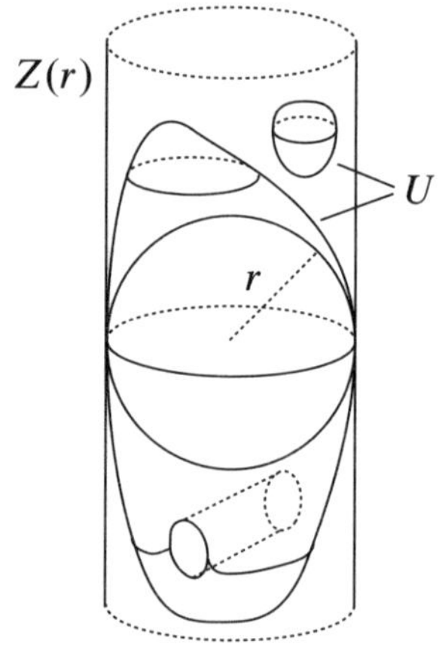

Figure VII.1. Illustration of example (1).

Theorem VII.5 (Gromov's non-squeezing theorem, 1985). *In the symplectic standard space* $(\mathbb{R}^{2n}, \omega_0)$ *there exists a symplectic embedding*

$$\varphi\colon B(R) \to Z(r)$$

if and only if $r \geq R$.

Proof [Monotonicity and normalization]. If $r \geq R$, then the inclusion map $B(R) \hookrightarrow Z(r)$ is a symplectic embedding. Conversely, if $\varphi\colon B(R) \to Z(r)$ is a symplectic embedding and if c is a symplectic capacity (whose existence will be proved later), then the monotonicity and Proposition VII.3 lead to $\pi R^2 = c(B(R)) \leq c(Z(r)) = \pi r^2$, so that $r \geq R$, as claimed in the theorem. □

In view of the monotonicity and the example above the following statement is obvious.

Theorem VII.6. *Let E and F be two ellipsoids in* $(\mathbb{R}^{2n}, \omega_0)$. *If there exists a symplectic embedding* $\varphi\colon E \to F$, *then* $r_1(E) \leq r_1(F)$.

Theorem VII.7. *In* $(\mathbb{R}^2, \omega_0) \times (\mathbb{R}^2, \omega_0)$ *there exists a symplectic diffeomorphism*

$$\varphi\colon B(r_1) \times B(r_2) \to B(s_1) \times B(s_2)$$

between products of open 2-discs satisfying $r_1 \leq r_2$ *and* $s_1 \leq s_2$, *if and only if* $r_1 = s_1$ *and* $r_2 = s_2$.

Proof. We denote by $B^m(r)$ the open ball in $\mathbb{R}^m$ centered at the origin. If φ is the symplectic diffeomorphism of the theorem, we consider the following composition of mappings:

$$B^4(r_1) \xrightarrow{i} B^2(r_1) \times B^2(r_2) \xrightarrow{\varphi} B^2(s_1) \times B^2(s_2) \xrightarrow{j} B^2(s_1) \times \mathbb{R}^2 = Z(s_1),$$

where i and j are the inclusion mappings. The composition $j \circ \varphi \circ i \colon B^4(r_1) \to Z(s_1)$ is a symplectic embedding so that $s_1 \geq r_1$ in view of Theorem VII.5. The same argument applied to the inverse map φ^{-1} leads to $r_1 \geq s_1$, so that $r_1 = s_1$. Because φ is volume preserving, it follows that $r_2 = s_2$, as claimed. □

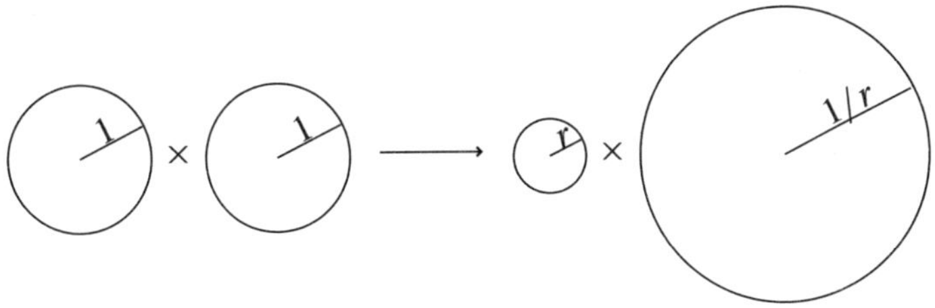

Figure VII.2. In view of Theorem VII.7 such a symplectic embedding is not possible.

The previous theorem shows clearly that volume preserving diffeomorphisms behave completely differently from symplectic diffeomorphisms. For every $r > 0$ there exists a (linear) volume preserving diffeomorphism

$$B(1) \times B(1) \to B(r) \times B(1/r).$$

The volume $\mathrm{vol}(B(r) \times B(1/r)) = \pi^2$ is independent of r. In contrast, the symplectic capacity function satisfies $c\big(B(r) \times B(1/r)\big) = \pi r^2 \to 0$ as $r \to 0$.

We have already seen above that every open ball $B(R)$ can be symplectically embedded into the *isotropic* cylinder $Z_{\mathrm{iso}}(r) = \{z \in \mathbb{R}^{2n} \mid x_1^2 + x_2^2 < r^2\} \subset (\mathbb{R}^{2n}, \omega_0)$. As a consequence its capacity is easily computed.

Proposition VII.8. *$c(Z_{\mathrm{iso}}(r), \omega_0) = \infty$ for every $r > 0$ and for every symplectic capacity c.*

Proof [Monotonicity]. Take the symplectic diffeomorphism $\varphi(x, y) = (\varepsilon x, \varepsilon^{-1} y)$ for $(x, y) \in \mathbb{R}^n \times \mathbb{R}^n$ where $\varepsilon > 0$. If $R > 0$, then for ε sufficiently small

$$\varphi \colon B(R) \to Z_{\mathrm{iso}}(r)$$

is a symplectic embedding and hence, in view of the monotonicity property of the capacity,

$$\pi R^2 = c(B(R)) \leq c(Z_{\mathrm{iso}}(r)).$$

This holds true for every $R > 0$ and so $c(Z_{\mathrm{iso}}(r)) = \infty$, as claimed. □

If one symplectic capacity has been found, it is easy to construct new ones. One example is the *Gromov width* $G(M, \omega)$ of a symplectic manifold which will be introduced next.

We consider the symplectic manifold (M, ω) of dimension $2n$. Given an open subset $U \subset M$ there exists in view of Darboux's theorem a symplectic embedding

$$\varphi\colon (B(r), \omega_0) \to (U, \omega),$$

of an open ball in $\mathbb{R}^{2n}$ having a sufficiently small radius $r > 0$. We now look for the largest open ball that can be symplectically embedded into the given manifold.

Definition. The *Gromov width* $G(M, \omega)$ is defined as

$$G(M, \omega) = \sup\{\pi r^2 \mid \text{there exists a symplectic embedding } (B(r), \omega_0) \to (M, \omega)\}.$$

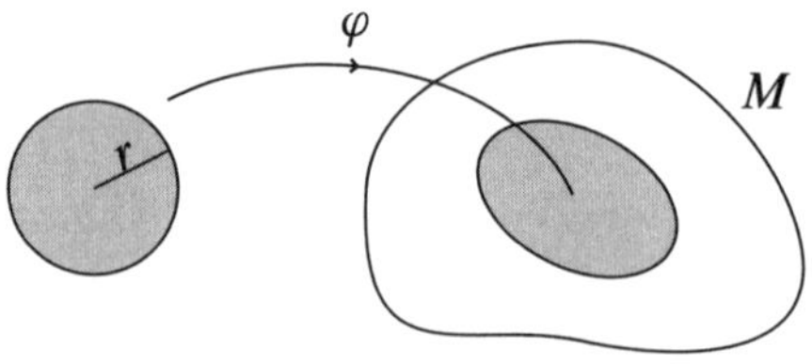

Figure VII.3. Embedding of the open ball $B(r)$, illustrating the definition of the Gromov width.

Theorem VII.9. *The Gromov width $G(M, \omega)$ is a symplectic capacity. It satisfies*

$$G(M, \omega) \leq c(M, \omega)$$

for every capacity c and every symplectic manifold (M, ω).

Proof [Gromov]. We shall verify that the function G satisfies the axioms (A1)–(A3) of a capacity function.

(A1) Let $\psi\colon (M, \omega) \to (N, \tau)$ be a symplectic embedding. For every symplectic embedding $\varphi\colon B(r) \to M$ into M the composition

$$\psi \circ \varphi\colon B(r) \to N$$

is a symplectic embedding into M. In view of the definition of the function G it therefore follows that $G(N, \tau) \geq G(M, \omega)$ and hence the axiom (A1) is verified.

(A2) Let $0 \neq \alpha \in \mathbb{R}$. We show that to every symplectic embedding

$$\varphi\colon (B(r), \omega_0) \to (M, \alpha\omega)$$

there belongs a symplectic embedding

$$\hat{\varphi}\colon \Big(B\Big(\frac{r}{\sqrt{|\alpha|}}\Big), \omega_0\Big) \to (M, \omega)$$

and conversely. Then the property $G(M, \alpha\omega) = |\alpha|\, G(M, \omega)$ of axiom (A2) follows.

If $\varphi\colon (B(r), \omega_0) \to (M, \alpha\omega)$ is a symplectic embedding, then $\varphi^*(\alpha\omega) = \alpha\varphi^*\omega = \omega_0$ and so

$$\varphi^*\omega = \frac{1}{\alpha}\omega_0.$$

Abbreviating $\delta = \frac{r}{\sqrt{|\alpha|}}$, the map

$$\psi\colon (B(\delta), \omega_0) \to \Big(B(r), \frac{1}{\alpha}\omega_0\Big), \quad \psi(x) := \sqrt{|\alpha|}x$$

satisfies

$$\psi^*\Big(\frac{1}{\alpha}\omega_0\Big) = \frac{1}{\alpha}\psi^*\omega_0 = \frac{\sqrt{|\alpha|}^2}{\alpha}\omega_0 = \operatorname{sign}(\alpha)\,\omega_0.$$

If $\operatorname{sign}(\alpha) = +1$, the composition

$$\hat{\varphi} = \varphi \circ \psi\colon (B(\delta), \omega_0) \to (M, \omega)$$

is a symplectic embedding, because

$$\hat{\varphi}^*\omega = \psi^* \circ \varphi^*\omega = \psi^*\Big(\frac{1}{\alpha}\omega_0\Big) = \operatorname{sign}(\alpha)\omega_0 = \omega_0.$$

If $\operatorname{sign}(\alpha) = -1$, we take the symplectic diffeomorphism

$$\psi_0\colon (B(\delta), \omega_0) \to (B(\delta), -\omega_0)$$

defined by $\psi_0(x, y) = (-x, y)$ for $(x, y) \in \mathbb{R}^n \times \mathbb{R}^n$, so that the composition

$$\hat{\varphi} = \varphi \circ \psi \circ \psi_0\colon (B(\delta), \omega_0) \to (M, \omega)$$

is a symplectic embedding as we wanted to show.

(A3) In order to confirm the normalization axiom, we first verify that $G(B(1), \omega_0) = \pi$. If $\varphi\colon B(r) \to B(1)$ is a symplectic embedding, it follows from the volume preservation that $r \le 1$ and hence $G(B(1)) \le \pi$. Since the inclusion map $B(1) \hookrightarrow B(1)$ is a symplectic embedding, $G(B(1)) = \pi$, as we wanted to show. It remains to prove that $G(Z(1), \omega_0) = \pi$. If $\varphi\colon B(r) \to Z(1)$ is a symplectic embedding, then, in view of the non-squeezing theorem, $r \le 1$ (here we have made use of the existence of a capacity!) and hence $G(Z(1), \omega_0) \le \pi$. The equality follows from the inclusion mapping. So far we have verified that the

Gromov width G is a symplectic capacity function and it remains to prove that it is the smallest capacity.

Take any symplectic capacity function c and let $\varphi\colon B(r) \to M$ be a symplectic embedding. Then, by the monotonicity property and Proposition VII.3,

$$\pi r^2 = c(B(r), \omega_0) \leq c(M, \omega).$$

This holds true for all the symplectic embeddings $\varphi\colon B(r) \to M$ and hence also for the supremum. Consequently, $G(M, \omega) \leq c(M, \omega)$ and Theorem VII.9 is proved. □

In the following sections we shall establish the existence of a capacity function. We shall construct a very distinguished symplectic capacity c_0 related to the dynamics of Hamiltonian systems and introduced in 1990 by H. Hofer and E. Zehnder [51]. This dynamical capacity measures the minimal oscillation of special Hamiltonian functions needed to conclude the existence of a distinguished fast periodic solution of the associated Hamiltonian vector field. In dimension 2 it agrees with the total area. The construction is based on a variational principle for the classical action functional of mechanics for periodic solutions of Hamiltonian equations.

VII.2 The Hofer–Zehnder capacity c_0

We recall that if (M, ω) is a symplectic manifold, then the smooth function $H: M \to \mathbb{R}$ determines the Hamiltonian vector field X_H on M by the requirement that $\omega(X_H(x), v) = -dH(x)v$ for all tangent vectors $v \in T_x M$. A *periodic solution* $x(t)$ of X_H having period T is a solution of the equation

$$\dot{x}(t) = X_H(x(t)), \quad t \in \mathbb{R},$$

satisfying $x(0) = x(T)$. Consequently, $x(t + T) = x(t)$ for all $t \in \mathbb{R}$. We next introduce a distinguished class of Hamiltonian functions.

Definition. If (M, ω) is a symplectic manifold we denote by $\mathcal{H}(M, \omega)$ the set of smooth functions $H: M \to \mathbb{R}$ having the following properties.

There exist an open set $U \subset M$ and a compact set $K \subset M$ (these sets depend on H) satisfying $U \subset K \subset M \setminus \partial M$ and

(i) $H|_U \equiv 0$ is *constant*,

(ii) $H|_{M \setminus K} \equiv m(H)$ is *constant*,

(iii) $0 \leq H(x) \leq m(H)$ for all $x \in M$.

The constant $m(H) = \max H - \min H$ is the *oscillation* of the function H. A function $H \in \mathcal{H}(M, \omega)$ is called *admissible*, if the associated Hamiltonian equation $\dot{x} = X_H(x)$ does not admit a non-constant periodic solution of period $0 < T \leq 1$, so that a *periodic* solution is either

(i) constant, $x(t) = x(0)$ for all $t \in \mathbb{R}$, or has

(ii) minimal period $T > 1$.

We denote by $\mathcal{H}_a(M, \omega)$ the set of admissible functions in $\mathcal{H}(M, \omega)$.

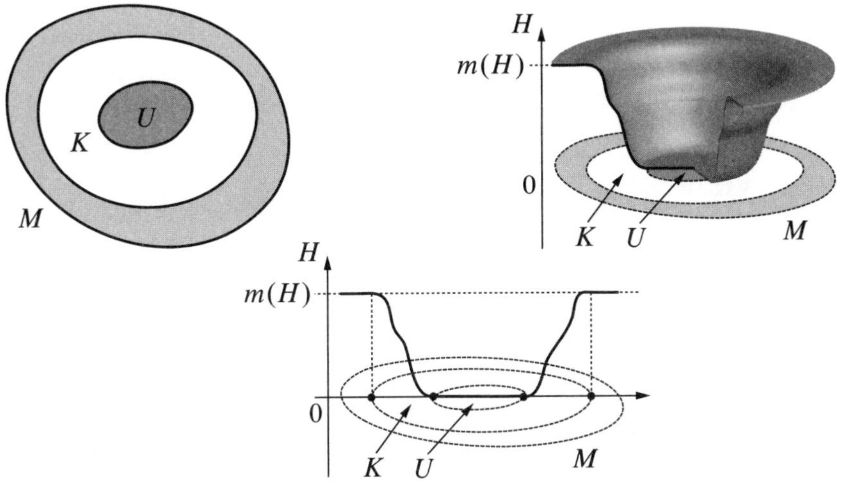

Figure VII.4. A function $H \in \mathcal{H}(M, \omega)$ with oscillation $m(H)$.

If $H \in \mathcal{H}(M, \omega) \setminus \mathcal{H}_a(M, \omega) \neq \emptyset$, then the Hamiltonian vector field $\dot{x} = X_H(x)$ possesses a non-constant periodic solution of period $0 < T \leq 1$. Since H vanishes on U and on $M \setminus K$, the non-constant periodic solution must be contained in $K \setminus U$.

If the manifold M is compact without boundary, one can choose $K = M$. In this case, $H \in \mathcal{H}(M, \omega)$ if the function H vanishes on an open set and satisfies $H \geq 0$.

Definition. The *symplectic capacity* c_0 of (M, ω) is defined as

$$c_0(M, \omega) = \sup\{m(H) \mid H \in \mathcal{H}_a(M, \omega)\}.$$

In the literature, this capacity c_0 is called Hofer–Zehnder capacity. We shall show that c_0 is, indeed, a symplectic capacity function.

The capacity function c_0 has the following *significance.* If $c_0(M, \omega) < \infty$, then for every function $H \in \mathcal{H}(M, \omega)$ satisfying $m(H) > c_0(M, \omega)$, the Hamiltonian vector field X_H on M possesses at least one non-constant periodic solution of period $0 < T \leq 1$. In fact, $c_0(M, \omega)$ is the infimum over all numbers having this property. The proof of the next theorem will take up the rest of the chapter.

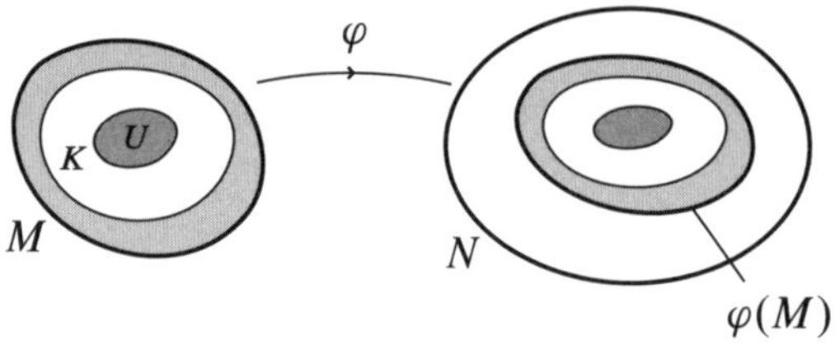

Figure VII.5. Illustrating the verification of the monotonicity axiom.

Theorem VII.10 (Hofer–Zehnder). *The map* $c_0 \colon (M, \omega) \mapsto c_0(M, \omega)$ *is a symplectic capacity.*

We shall illustrate the theorem with the open ball $B(r)$. From $c_0(B(r), \omega_0) = \pi r^2$ it follows for every function $H \in \mathcal{H}(B(r), \omega_0)$ satisfying $m(H) > \pi r^2$ that the Hamiltonian vector field X_H has a periodic solution of period $0 < T \leq 1$. If $r = 1$ we obtain, by way of the capacity c_0, a dynamical definition of the number π in terms of periodic solutions of Hamiltonian systems.

We point out that in dimension 2 the special capacity c_0 agrees with the area, so that $c_0(U, \omega_0) = \text{ area } U$, if $U \subset \mathbb{R}^2$ and

$$c_0(M, \omega) = \left| \int_M \omega \right|, \quad \dim M = 2.$$

This result is due to K. F. Siburg [104] and M. Y. Jiang [54], it can also be found in the book [52].

In order to prove Theorem VII.10, we have to verify that the function c_0 meets the axioms (A1)–(A3). It turns out that the monotonicity and the conformality of c_0 easily follow from the definition. The difficulty is the verification of the normalization axiom (A3), that is, the computation of the values of c_0 for the cylinder $Z(1)$ and the ball $B(1)$ equipped with the canonical symplectic structure. We begin with the monotonicity.

Lemma VII.11. *The map* c_0 *satisfies the monotonicity axiom* (A1).

Proof. If $\varphi \colon (M, \omega) \to (N, \tau)$ is a symplectic embedding, we have to prove that $c_0(M, \omega) \leq c_0(N, \tau)$. In view of the definition of c_0 it suffices to verify that for every function $H \in \mathcal{H}_a(M, \omega)$ on M, there exists a function $H' \in \mathcal{H}_a(N, \tau)$ on N having the same oscillation, that is, $m(H') = m(H)$.

We define the map $\varphi_* \colon \mathcal{H}(M, \omega) \to \mathcal{H}(N, \tau)$ by

$$\varphi_*(H)(x) := \begin{cases} H \circ \varphi^{-1}(x), & x \in \varphi(M), \\ m(H), & x \notin \varphi(M). \end{cases}$$

Since $K \subset M \setminus \partial M$ is compact, also its image $\varphi(K) \subset N \setminus \partial N$ is compact. Moreover, the set $\varphi(M \setminus \partial M)$ is open in N and the sets $\varphi(K)$ and $N \setminus \varphi(M \setminus \partial M)$ can be separated in N by open sets, so that the function $\varphi_* H$ is a smooth function on N. In addition, $\varphi(U) \subset N$ is open and we see that $\varphi_* H \in \mathcal{H}(N, \tau)$. By construction, $m(\varphi_*(H)) = m(H)$. It only remains to show that $\varphi_*(H)$ is admissible, if H is admissible. Since φ is a symplectic map,

$$\varphi^* X_{\varphi_*(H)} = X_{\varphi_*(H) \circ \varphi} = X_H,$$

in view of Proposition V.26. Consequently, the flows are conjugated in the sense that

$$\varphi \circ \varphi_H^t \circ \varphi^{-1} = \varphi_{\varphi_*(H)}^t.$$

In particular, all the corresponding periodic solutions have the same periods. □

Lemma VII.12. *The map c_0 satisfies the conformality axiom* (A2).

Proof. If $\alpha \neq 0$, we have to show that $c_0(M, \alpha\omega) = |\alpha|\, c_0(M, \omega)$. For this purpose we define the bijection $\psi \colon \mathcal{H}(M, \omega) \to \mathcal{H}(M, \alpha\omega)$ by

$$H \mapsto |\alpha| \cdot H =: H_\alpha.$$

Clearly,

$$m(H_\alpha) = |\alpha|\, m(H),$$

so that the lemma follows if we show that ψ also defines a bijection

$$\psi \colon \mathcal{H}_a(M, \omega) \to \mathcal{H}_a(M, \alpha\omega)$$

between the admissible functions. By the definition of the Hamiltonian vector fields,

$$(\alpha\omega)(X_{H_\alpha}, \cdot) = -dH_\alpha = -|\alpha|\, dH = |\alpha|\, \omega(X_H, \cdot)$$

so that

$$\omega\Big(\frac{\alpha}{|\alpha|} X_{H_\alpha}, \cdot\Big) = \omega(X_H, \cdot).$$

Due to the nondegeneracy of ω we obtain $\frac{\alpha}{|\alpha|} X_{H_\alpha} = X_H$ on M and so

$$X_{H_\alpha} = \pm X_H.$$

In both cases the vector fields have the same periodic orbits and the corresponding periods are the same. This completes the proof of Lemma VII.12. □

The next lemma is the first step towards the verification of axiom (A3).

Lemma VII.13. $c_0(B(1), \omega_0) \geq \pi$.

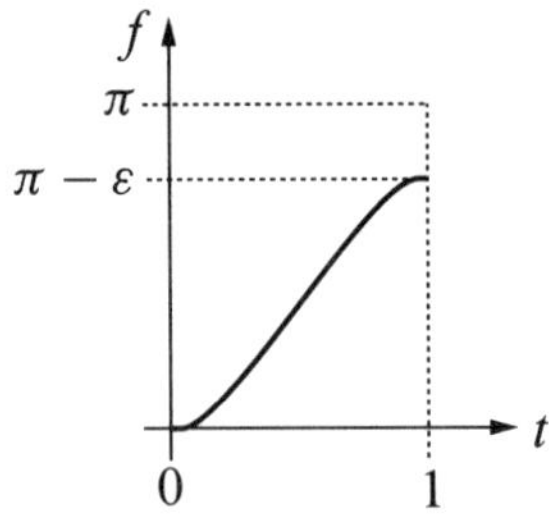

Figure VII.6. The function $f\colon [0,1] \to \mathbb{R}$.

Proof. We show that for every $0 < \varepsilon < \pi$ there exists an admissible function $H \in \mathcal{H}_a$ whose oscillation is equal to $m(H) = \pi - \varepsilon$ which then proves the lemma. For this purpose we first choose a C^∞-function $f\colon [0,1] \to \mathbb{R}$ having the following properties, illustrated by Figure VII.6,

(i) $f(t) = 0$ for t near 0,
(ii) $f(t) = \pi - \varepsilon$ for t near 1,
(iii) $0 \le f(t) \le \pi - \varepsilon$ for all t,
(iv) $0 \le f'(t) < \pi$ for all t.

The Hamiltonian function $H\colon B(1) \to \mathbb{R}$ is defined by

$$H(x) = f(|x|^2).$$

From the properties of f we conclude that $H \in \mathcal{H}(B(1), \omega_0)$ and

$$m(H) = \pi - \varepsilon,$$

so that the proof is complete, if we show that H belongs to the set $\mathcal{H}_a(B(1), \omega_0)$ of admissible functions. The associated Hamiltonian vector field X_H, explicitly given by

$$\dot{x} = J\nabla H(x) = 2f'(|x|^2)\, Jx,$$

has the function $G(x) = \frac{1}{2}|x|^2$ as an *integral*, because

$$dG(x)\, X_H(x) = \langle \nabla G(x), X_H(x)\rangle = 2f'(|x|^2)\, \langle x, Jx\rangle = 0.$$

Consequently, for every solution $x(t)$ the value of the function $2G(x(t)) = |x(t)|^2$ and hence also the value of the function $2f'(|x(t)|^2) =: a(x(0))$ is constant in time t along the solution. Therefore, every solution $x(t)$ solves the linear equation

$$\dot{x} = aJx,$$

for a real number $a \geq 0$, and so is equal to $x(t) = e^{atJ}x(0) = (\cos at)x(0) + (\sin at)Jx(0)$. We see that all the solutions are periodic! In the case $a = 0$, the solution is *constant* $= x(0)$, in the case $a > 0$ the solution has the minimal period $T = 2\pi/a$. From $0 \leq f' < \pi$, it follows that $a < 2\pi$ and hence $T > 1$. We have verified that the function H does, indeed, belong to the set $\mathcal{H}_a$ of admissible functions. □

We have demonstrated the estimate $c_0(B(1), \omega_0) \geq \pi$ by means of a single example. Since the inclusion map $B(1) \hookrightarrow Z(1)$ is a symplectic embedding, we obtain, in view of Lemma VII.11, the following estimate.

Corollary VII.14. $\pi \leq c_0(B(1)) \leq c_0(Z(1))$.

In order to verify the normalization axiom (A3), it remains to prove the estimate $c_0(Z(1)) \leq \pi$. This requires an *existence proof*, namely, the proof of the following theorem. If proved, Theorem VII.10 is also proved and c_0 is a symplectic capacity, as claimed.

Theorem VII.15. *If $H \in \mathcal{H}(Z(1))$ satisfies $m(H) > \pi$, then the Hamiltonian vector field X_H on $Z(1)$ possesses a non-constant periodic solution $x(t) \in Z(1)$ of period $T = 1$.*

It follows that $c_0(Z(1)) \leq \pi$, so that Theorem VII.10 is proved.

It remains to find a global periodic solution of *period* $T = 1$, for a Hamiltonian vector field X_H defined on the open cylinder $Z(1)$ where the Hamiltonian function H belongs to the set $\mathcal{H}(Z(1), \omega_0)$ and satisfies $m(H) > \pi$.

Our existence proof will be based on a variational principle for the action functional of classical mechanics and we start with some preparations.

Since the Hamiltonian function $H\colon Z(1) \to \mathbb{R}$ belongs to $\mathcal{H}(Z(1), \omega_0)$ it satisfies $H \equiv 0$ on an open set U and $H \equiv m(H)$ is constant outside of a compact set K that contains U. We have already proved that c_0 is a symplectic invariant, because the monotonicity axiom (A1) holds true for c_0. Therefore, we may assume that U is an open neighborhood of $0 \in \mathbb{R}^{2n}$ by means of a symplectic diffeomorphism of $\mathbb{R}^{2n}$ having compact support in the cylinder $Z(1)$.

We prefer to work with a Hamiltonian function defined on the whole space $\mathbb{R}^{2n}$ and not merely on the cylinder $Z(1)$ and we therefore *extend* the function H to a smooth function $\bar{H}\colon \mathbb{R}^{2n} \to \mathbb{R}$. In principle, this is not a problem, we could, for instance, extend H by the constant $m(H)$. For technical reasons, this is not good enough and we shall choose a more clever extension which, far away from the origin, looks like a specific quadratic form.

Since the set $K \subset Z(1)$ is compact, there exists an ellipsoid $E = E_N \subset Z(1)$ containing the compact set $K \subset E$, so that the Hamiltonian function H also belongs to the set $\mathcal{H}(E, \omega_0)$. Indeed, we choose

$$E = \{z \in \mathbb{R}^{2n} \mid q(z) < 1\}$$

for $z = (x, y) \in \mathbb{R}^n \times \mathbb{R}^n$, with the quadratic function q defined by

$$q(z) = x_1^2 + y_1^2 + \frac{1}{N^2} \sum_{j=2}^{n} (x_j^2 + y_j^2)$$

and choose the integer N sufficiently large.

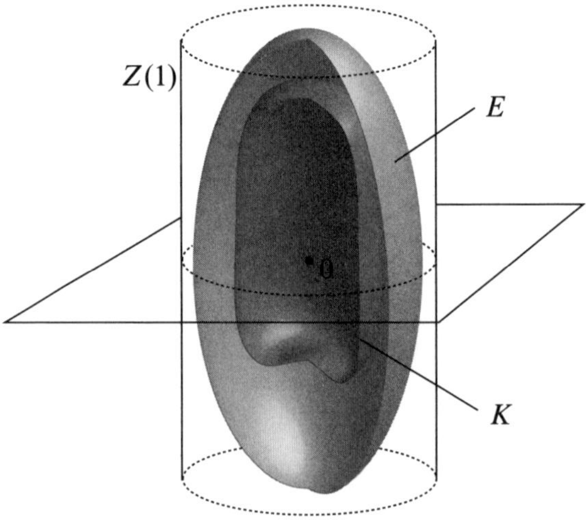

Figure VII.7. The ellipsoid $E = E_N$ around the compact set $K \subset Z(1)$.

We have to keep in mind that q is *homogeneous* of degree 2, so that

$$q(\lambda z) = \lambda^2 q(z)$$

for all $\lambda \in \mathbb{R}$ and $z \in \mathbb{R}^{2n}$. Differentiating this equation in the variable λ at the point $\lambda = 1$ we obtain the identity

$$\langle \nabla q(z), z \rangle = 2q(z), \quad z \in \mathbb{R}^{2n}.$$

By assumption, the function H possesses the oscillation $m(H) > \pi$ and we choose a small constant $\varepsilon > 0$ such that

$$m(H) > \pi + \varepsilon.$$

Next, we choose a C^∞-function $f : \mathbb{R} \to \mathbb{R}$, illustrated in Figure VII.8, having the following properties:

(i) $f(s) = m(H)$ for $s \leq 1$,

(ii) $f(s) \geq (\pi + \varepsilon)s$ for all $s \in \mathbb{R}$,

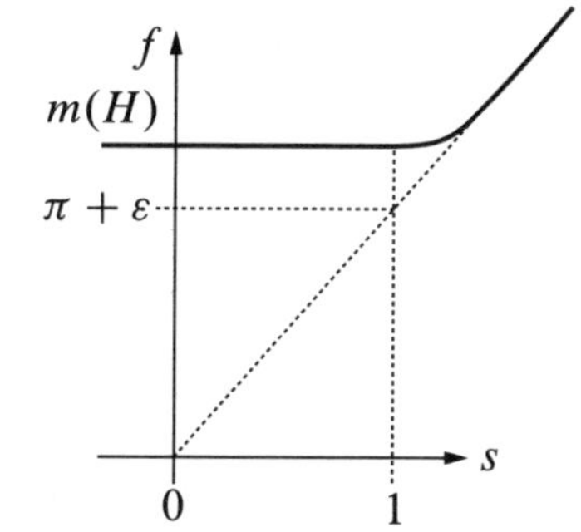

Figure VII.8. The function $f : \mathbb{R} \to \mathbb{R}$.

(iii) $f(s) = (\pi + \varepsilon)s$ for s large,
(iv) $0 < f'(s) \leq \pi + \varepsilon$ for $s > 1$.

The *extension* $\bar{H}$ of the Hamiltonian H is finally defined as the function

$$\bar{H}(z) = \begin{cases} H(z), & z \in E, \\ f(q(z)), & z \notin E. \end{cases}$$

The function $\bar{H}$ is a non-negative smooth function on $\mathbb{R}^{2n}$ which is *quadratic at infinity* in the sense that

$$\bar{H}(z) = (\pi + \varepsilon)q(z), \quad |z| \geq R$$

for a sufficiently large R.

We are looking for periodic solutions of the extended Hamiltonian vector field $X_{\bar{H}}$ which are contained in the ellipsoid E, where the vector field $X_{\bar{H}}$ agrees with the original vector field X_H. For this purpose the following *a-priori statement* is crucial. It localizes the distinguished periodic solutions we are looking for.

Lemma VII.16. *If $x(t)$ is a periodic solution of the Hamiltonian vector field $\dot{x} = X_{\bar{H}}(x)$ of period* 1 *which satisfies*

$$\Phi(x) := \int_0^1 \frac{1}{2}\langle -J\dot{x}(t), x(t)\rangle - \bar{H}(x(t))\, dt > 0,$$

*then $x(t)$ is **non-constant** and lies in the ellipsoid E for all t. In particular $x(t)$ is a non-constant* 1*-periodic solution of the original system $\dot{x} = X_H(x)$ in the open cylinder $Z(1)$.*

Proof. If $x(t) = x^*$ is a constant solution, then $\Phi(x^*) \leq 0$, because $\dot{x} = 0$ and $\bar{H}(x) \geq 0$. It remains to prove for a non-constant periodic solution $x(t)$ of $\dot{x} = X_{\bar{H}}(x)$ that $\Phi(x) \leq 0$, if $x(t)$ is not contained in E for all t.

The Hamiltonian vector field $X_{\bar{H}}$ *vanishes* on the boundary $\partial E = \{z \mid q(z) = 1\}$, because $f'(1) = 0$. Therefore, if a solution $x(t)$ satisfies $x(t_0) \notin E$ for some $t_0 \in \mathbb{R}$, then $x(t) \notin E$ for *all* t and there $x(t)$ solves the equation

$$\dot{x} = X_{\bar{H}}(x) = J\nabla\bar{H}(x) = f'(q(x))\,J\nabla q(x), \quad x \subset \mathbb{R}^{2n} \setminus \bar{E}.$$

However, outside of the ellipsoid E the function q is an *integral* of the vector field $X_{\bar{H}}$, since here

$$\langle \nabla q, J\nabla\bar{H}\rangle = \langle \nabla q, J\nabla q\rangle \cdot f'(q) = 0.$$

Consequently along the solution $x(t)$ in $\bar{E}^c = \{q > 1\}$, the function

$$q(x(t)) = q(x(0)) =: \tau > 1$$

is independent of t. Recalling that $\langle \nabla q(x), x\rangle = 2q(x)$ for all $x \in \mathbb{R}^{2n}$, and using $J^2 = -\mathbb{1}$, and the properties of the function f, we compute

$$\begin{aligned}
\Phi(x) &= \int_0^1 \frac{1}{2}\langle -J\dot{x}(t), x(t)\rangle - \bar{H}(x(t))\,dt \\
&= \int_0^1 \frac{1}{2} f'(\tau)\,\langle \nabla q(x), x\rangle\,dt - \int_0^1 f(q(x))\,dt \\
&= \int_0^1 \frac{1}{2} f'(\tau) 2q(x(t))\,dt - f(\tau) \\
&= f'(\tau)\tau - f(\tau) \\
&\leq (\pi + \varepsilon)\tau - (\pi + \varepsilon)\tau \\
&= 0.
\end{aligned}$$

Hence $\Phi(x) \leq 0$ for all solutions x not contained in E and the lemma is proved. □

In view of this lemma our task is now the following.

- *Find a* 1*-periodic solution of* $\dot{x} = X_{\bar{H}}(x)$ *in* $\mathbb{R}^{2n}$ *satisfying* $\Phi(x) > 0$!

To simplify the notation, we shall in the following again write H instead of $\bar{H}$,

$$H \text{ instead of } \bar{H}.$$

In order to establish the existence of a periodic solution we shall make use of a well-known variational principle for which the critical points are the required periodic solutions. In order to introduce this principle we proceed at first on an informal level and consider the loop space of smooth closed curves

$$\Omega = C^\infty(S^1, \mathbb{R}^{2n}),$$

where $S^1 = \mathbb{R}/\mathbb{Z}$. On this loop space we define the functional $\Phi\colon \Omega \to \mathbb{R}$ by setting

$$\Phi(x) := \int_0^1 \frac{1}{2}\langle -J\dot{x}(t), x(t)\rangle - H(x(t))\,dt.$$

The functional Φ is the *action functional of classical mechanics*.

Claim. The critical points of the functional Φ are precisely those loops that solve the Hamiltonian equation $\dot{x} = X_H(x)$, so that the critical points are the 1-periodic solutions of X_H.

Proof. The derivative of the functional Φ at the loop $x \in \Omega$ in the direction of $y \in \Omega$ is easily computed,

$$\begin{aligned}\Phi'(x)y &:= \frac{d}{d\varepsilon}\Phi(x+\varepsilon y)\Big|_{\varepsilon=0}\\ &= \int_0^1 \frac{1}{2}\langle -J\dot{y}(t), x(t)\rangle + \frac{1}{2}\langle -J\dot{x}(t), y(t)\rangle - \langle \nabla H(x(t)), y(t)\rangle\,dt.\end{aligned}$$

By partial integration we obtain

$$\int_0^1 \langle -J\dot{y}, x\rangle = \int_0^1 \langle Jy, \dot{x}\rangle = \int_0^1 \langle y, J^T\dot{x}\rangle = \int_0^1 \langle y, -J\dot{x}\rangle,$$

because the boundary terms cancel each other due to the periodicity of x and y. Therefore,

$$\Phi'(x)y = \int_0^1 \langle -J\dot{x} - \nabla H(x), y\rangle\,dt.$$

If x is a critical point of Φ, that is, if

$$\Phi'(x)(y) = 0 \quad \text{for all } y \in \Omega,$$

then, necessarily $-J\dot{x}(t) - \nabla H(x(t)) = 0$ for all t and hence

$$\dot{x}(t) = J\nabla H(x(t)) = X_H(x(t)) \quad \text{for all } t \in \mathbb{R} \quad \text{and} \quad x(0) = x(1).$$

We see that the critical points are precisely those loops that are 1-periodic solutions of the Hamiltonian equation, as claimed. □

From the potentially extremely complicated set of solutions of the Hamiltonian system the variational principle selects precisely the 1-periodic solutions. We are dealing with a *global boundary value problem* with *periodic* boundary conditions.

Unfortunately, we are confronted with a difficulty. The functional Φ is neither bounded below nor bounded above, so that the so-called direct methods of the calculus of variation based on minimizing sequences are *not applicable*.

Example. We take a sequence of loops $x_j \in \Omega$, defined by

$$x_j(t) = e^{j2\pi Jt}x_0 = (\cos j2\pi t)x_0 + (\sin j2\pi t)Jx_0$$

for a constant $x_0 \in \mathbb{R}^{2n}$ satisfying $|x_0| = 1$, so that $\|x_j\|_{L^2} = 1$ for all $j \in \mathbb{Z}$. For the first term of the action functional we compute

$$\int_0^1 \frac{1}{2}\langle -J\dot{x}_j(t), x_j(t)\rangle\, dt = \int_0^1 \frac{1}{2}\langle -J^2 j2\pi e^{j2\pi Jt}x_0, e^{j2\pi Jt}x_0\rangle\, dt = \pi j,$$

while the second term is bounded,

$$\left|\int_0^1 H(x_j(t))\, dt\right| \le \sup_{|z|=1} |H(z)|.$$

Hence, $\Phi(x_j) \to \infty$ as $j \to \infty$ and $\Phi(x_j) \to -\infty$ as $j \to -\infty$.

In view of these examples it is useless to search for global maxima or minima and the task is to find saddle points. Only relatively recently, P. Rabinowitz [89] and [90] demonstrated that our degenerate variational principle can be used very effectively for existence proofs. For this purpose he designed minimax arguments adapted to the structure of the functional. Before we start with the technical details for the functional Φ, it is helpful to describe the minimax idea first in an abstract setting.

VII.3 Minimax principles

We assume E to be a real *Hilbert space* having the scalar product $\langle\cdot,\cdot\rangle$ and the induced norm $\|\cdot\| = \langle\cdot,\cdot\rangle^{1/2}$ and let $f\colon E \to \mathbb{R}$ be a C^1-function. We are looking for *critical points* of the function f. By definition these are points $x^* \in E$ satisfying

$$df(x^*) = 0,$$

so that $df(x^*)y = 0$ for all $y \in E$. The derivative $df(x^*)$ is an element of the dual space $E^* = \mathcal{L}(E, \mathbb{R})$ which is the space of continuous linear functionals equipped with the supremum norm. In view of a well-known theorem by F. Riesz there exists a distinguished linear *surjective isometry*

$$I\colon E^* \to E, \quad e^* \mapsto I(e^*)$$

such that the linear functional $e^* \in E^*$ has a unique representation as the scalar product

$$e^*(y) = \langle I(e^*), y\rangle \quad \text{for all } y \in E$$

and $\|I(e^*)\| = \|e^*\|_{E^*}$ holds true. Therefore, there exists for every $x \in E$ a unique $v(x) \in E$ satisfying

$$df(x)y = \langle v(x), y\rangle \quad \text{for all } y \in E$$

and $\|df(x)\|_{E^*} = \|v(x)\|$, namely $v(x) = I(df(x))$. The point $v(x) \in E$ is called the *gradient* of the function f at the point x with respect to the scalar product $\langle\cdot,\cdot\rangle$ and denoted by $v(x) = \nabla f(x)$, so that

$$df(x)y = \langle \nabla f(x), y\rangle \quad \text{for all } y \in E.$$

The critical points of the function f are the zeroes of the *gradient vector field*

$$x \mapsto \nabla f(x)$$

on E and we can interpret the critical points of f dynamically as equilibrium points of the gradient equation

$$\dot{x} = -\nabla f(x), \quad x \in E$$

which is an *ordinary differential equation* in the Hilbert space E.

In the following we assume that the gradient equation generates a unique global flow φ^t satisfying

$$\begin{cases} \frac{d}{dt}\varphi^t(x) = -\nabla f(\varphi^t(x)), & t \in \mathbb{R}, \\ \varphi^0(x) = x, \quad x \in E, \end{cases}$$

by requiring that the initial value problem can be uniquely solved for all times and for all initial conditions $x \in E$. Such a global flow does exist, for example, if the gradient of the function f satisfies a global Lipschitz estimate $\|\nabla f(x) - \nabla f(y)\| \le M\,\|x - y\|$ for a constant M and all $x, y \in E$, as we have seen in Chapter IV.

From Chapter IV we recall the crucial property of the gradient flow. The function $t \mapsto f(\varphi^t(x))$ *decreases along non-constant solutions*, in view of $\frac{d}{ds} f(\varphi^s(x)) = -\|\nabla f(\varphi^s(x))\|^2$, from which one obtains by integration the identity

$$f(\varphi^t(x)) = f(x) - \int_0^t \|\nabla f(\varphi^s(x))\|^2\, ds.$$

A differentiable function $f\colon E \to \mathbb{R}$ does not need to have critical points, even if it is bounded below, because the space E is non-compact. The function $f(x) = e^{-x}$ on $\mathbb{R}$ satisfies $\|\nabla f(x_j)\| \to 0$ for every sequence $x_j \to \infty$, and the sequence $f(x_j)$ is bounded, however the sequence does not have a convergent subsequence.

We introduce the so-called P.S.-condition, which goes back to R. S. Palais and S. Smale. It is one of the classical compactness conditions of nonlinear functional analysis which, in many cases, guarantees critical points.

Definition. A function $f \in C^1(E, \mathbb{R})$ *satisfies the P.S.-condition*, if every sequence $x_j \in E$ satisfying

$$\nabla f(x_j) \to 0 \text{ in } E \quad \text{and} \quad |f(x_j)| \le c \quad \text{for all } j$$

possesses a convergent subsequence where c is a constant.

Then, the limit x^* of such a subsequence is necessarily a *critical point* of f,

$$\nabla f(x^*) = 0,$$

because the function f is of class C^1.

Definition. Let $\mathfrak{F}$ be a family of subsets $F \subset E$ and $f \in C^1(E, \mathbb{R})$. The *minimax* $c(f, \mathfrak{F})$ of the pair $(f, \mathfrak{F})$ is defined as

$$c(f, \mathfrak{F}) := \inf_{F \in \mathfrak{F}} \Big(\sup_{x \in F} f(x) \Big) \in \mathbb{R} \cup \{\pm\infty\}.$$

We shall apply the strategy of the following proposition later on in order to guarantee a critical point of the action functional.

Proposition VII.17 (Minimax principle). *We require that the function $f \in C^1(E, \mathbb{R})$ and the family $\mathfrak{F}$ meet the following conditions.*

(i) *f satisfies the P.S.-condition.*
(ii) *$\dot{x} = -\nabla f(x)$ defines a unique global flow $\varphi^t(x)$ on the Hilbert space E.*
(iii) *The family $\mathfrak{F}$ is positively invariant under the flow, i.e., for all $t \ge 0$ we conclude from $F \in \mathfrak{F} \Longrightarrow \varphi^t(F) \in \mathfrak{F}$.*
(iv) *The minimax is finite, $-\infty < c(f, \mathfrak{F}) < \infty$.*

Then the real number $c(f, \mathfrak{F})$ is a **critical** *value of f, i.e., there exists a point $x^* \in E$ satisfying*

$$\nabla f(x^*) = 0 \quad \textit{and} \quad f(x^*) = c(f, \mathfrak{F}).$$

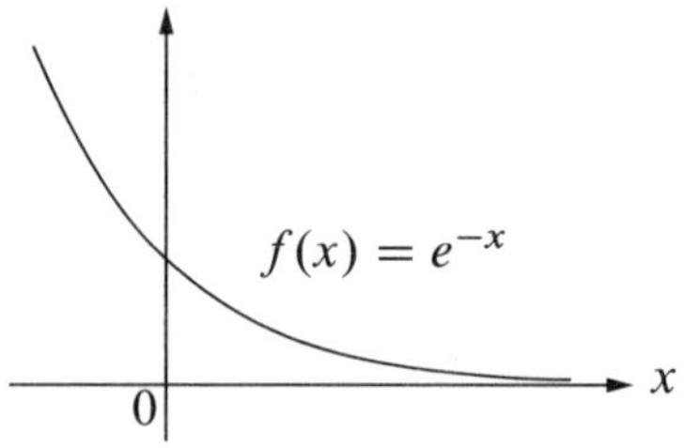

Figure VII.9. The function $f(x) = e^{-x}$ on $E = \mathbb{R}$ does not have a critical point, even though $\nabla f(x) \to 0$ for $x \to \infty$.

Proof. We abbreviate the minimax by $c := c(f, \mathfrak{F})$. It suffices to prove the following claim.

Claim. For every $\varepsilon > 0$ there exists a point $x \in E$ satisfying

$$c - \varepsilon \leq f(x) \leq c + \varepsilon \quad \text{and} \quad \|\nabla f(x)\| \leq \varepsilon.$$

Indeed, if this holds true, the proposition follows, since taking the sequence $\varepsilon_j = \frac{1}{j} \to 0$ we obtain a sequence $x_j \in E$ satisfying

$$c - \frac{1}{j} \leq f(x_j) \leq c + \frac{1}{j} \quad \text{and} \quad \|\nabla f(x_j)\| \leq \frac{1}{j}.$$

Because f satisfies the P.S.-condition, there exists a convergent subsequence of x_j. Its limit x^* satisfies $\nabla f(x^*) = 0$ and $f(x^*) = c(f, \mathfrak{F})$, because the functions f and ∇f are continuous.

Proof of the claim. In order to prove the claim, we proceed by contradiction and assume that there exists an $\varepsilon > 0$ satisfying

$$\|\nabla f(x)\| > \varepsilon \quad \text{for all } x \in \{x \mid c - \varepsilon \leq f(x) \leq c + \varepsilon\}.$$

Using the definition of c we find a set $F \in \mathfrak{F}$ satisfying $\sup_{x \in F} f(x) \leq c + \varepsilon$. If $x \in F$, then $f(x) \leq c + \varepsilon$ and we shall show that

$$(*) \quad f(\varphi^{t^*}(x)) \leq c - \varepsilon \quad \text{for } t^* = 2\varepsilon^{-1}.$$

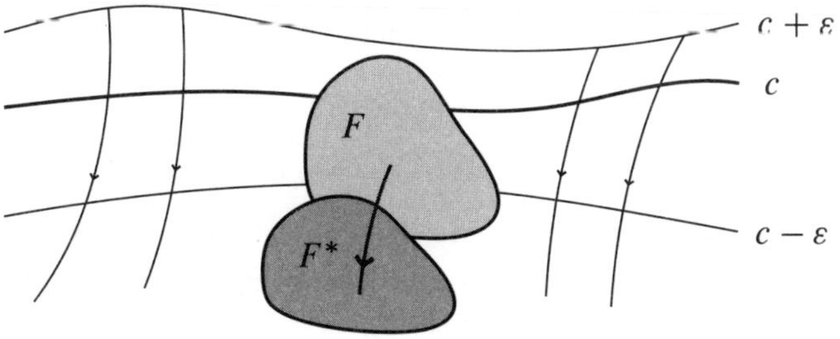

Figure VII.10. The proof of the minimax principle by contradiction.

This produces the desired contradiction, since for the set $F^* = \varphi^{t^*}(F) \in \mathfrak{F}$ it follows that

$$\sup_{x \in F^*} f(x) \leq c - \varepsilon,$$

which is in contradiction to the definition of the minimax c. It remains to prove the estimate $(*)$. If $f(\varphi^t(x)) \le c-\varepsilon$ for some time t in the interval $0 \le t \le t^*$, then, indeed, $f(\varphi^{t^*}(x)) \le c-\varepsilon$, because $f(\varphi^t(x))$ is decreasing if t increases. If, on the other hand

$$(**) \quad f(\varphi^t(x)) > c-\varepsilon \quad \text{for all } 0 \le t \le t^*,$$

then it follows from our assumption that $\|\nabla f(\varphi^t(x))\| > \varepsilon$ for all $0 \le t \le t^*$. Using the identity $f(\varphi^t(x)) = f(x) - \int_0^t \|\nabla f(\varphi^s(x))\|^2\, ds$ we obtain the estimate

$$f(\varphi^{t^*}(x)) \le c+\varepsilon-\varepsilon^2 t^* = c-\varepsilon,$$

in contradiction to the estimate $(**)$. Thus, the estimate $(*)$ holds true and the minimax principle is proved. □

Example. We assume that the function f on E satisfies the P.S.-condition and that the corresponding gradient equation generates a unique global flow, as specified in the assumptions (i) and (ii) of the proposition. In addition, we require that the function f is *bounded below*, so that $-\infty < \alpha \le f(x)$ for all $x \in E$. As a minimax family $\mathcal{F}$ of sets we choose the family of all points, namely $\mathcal{F} = \{x\}_{x\in E}$. Then

$$\alpha \le c(f, \mathcal{F}) := \inf_{x\in E} f(x) < \infty$$

and therefore $c(f, \mathcal{F}) \in \mathbb{R}$. Since there is nothing to prove for (iii), we conclude from the minimax principle the existence of a point $x^* \in E$ satisfying

$$f(x^*) = \inf_E f \quad \text{and} \quad \nabla f(x^*) = 0.$$

We have proved that a *global minimum* of the function f exists!

As a further illustration we formulate the mountain pass lemma.

Definition. A subset $R \subset E$ is called a *mountain range for the function* f,

(i) if the restriction $f|_R$ is bounded below, so that $\inf_{x\in R} f(x) =: \alpha > -\infty$,

(ii) if the set $E \setminus R$ has at least two connected components and if in every component there exists a point x satisfying $f(x) < \alpha$.

Proposition VII.18 (Mountain pass lemma). *We assume that a function $f: E \to \mathbb{R}$ is continuously differentiable and satisfies the P.S.-condition and that, moreover, $\dot{x} = -\nabla f(x)$ generates a unique global flow. If $R \subset E$ is a mountain range for f, then f has a critical value $c \in \mathbb{R}$ satisfying*

$$c \ge \inf_R f.$$

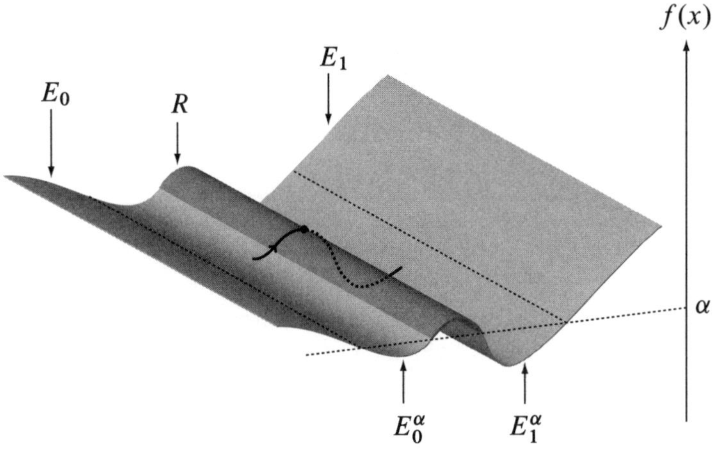

Figure VII.11. A mountain range R and the connected components E_0, E_1 of $E \setminus R$.

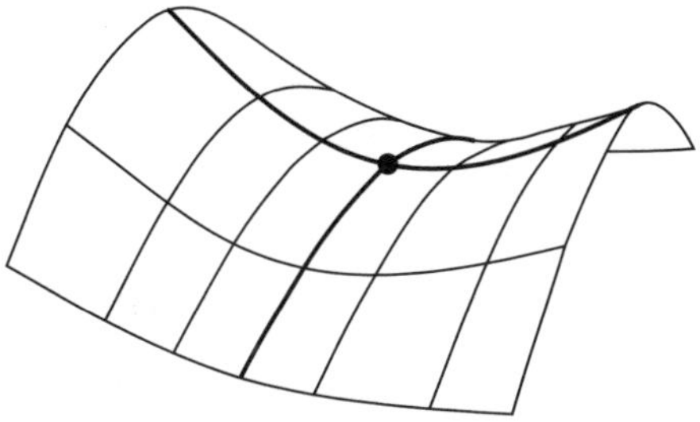

Figure VII.12. Saddle point on a path $\gamma \in \Gamma$ over the mountain range.

Proof [Minimax principle]. If we choose two different connected components E_0 and E_1 of $E \setminus R$ according to the definition (ii) of a mountain range and if we define for $j = 0, 1$ and $\alpha = \inf_{x \in R} f(x)$ the sublevel sets

$$E_j^\alpha = \{x \in E_j \mid f(x) < \alpha\},$$

then $E_j^\alpha \neq \emptyset$. Next, we consider the set Γ of paths connecting the two valleys, defined by

$$\Gamma = \{\gamma\colon [0,1] \to E \text{ continuous} \mid \gamma(0) \in E_0^\alpha \text{ and } \gamma(1) \in E_1^\alpha\}$$

and define the family $\mathcal{F}$ of subsets of E as

$$\mathcal{F} = \{\gamma([0,1]) \mid \gamma \in \Gamma\}.$$

The sets $\gamma([0,1])$ are compact, because the curves are continuous. Since $\gamma(0)$ and $\gamma(1)$ belong to different components of $E \setminus R$, there exists a parameter value $t_0 \in (0,1)$ satisfying $\gamma(t_0) \in R$, so that

$$f(\gamma(t_0)) \geq \alpha.$$

Therefore, $\alpha \leq \sup_{x \in \gamma([0,1])} f(x) < \infty$ for all paths $\gamma \in \Gamma$, so that

$$-\infty < \alpha \leq c(f, \mathcal{F}) = \inf_{\gamma \in \Gamma} \sup_{x \in \gamma([0,1])} f(x) < \infty.$$

Consequently, the minimax is finite. It remains to prove that $\mathcal{F}$ is positively invariant under the flow φ^t of the vector field $-\nabla f$. Due to the monotonicity,

$$f(\varphi^t(x)) \leq f(\varphi^0(x)) = f(x), \quad x \in E,\ t \geq 0.$$

Hence, in view of the definition of R, the flow cannot leave the sublevel sets of the components $E \setminus R$ so that $\varphi^t \circ \gamma \in \Gamma$ for $\gamma \in \Gamma$ and $t \geq 0$. Now, we can apply the minimax principle from which the proposition follows. □

Our next task is to extend our action functional Φ to a continuously differentiable function defined on a suitable Hilbert space E which contains the set $C^\infty(S^1, \mathbb{R})$ of smooth loops as a linear subspace. Moreover, we have to look for a suitable family $\mathcal{F}$, so that we can perform the minimax strategy.

VII.4 The functional analysis of the action functional

We return to the action functional $\Phi\colon \Omega = C^\infty(S^1, \mathbb{R}^{2n}) \to \mathbb{R}$, defined by

$$\Phi(x) := \int_0^1 \frac{1}{2} \langle -J\dot{x}(t), x(t) \rangle - H(x(t))\, dt,$$

where the smooth function $H\colon \mathbb{R}^{2n} \to \mathbb{R}$ has the properties

(i) $H \equiv 0$ near $z = 0$ and

(ii) $H(z) = Q(z)$ for $|z|$ large.

The quadratic form Q is defined by

$$Q(z) = (\pi + \varepsilon)\, q(z)$$

for a small $\varepsilon > 0$. We shall extend the function Φ onto a Hilbert space $E \supset \Omega$ and shall translate the pointwise properties (i) and (ii) of H into the qualitative behavior of the extended function $\Phi\colon E \to \mathbb{R}$. In order to find a suitable Hilbert space E, we consider the dominant part of Φ, namely

$$a(x, y) := \int_0^1 \frac{1}{2} \langle -J\dot{x}(t), y(t) \rangle\, dt.$$

We represent the loops $x, y \in C^\infty(S^1, \mathbb{R}^{2n})$ by their Fourier series

$$x(t) = \sum_{j\in\mathbb{Z}} e^{j2\pi tJ} x_j, \quad x_j \in \mathbb{R}^{2n}.$$

Since x belongs to $C^\infty(S^1)$, its Fourier series does converge not only in L^2, but also in the supremum norm and even in every C^k-norm. Inserting the Fourier series of the loops $x(t)$ and $y(t)$ into $a(x, y)$, recalling that $J^2 = -1$ and observing that $\int_0^1 \langle e^{j2\pi tJ} x_j, e^{k2\pi tJ} x_k\rangle\, dt = \delta_{jk}\langle x_j, x_k\rangle$, one obtains

$$\begin{aligned} 2a(x, y) &= 2\pi \sum_{j\in\mathbb{Z}} j\,\langle x_j, y_j\rangle \\ &= 2\pi \sum_{j>0} |j|\,\langle x_j, y_j\rangle - 2\pi \sum_{j<0} |j|\,\langle x_j, y_j\rangle. \end{aligned}$$

This representation shows that $a(x, y)$ can be extended to the Sobolev space $H^{\frac{1}{2}}(S^1, \mathbb{R}^{2n})$ as a continuous bilinear form.

Definition. The Sobolev space $H^s = H^s(S^1, \mathbb{R}^{2n})$ for a real number $s \ge 0$ is defined as

$$H^s = \big\{x \in L^2(S^1, \mathbb{R}^{2n}) \mid \textstyle\sum_{j\in\mathbb{Z}\setminus\{0\}} |j|^{2s}|x_j|^2 < \infty\big\},$$

where $x_j \in \mathbb{R}^{2n}$ denote the coefficients of the Fourier series $x(t) = \sum_{j\in\mathbb{Z}} e^{j2\pi tJ} x_j$ which converges in L^2. The space H^s is a Hilbert space equipped with the scalar product

$$\langle x, y\rangle_s := \langle x_0, y_0\rangle + 2\pi \sum_{j\in\mathbb{Z}\setminus\{0\}} |j|^{2s}\langle x_j, y_j\rangle$$

and the norm $\|x\|_s^2 = \langle x, x\rangle_s$.

The norm $\|\cdot\|_0$ for $s = 0$ is equivalent to the L^2 norm, since, according to the theorem of Plancherel, $\|x\|_{L^2}^2 = \int_0^1 |x(t)|^2\, dt = \sum_{j\in\mathbb{Z}} |x_j|^2$ and hence $\|\cdot\|_{L^2} \le \|\cdot\|_0 \le \sqrt{2\pi}\,\|\cdot\|_{L^2}$. In the following the Sobolev space $H^{\frac{1}{2}}$ will be the desired Hilbert space E we shall work on and so we introduce the abbreviations

$$\begin{aligned} E &:= H^{\frac{1}{2}}(S^1, \mathbb{R}^{2n}), \\ \langle x, y\rangle &:= \langle x, y\rangle_{\frac{1}{2}}, \\ \|x\|^2 &:= \langle x, x\rangle_{\frac{1}{2}}. \end{aligned}$$

There exists an *orthogonal splitting of* E,

$$E = E^- \oplus E^0 \oplus E^+,$$

into the closed subspaces consisting of the elements $x \in E$ which have non-vanishing Fourier coefficients only for $j < 0$, $j = 0$ or $j > 0$. The corresponding *orthogonal projectors* are denoted by P^+, P^0 and P^-. Therefore, every $x \in E$ has the *unique decomposition*

$$x = x^- + x^0 + x^+ \in E^- \oplus E^0 \oplus E^+$$

where $x^+ = P^+ x$, $x^0 = P^0 x$ and $x^- = P^- x$.

Definition. For $x, y \in E$ we define the continuous bilinear form a by

$$\begin{aligned} a(x,y) &= \frac{1}{2}\langle x^+, y^+\rangle - \frac{1}{2}\langle x^-, y^-\rangle \\ &= \frac{1}{2}\langle (P^+ - P^-)x, y\rangle. \end{aligned}$$

The bilinear form a is an extension of the original bilinear form a defined on the subspace Ω of smooth loops. Indeed, for two smooth loops $x, y \in \Omega \subset E$ we have the representation

$$a(x,y) = \int_0^1 \frac{1}{2}\langle -J\dot{x}, y\rangle\, dt.$$

Definition. The corresponding quadratic form $a\colon E \to \mathbb{R}$ is defined by

$$a(x) = a(x,x) = \frac{1}{2}\|x^+\|^2 - \frac{1}{2}\|x^-\|^2, \quad x = x^+ + x^0 + x^-.$$

The function $a\colon E \to \mathbb{R}$ is continuously differentiable and its derivative $da(x)$ in the point $x \in E$ is equal to

$$\begin{aligned} da(x)(y) &= \langle x^+, y^+\rangle - \langle x^-, y^-\rangle \\ &= \langle (P^+ - P^-)x, y\rangle. \end{aligned}$$

By definition of the gradient, $da(x)(y) = \langle \nabla a(x), y\rangle$ for all $y \in E$, so that the gradient of the function a is equal to

$$\nabla a(x) = (P^+ - P^-)x = x^+ - x^- \in E.$$

The map $x \mapsto \nabla a(x)$ on E is linear and continuous.

Next, we shall prove some useful *properties of the spaces* H^s. First we note that these spaces decrease according to

$$H^t \subset H^s \subset H^0 \quad \text{for } t \geq s \geq 0,$$

while the norms increase,

$$\|x\|_t \geq \|x\|_s \geq \|x\|_0, \quad \text{if } x \in H^t.$$

From these estimates one reads off that the *inclusion map*

$$I: H^t \to H^s, \quad t \geq s$$

is a bounded linear operator and therefore also continuous.

Remark. The loop space $\Omega = C^\infty(S^1, \mathbb{R}^{2n})$ is contained in the Sobolev space H^s and is *dense*, for every $s \geq 0$.

Proof. The Fourier coefficients $x_j \in \mathbb{R}^{2n}$ of a smooth loop $x \in \Omega$ satisfy the following estimates. For every integer $k \geq 0$ there exists a constant c_k such that

$$|x_j| \leq c_k \frac{1}{|j|^k} \quad \text{for all } j \in \mathbb{Z}.$$

Hence $\|x\|_s < \infty$ for every s. Since already the subset of Ω consisting of *trigonometric polynomials* $x(t) = \sum_{j=-N}^{N} e^{j2\pi tJ} x_j$ with $N \in \mathbb{N}$ and $x_j \in \mathbb{R}^{2n}$ is dense in H^s for $s \geq 0$, the claim follows. □

Definition. A map between metric spaces is called *compact*, if it maps bounded sets onto relatively compact sets, i.e., onto sets whose closure is compact. Equivalently, the image of a bounded sequence possesses a convergent subsequence.

Since compact sets are bounded, the compact maps map bounded sets onto bounded sets. In particular, compact *linear* maps between normed spaces are continuous maps.

Proposition VII.19. *If $t > s \geq 0$, the inclusion map $I: H^t \to H^s$ is compact.*

Proof. If we define the map $P_N: H^t \to H^s$ for the integer $N \in \mathbb{N}$ by

$$(P_N x)(t) := \sum_{|j| \leq N} e^{j2\pi tJ} x_j \quad \text{for } x(t) = \sum_{j \in \mathbb{Z}} e^{j2\pi tJ} x_j,$$

then $P_N \in \mathcal{L}(H^t, H^s)$ is a linear and bounded map. The image under the map P_N is a finite dimensional space in which relatively compact sets and bounded sets are the same. Therefore P_N is a compact map. The set of compact linear operators is closed with respect to the operator norm, as can be verified easily. The proposition therefore follows if we show that

$$P_N \to I \text{ in } \mathcal{L}(H^t, H^s), \quad N \to \infty.$$

For this purpose we compute

$$\begin{aligned}\|(P_N - I)x\|_s^2 &= \Big\| \sum_{|j|>N} e^{j2\pi tJ} x_j \Big\|_s^2 \\ &= 2\pi \sum_{|j|>N} |j|^{2s} |x_j|^2 \\ &= 2\pi \sum_{|j|>N} |j|^{2(s-t)} |j|^{2t} |x_j|^2 \\ &\le N^{2(s-t)} 2\pi \sum_{|j|>N} |j|^{2t} |x_j|^2 \\ &\le N^{2(s-t)} \|x\|_t^2.\end{aligned}$$

Because $t > s$,

$$\|P_N - I\|_{\mathcal{L}(H^t, H^s)} \le N^{(s-t)} \to 0$$

as $N \to \infty$ and the proposition is proved. □

Not all the elements of the space $E = H^{\frac{1}{2}}$ can be represented by *continuous* functions. Yet for $s > \frac{1}{2}$ the Sobolev space $H^s \subset C(S^1, \mathbb{R}^{2n})$ is continuously embedded in the space of continuous functions equipped with the supremum norm.

Proposition VII.20. *If $s > \frac{1}{2}$, then*

$$x \in H^s \Longrightarrow x \in C(S^1, \mathbb{R}^{2n}),$$

that is, there exists a continuous representative of $x \in H^s \subset L^2$. Moreover, there exists a constant $c > 0$ such that

$$\sup_{t \in S^1} |x(t)| \le c\|x\|_s \quad \textit{for all } x \in H^s.$$

Proof. We shall show that the Fourier series

$$x(t) = \sum_{j \in \mathbb{Z}} e^{j2\pi tJ} x_j$$

of $x \in H^s$ does not only converge in L^2, but also in the supremum norm, so that the limit function is, indeed, a continuous function. For $s > \frac{1}{2}$ we compute, using

Hölder's inequality,

$$\begin{aligned}\sum_{j\neq 0} |e^{j2\pi tJ} x_j| &= \sum_{j\neq 0} |x_j| \\ &= \sum_{j\neq 0} |j|^{-s}|j|^{s}|x_j| \\ &\leq \Big[\sum_{j\neq 0} |j|^{-2s}\Big]^{\frac{1}{2}}\Big[\sum_{j\neq 0} |j|^{2s}|x_j|^2\Big]^{\frac{1}{2}} \\ &= c\|x\|_s,\end{aligned}$$

where $c^2 := \sum_{j\neq 0} |j|^{-2s} < \infty$ converges in view of the assumption $2s > 1$. □

Similarly, one proves for $s > \frac{1}{2} + r$, where $r \in \mathbb{N}$ is a positive integer, the implication

$$x \in H^s \Longrightarrow x \in C^r(S^1, \mathbb{R}^{2n}),$$

and, moreover, the existence of a constant $c > 0$ such that

$$\|x\|_{C^r(S^1,\mathbb{R}^{2n})} \leq c\|x\|_s \quad \text{for all } x \in H^s.$$

In view of Proposition VII.19 the embedding

$$j\colon H^{\frac{1}{2}} \to L^2 \cong H^0$$

is a *compact* linear operator and we consider its adjoint operator

$$j^*\colon L^2 \to H^{\frac{1}{2}},$$

which satisfies

$$(j(x), y)_{L^2} = \langle x, j^*(y)\rangle_{\frac{1}{2}} \quad \text{for all } x \in H^{\frac{1}{2}} \text{ and all } y \in L^2.$$

In order to recall the definition of the adjoint operator we take an element $y \in L^2$ and observe that the functional $f_y\colon x \mapsto (j(x), y)_{L^2}$ is continuous on the space $H^{\frac{1}{2}}$, because, by Cauchy–Schwarz,

$$|(j(x), y)_{L^2}| \leq \|j(x)\|_{L^2}\|y\|_{L^2} \leq \|x\|_{\frac{1}{2}}\|y\|_{L^2},$$

for all $x \in H^{\frac{1}{2}}$, so that $\|f_y\|_{\left(H^{\frac{1}{2}}\right)^*} \leq \|y\|_{L^2}$. Thus a well-known theorem of F. Riesz guarantees a unique element $y^* \in H^{\frac{1}{2}}$ such that the continuous functional is represented by the scalar product as

$$f_y(x) = (j(x), y)_{L^2} = \langle x, y^*\rangle_{\frac{1}{2}} \quad \text{for all } x \in H^{\frac{1}{2}},$$

moreover,

$$\|y^*\|_{\frac{1}{2}} = \|f_y\| \le \|y\|_{L^2}.$$

The adjoint operator $j^* \in \mathcal{L}(L^2, H^{\frac{1}{2}})$ is now defined by

$$j^*(y) = y^*$$

and satisfies $\|j^*(y)\|_{\frac{1}{2}} \le \|y\|_{L^2}$ for all $y \in L^2$.

Proposition VII.21. *$j^*(L^2) \subset H^1$ and*

$$\|j^*(y)\|_1 \le \|y\|_{L^2}$$

for all $y \in L^2$.

Proof. We write the definition of the adjoint operator explicitly and obtain, by the theorem of Plancherel,

$$\begin{aligned}\sum_{k\in\mathbb{Z}} \langle x_k, y_k\rangle &= (j(x), y)_{L^2} \\ &= \langle x, j^*(y)\rangle_{\frac{1}{2}} \\ &= \langle x_0, j^*(y)_0\rangle + 2\pi \sum_{k\in\mathbb{Z}} |k|\,\langle x_k, j^*(y)_k\rangle\end{aligned}$$

for *all* $x \in H^{\frac{1}{2}} \subset L^2$ and $y \in L^2$. Hence,

$$\begin{aligned} j^*(y)_0 &= y_0, \\ j^*(y)_k &= \frac{1}{2\pi|k|} y_k, \quad k \ne 0. \end{aligned}$$

Therefore,

$$j^*(y)(t) = y_0 + \sum_{k\ne 0} e^{k2\pi t J} \frac{1}{2\pi|k|} y_k, \quad y \in L^2,$$

and by definition of the norm in H^1 we obtain the estimate $\|j^*(y)\|_1 \le \|y\|_{L^2}$, as claimed. □

Proposition VII.22. *The adjoint operator $j^*\colon L^2 \to H^{\frac{1}{2}}$ is a compact linear operator.*

Proof. In view of Proposition VII.21 the map factorizes through the Sobolev space H^1, so that

$$j^*\colon L^2 \to H^1 \xrightarrow{I} H^{\frac{1}{2}},$$

where, by Proposition VII.19, the embedding $I\colon H^1 \to H^{\frac{1}{2}}$ is compact. The composition of a continuous (hence bounded) operator with a compact operator is again a compact operator, so that the proposition is proved. □

Next we extend the *second part* of the functional Φ which is given by

$$b(x) := \int_0^1 H(x(t))\,dt, \quad x \in \Omega.$$

Since $H(z) \equiv 0$ near $z = 0$ and $H(z) = Q(z)$ for $|z|$ large, we have for a sufficiently large constant M the estimates

$$\begin{aligned}|H(z)| &\le M|z|^2,\\ |d^2H(z)| &\le M\end{aligned}$$

for all $z \in \mathbb{R}^{2n}$. Therefore, the function b is defined for all $x \in L^2(S^1, \mathbb{R}^{2n})$ and, in particular, for every $x \in E \subset L^2$. If we consider b as a function of $x \in L^2$, we denote it by $\hat{b}$, so that

$$\hat{b}(x) = \int_0^1 H(x(t))\,dt, \quad x \in L^2.$$

Using the inclusion $j\colon E \to L^2$, the function $b\colon E \to \mathbb{R}$ is the *composition*

$$b(x) = \hat{b}(j(x)), \quad x \in E.$$

We distinguish between b and $\hat{b}$, because the compactness of the embedding j will play a role later on.

Claim. We claim that the function $\hat{b}\colon L^2 \to \mathbb{R}$ is differentiable and its derivative is the bounded linear map $d\hat{b}(x)(h) = \int_0^1 \langle \nabla H(x), h\rangle\,dt = (\nabla H(x), h)_{L^2}$. Therefore, the L^2-gradient is equal to $\nabla\hat{b}(x) = \nabla H(x)$.

Proof. We start with *pointwise* identities. Using the chain rule we compute

$$\begin{aligned}H(z+\xi) - H(z) &= \int_0^1 \frac{d}{dt} H(z + t\xi)\,dt\\ &= \int_0^1 \langle \nabla H(z+t\xi), \xi\rangle\,dt\\ &= \langle \nabla H(z), \xi\rangle + \int_0^1 \langle \nabla H(z+t\xi) - \nabla H(z), \xi\rangle\,dt,\end{aligned}$$

and hence,

$$(*) \quad H(z+\xi) = H(z) + \langle \nabla H(z), \xi\rangle + \int_0^1 \langle \nabla H(z+t\xi) - \nabla H(z), \xi\rangle\,dt$$

for all $z, \xi \in \mathbb{R}^{2n}$. Due to the bound $|d^2H(z)| \le M$ on $\mathbb{R}^{2n}$ it follows by means of the mean value theorem that

$$\begin{aligned} |\langle \nabla H(z+t\xi) - \nabla H(z), \cdot \rangle| &= |dH(z+t\xi) - dH(z)| \\ &\le \sup_{a \in \mathbb{R}^{2n}} |d^2H(a)|\,|t\xi| \\ &\le M|\xi|. \end{aligned}$$

Therefore, the integral in $(*)$ can be estimated for all $z \in \mathbb{R}^{2n}$ by $M|\xi|^2$. If $x \in L^2$, then also the composition

$$\nabla H(x) = \nabla H(x(t))$$

is an element in L^2, because $|\nabla H(z)| \le C|z|$ for all $z \in \mathbb{R}^{2n}$. Now, if $x, h \in L^2$, we insert $z = x(t)$ and $\xi = h(t)$ into the identity $(*)$ and integrate over t to obtain

$$\hat{b}(x+h) = \hat{b}(x) + \int_0^1 \langle \nabla H(x), h \rangle \, dt + R(x,h)$$

with a rest term R satisfying the estimate

$$|R(x,h)| \le M\|h\|_{L^2}^2.$$

Recalling the definition of a derivative, the function $\hat{b}\colon L^2 \to \mathbb{R}$ is differentiable at the point $x \in L^2$ and its derivative $d\hat{b}(x) \in \mathcal{L}(L^2, \mathbb{R})$ is given by the formula

$$\begin{aligned} d\hat{b}(x)(h) &= \int_0^1 \langle \nabla H(x), h \rangle \, dt \\ &= (\nabla H(x), h)_{L^2}. \end{aligned}$$

From the definition of the gradient in L^2 we deduce that $\nabla \hat{b}(x) = \nabla H(x)$, as claimed. □

We recall that, using the continuous embedding $j\colon E \to L^2$, the map $b\colon E \to \mathbb{R}$ is defined as the composition

$$b(x) = \hat{b}(j(x)), \quad x \in E.$$

By the chain rule its derivative is equal to

$$\begin{aligned} db(x)(y) &= d\hat{b}(j(x))\, j(y) \\ &= \big(\nabla \hat{b}(j(x)), j(y)\big)_{L^2} \\ &= \big\langle j^* \nabla \hat{b}(j(x)), y \big\rangle_{\frac{1}{2}} \end{aligned}$$

for all $x, y \in E$. Hence, by definition of the gradient in the Hilbert space E,

$$\nabla b(x) = j^* \nabla \hat{b}(j(x)) = j^* \nabla H(j(x)).$$

Proposition VII.23. *The map $b\colon E \to \mathbb{R}$ is continuously differentiable. The gradient $\nabla b\colon E \to E$ is continuous and compact. Moreover, for all $x, y \in E$ we have the estimates*

$$\|\nabla b(x) - \nabla b(y)\|_{\frac{1}{2}} \le M\|x - y\|_{\frac{1}{2}},$$
$$|b(x)| \le M\|x\|_{L^2}^2$$

with a constant $M > 0$.

Proof. From the estimate $M \ge \sup_{z\in\mathbb{R}^{2n}} |d^2H(z)|$ we obtain by the mean value theorem the estimate

$$\|\nabla H(x) - \nabla H(y)\|_{L^2} \le M\|x - y\|_{L^2} \quad \text{for all } x, y \in L^2.$$

Therefore, the L^2-gradient $\nabla H\colon L^2 \to L^2$ maps bounded sets onto bounded sets. Since $j^*\colon L^2 \to E$ is compact, the E-gradient $\nabla b = j^* \circ \nabla H \circ j\colon E \to E$ is a compact map. Moreover, using $\|\cdot\|_{L^2} \le \|\cdot\|_{\frac{1}{2}} \le \|\cdot\|_1$ and Proposition VII.21, we can estimate

$$\begin{aligned}\|\nabla b(x) - \nabla b(y)\|_{\frac{1}{2}} &\le \|j^*(\nabla H(x) - \nabla H(y))\|_{\frac{1}{2}}\\ &\le \|j^*(\nabla H(x) - \nabla H(y))\|_1\\ &\le \|\nabla H(x) - \nabla H(y)\|_{L^2}\\ &\le M\|x - y\|_{L^2}\\ &\le M\|x - y\|_{\frac{1}{2}}.\end{aligned}$$

For M large enough the estimate $|H(z)| \le M|z|^2$ holds true for all $z \in \mathbb{R}^{2n}$, so that $|b(x)| \le \int_0^1 |H(x(t))|\,dt \le M\int_0^1 |x(t)|^2\,dt = M\|x\|_{L^2}^2$, as claimed in the proposition. □

Summarizing the discussion so far, we have extended the action functional $\Phi\colon \Omega \to \mathbb{R}$ from the space of smooth loops to the function $\Phi\colon E \to \mathbb{R}$, still denoted by the same letter, and defined on the Hilbert space $E \supset \Omega$ by

$$\Phi(x) = a(x) - b(x), \quad x \in E.$$

The function $\Phi\colon E \to \mathbb{R}$ is continuously differentiable and its E-gradient is equal to

$$\nabla\Phi(x) = x^+ - x^- - \nabla b(x) \in E$$

for $x = x^+ + x^0 + x^- \in E$. By Proposition VII.23 this gradient is globally *Lipschitz-continuous*. We are interested in classical solutions of the Hamiltonian equation. It is, therefore, crucial to observe that a critical point of the function $\Phi\colon E \to \mathbb{R}$ is not simply an element in E, which might not even be a continuous function, but is actually a smooth periodic solution of the Hamiltonian vector field X_H having period equal to 1. Indeed, the following regularity theorem holds true.

Proposition VII.24 (Regularity theorem). *If $x \in E$ is a critical point of $\Phi\colon E \to \mathbb{R}$, so that*

$$\nabla\Phi(x) = 0,$$

then x belongs to $C^\infty(S^1, \mathbb{R}^{2n})$ and solves the Hamiltonian equation

$$\dot{x}(t) = J\nabla H(x(t)), \quad t \in \mathbb{R}.$$

Proof [Fourier series, Proposition VII.20]. (1) We first prove that *x is a continuous function.* If $x \in E$ satisfies $\nabla\Phi(x) = \nabla a(x) - \nabla b(x) = 0$, then

$$\langle \nabla a(x), v\rangle = \langle \nabla b(x), v\rangle \quad \text{for all } v \in E,$$

and hence, using the explicit expressions for the gradients,

$$\langle (P^+ - P^-)x, v\rangle = \langle j^*\nabla H(j(x)), v\rangle = \big(\nabla H(j(x)), j(v)\big)_{L^2}.$$

Because $x = x(t)$ and $v = v(t)$ are elements of L^2,

$$(*) \quad \langle (P^+ - P^-)x, v\rangle = \int_0^1 \langle \nabla H(x), v\rangle\, dt.$$

Next, we develop $x \in L^2$ and $\nabla H(x) \in L^2$ into their Fourier series

$$x = \sum_{k\in\mathbb{Z}} e^{k2\pi tJ} x_k, \quad \nabla H(x) = \sum_{k\in\mathbb{Z}} e^{k2\pi tJ} a_k,$$

where the *Fourier coefficients* are defined by

$$x_k = \int_0^1 e^{-k2\pi tJ} x(t)\, dt, \quad a_k = \int_0^1 e^{-k2\pi tJ} \nabla H(x(t))\, dt.$$

If v_k are the Fourier coefficients of v, we deduce from $(*)$ the identity

$$(**) \quad \sum_{k\in\mathbb{Z}} \langle 2\pi k x_k, v_k\rangle = \sum_{k\in\mathbb{Z}} \langle a_k, v_k\rangle,$$

by using the definition of the scalar product $\langle\cdot,\cdot\rangle_{\frac{1}{2}}$ and the theorem of Plancherel. Inserting the *test functions* $v(t) := e^{k2\pi tJ}\bar{v} \in E$ for $\bar{v} \in \mathbb{R}^{2n}$ and $k \in \mathbb{Z}$ into the identity $(**)$, we obtain $\langle 2\pi k x_k, \bar{v}\rangle = \langle a_k, \bar{v}\rangle$ for all $\bar{v} \in \mathbb{R}^{2n}$ and conclude that

$$2\pi k x_k = a_k, \quad k \in \mathbb{Z}.$$

In particular,

$$0 = a_0 = \int_0^1 \nabla H(x(t))\, dt.$$

Consequently, by Plancherel' s theorem,

$$\sum |k|^2|x_k|^2 \le \sum |a_k|^2 = \|\nabla H(x))\|_{L^2}^2 < \infty,$$

so that x is an element of H^1 and hence by Proposition VII.20 a continuous function.

(2) Next, we prove that x *is continuously differentiable.* In view of step (1) the function $\nabla H(x(t))$ is continuous, since H is smooth. The continuously differentiable function $\xi\colon [0,1] \to \mathbb{R}^{2n}$, defined by

$$\xi(t) = \int_0^t J\nabla H(x(s))\, ds,$$

is periodic of period 1, hence belongs to the space $C^1(S^1, \mathbb{R}^{2n})$. Indeed,

$$\xi(1) = J\int_0^1 \nabla H(x(s))\, ds = Ja_0 = 0$$

so that $\xi(1) = \xi(0) = 0$. The derivative of ξ satisfies

$$\dot{\xi}(t) = J\nabla H(x(t)), \quad t \in \mathbb{R}.$$

In order to determine the Fourier coefficients ξ_k of ξ for $k \neq 0$ we compute, using partial integration and $\xi(1) = 0 = \xi(0)$,

$$\begin{aligned}
\xi_k &= \int_0^1 e^{-k2\pi tJ}\xi(t)\, dt \\
&= (-k2\pi J)^{-1}e^{-k2\pi tJ}\xi(t)\Big|_0^1 - \int_0^1 (-k2\pi J)^{-1}e^{-k2\pi tJ}\dot{\xi}(t)\, dt \\
&= -\int_0^1 (-k2\pi J)^{-1}e^{-k2\pi tJ} J\nabla H(x(t))\, dt \\
&= \frac{1}{k2\pi}\int_0^1 e^{-k2\pi tJ}\nabla H(x(t))\, dt \\
&= \frac{1}{k2\pi}a_k \\
&= x_k
\end{aligned}$$

for all integers $k \neq 0$. Due to the uniqueness of the Fourier coefficients, $\xi(t) = x(t) + c$ and so, using $\xi(0) = 0$,

$$\xi(t) = x(t) - x(0).$$

We see that the function x is of class C^1.

(3) Finally, we prove that x *is a smooth function.* In view of $\dot{x} = \dot{\xi}$, the function x solves the Hamiltonian equation

$$\dot{x}(t) = J\nabla H(x(t)).$$

The right-hand side of the Hamiltonian equation is of class C^1. Therefore, also the derivative on the left-hand side is of class C^1, so that x is of class C^2 and hence the right-hand side is of class C^2, so that x is of class C^3, and so on. Iteratively using the Hamiltonian equation it follows that $x \in C^\infty(S^1, \mathbb{R}^{2n})$ is a smooth, 1-periodic solution of the Hamiltonian equation and the proof of Proposition VII.24 is complete. □

Next we show that the extended functional $\Phi\colon E \to \mathbb{R}$ satisfies the *P.S.-condition.* Here we shall use the dynamical behavior of the Hamiltonian system X_H far away from $0 \in \mathbb{R}^{2n}$ whose Hamiltonian function

$$H(z) = Q(z) = (\pi + \varepsilon)q(z), \quad |z| \geq R$$

is a quadratic form, for a large R. The corresponding Hamiltonian vector field $\dot{z} = X_Q(z) = J\nabla Q(z)$ for $z \in \mathbb{R}^{2n}$ decomposes into n *linear oscillators* in the symplectic planes $\operatorname{span}\{e_j, f_j\}$, so that explicitly, in the symplectic coordinates $z = (x, y)$,

$$\begin{aligned} \dot{x}_j &= c_j y_j, \\ \dot{y}_j &= -c_j x_j \end{aligned}$$

for $1 \leq j \leq n$. Recalling

$$q(z) = x_1^2 + y_1^2 + \frac{1}{N^2}\sum_{n=2}^{n}(x_j^2 + y_j^2)$$

for $z = (x, y) \in \mathbb{R}^n \times \mathbb{R}^n$, with a large integer N, we read off that the constants are equal to

$$\begin{aligned} c_1 &= 2(\pi + \varepsilon), \\ c_j &= 2(\pi + \varepsilon)/N^2, \quad 2 \leq j \leq n. \end{aligned}$$

The minimal periods of the oscillators are $T_j = 2\pi/c_j$. Since ε is positive and small, *we see that non-constant periodic solutions of period* 1 *do not exist*! This fact is one of the reasons for having chosen our special extension of the Hamiltonian function from the ellipsoid onto the whole space $\mathbb{R}^{2n}$. It will be crucial in the proof of the following proposition.

Proposition VII.25. *Every sequence x_k in E satisfying $\nabla\Phi(x_k) \to 0$ in E possesses a convergent subsequence. In particular $\Phi\colon E \to \mathbb{R}$ satisfies the P.S.-condition.*

Proof. Explicitly, the condition $\nabla\Phi(x_k) \to 0$ in E reads as

$$(*) \quad x_k^+ - x_k^- - \nabla b(x_k) \to 0 \quad \text{in } E.$$

(1) *We first assume that the sequence x_k is bounded in E.*

Since ∇b is compact (Proposition VII.23), a subsequence of $\nabla b(x_k)$ converges in E. Due to $(*)$ a subsequence of $x_k^+ - x_k^-$ also converges in E. Because the sequence of constants $x_k^0 \in \mathbb{R}^{2n}$ is bounded, it possesses a convergent subsequence, too. All in all, a subsequence of the sequence $x_k = x_k^+ + x_k^0 + x_k^-$ converges in E and the proposition is proved in this case.

(2) *If x_k is not bounded in E,* then there exists a subsequence satisfying

$$\|x_k\| \to \infty.$$

We shall show indirectly that this case never happens, so that step (1) demonstrates the proposition. The sequence

$$y_k = \frac{x_k}{\|x_k\|}$$

satisfies $\|y_k\| = 1$. The following claim leads to the desired contradiction.

Claim. There exists a convergent subsequence

$$y_k \to y \quad \text{in } E$$

whose limit $y \in E$ satisfies $\|y\| = 1$ and

$$(P^+ - P^-)y - \nabla b_\infty(y) = 0,$$

where $b_\infty(y) := \int_0^1 Q(y)\,dt$.

Postponing the proof of the claim we first use it to finish the proof of Proposition VII.25. One concludes, as in the proof of Proposition VII.24, that the limit y belongs to $C^\infty(S^1, \mathbb{R}^{2n})$ and solves the Hamiltonian equation

$$\dot{y}(t) = X_Q(y(t)) = J\nabla Q(y(t)).$$

Since $\|y\| = 1$, the solution y is a non-constant 1-periodic solution of the Hamiltonian vector field $\dot{z} = X_Q(z)$. However, in view of the above discussion such a solution cannot exist, so that only case (1) occurs.

Proof of the claim. We multiply the sequence in $(*)$ by $\frac{1}{\|x_k\|}$ and obtain by means of the explicit representation of the gradient $\nabla b(x_k)$ the converging sequence

$$(**) \quad \frac{\nabla\Phi(x_k)}{\|x_k\|} = (P^+ - P^-)y_k - j^*\left(\frac{\nabla H(x_k)}{\|x_k\|}\right) \to 0 \quad \text{in } E.$$

Since $|\nabla H(z)| \leq M|z|$ for a constant $M > 0$,

$$\frac{\|\nabla H(x_k)\|_{L^2}}{\|x_k\|} \leq M \frac{\|x_k\|_{L^2}}{\|x_k\|} \leq CM \frac{\|x_k\|}{\|x_k\|} = CM,$$

where $C < \infty$ is the norm of the embedding $j\colon E \to L^2$. So, this sequence is bounded in L^2. Since the adjoint operator $j^*\colon L^2 \to E$ is a compact operator (Proposition VII.19), the sequence

$$j^*\left(\frac{\nabla H(x_k)}{\|x_k\|}\right)$$

possesses a convergent subsequence in E. Consequently, due to the convergence in $(**)$, the sequence $(P^+ - P^-)y_k$ has a convergent subsequence. Moreover, the sequence $P^0 y_k$ of constants is bounded and so possesses a convergent subsequence. All in all, there is a converging subsequence

$$y_k \to y \quad \text{in } E$$

whose limit satisfies $\|y\| = 1$, since all the elements y_k have norms equal to 1. Because the embedding $E \subset L^2$ is continuous, the convergence

$$y_k \to y \quad \text{in } L^2$$

follows. The functions H and Q agree outside of a ball, so that

$$|\nabla H(z) - \nabla Q(z)| \leq K \quad \text{for all } z \in \mathbb{R}^{2n}$$

with a sufficiently large constant $K > 0$. In addition, the map $y \mapsto \nabla Q(y)\colon L^2 \to L^2$ is linear and continuous, because Q is a quadratic form. Using these ingredients we obtain the estimate

$$\begin{aligned}\left\|\frac{\nabla H(x_k)}{\|x_k\|} - \nabla Q(y)\right\|_{L^2} &\leq \frac{\|\nabla H(x_k) - \nabla Q(x_k)\|_{L^2}}{\|x_k\|} + \left\|\frac{\nabla Q(x_k)}{\|x_k\|} - \nabla Q(y)\right\|_{L^2} \\ &\leq \frac{K}{\|x_k\|} + \|\nabla Q(y_k) - \nabla Q(y)\|_{L^2} \to 0\end{aligned}$$

from which we conclude the convergence

$$\frac{\nabla H(x_k)}{\|x_k\|} \to \nabla Q(y) \quad \text{in } L^2.$$

Since $j^*\colon L^2 \to E$ is continuous, the convergence $j^*\left(\frac{\nabla H(x_k)}{\|x_k\|}\right) \to j^*(\nabla Q(y))$ in E follows. From $(**)$ we now conclude that the limit y solves the equation

$$(P^+ - P^-)y - j^*(\nabla Q(y)) = 0.$$

Now, arguing as in the computation of the gradient ∇b, we find that $\nabla b_\infty(y) = j^*(\nabla Q(y))$ is the E-gradient of the function b_∞. We have confirmed the claim and so the proposition is proved. □

The vector field $\nabla\Phi\colon E \to E$ is *globally Lipschitz-continuous*. Indeed, $\nabla\Phi = \nabla a - \nabla b$ where $\nabla a = (P^+ - P^-)$ is a bounded linear operator and where ∇b is Lipschitz-continuous by Proposition VII.23. Thus, the gradient equation

$$\dot{x} = -\nabla\Phi(x), \quad x \in E$$

generates a *global flow* $\varphi^t(x)$ which is defined for all $x \in E$ and $t \in \mathbb{R}$. The map

$$\varphi\colon \mathbb{R} \times E \to E, \quad (t,x) \to \varphi^t(x) =: x \cdot t$$

maps bounded sets onto bounded sets and possesses, in addition, the following *compactness property*.

Proposition VII.26 (Representation formula for the flow). *The flow $\varphi^t(x)$ of the gradient equation $\dot{x} = -\nabla\Phi(x)$ on E has the representation*

$$\varphi^t(x) = e^t x^- + x^0 + e^{-t}x^+ + K(t,x)$$

for all $t \in \mathbb{R}$ and all $x = x^- + x^0 + x^+ \in E$, where the rest term $K\colon \mathbb{R}\times E \to E$ is a compact map, that is, it maps bounded sets onto relatively compact sets.

Proof [Duhamel's variation of the constants formula]. The flow $\varphi^t(x)$ of a Lipschitz-continuous ordinary differential equation in E of the form

$$\dot{x} = Ax + f(x)$$

where $A \in \mathcal{L}(E)$ satisfies the formula

$$\varphi^t(x) = e^{tA}x + \int_0^t e^{(t-s)A} f(\varphi^s(x))\, ds.$$

Indeed, the right hand side, denoted by $\xi(t)$, solves the equation $\dot{\xi} = A\xi(t) + f(\varphi^t(x))$ and the initial condition $\xi(0) = x$, as can easily be verified. Because the flow $\varphi^t(x)$ solves the same initial value problem, the uniqueness of the solution implies that $\xi(t) = \varphi^t(x)$. Applied to our gradient equation

$$\dot{x} = -\nabla\Phi(x) = (P^- - P^+)x + \nabla b(x)$$

we obtain for the flow $\varphi^t(x)$ the desired representation formula with the rest term K defined by

$$K(t,x) := \int_0^t (e^{t-s}P^- + P^0 + e^{-t+s}P^+)\nabla b(\varphi^s(x))\, ds.$$

We have used that if $x = x^- + x^0 + x^+ \in E$, then, in view of the orthogonality of the projectors,

$$\begin{aligned} e^{t(P^- - P^+)}x &= e^{t(P^- - P^+)}(P^- + P^0 + P^+)x \\ &= (e^t P^- + P^0 + e^{-t}P^+)x \\ &= e^t x^- + x^0 + e^{-t}x^+. \end{aligned}$$

It remains to verify that $K(t,x)$ is a compact map. Recalling

$$\nabla b(x \cdot s) = j^* \nabla H(j(x \cdot s))$$

where $j^* \in \mathcal{L}(L^2, E)$ and $x \cdot s = \varphi^s(x)$, we can write the map K as

$$K(t,x) = j^*\Big(\int_0^t (e^{t-s}P^- + P^0 + e^{-t+s}P^+)\nabla H(j(x \cdot s))\, ds\Big).$$

We see that the map K is the composition of a continuous, bounded map $\mathbb{R}\times E \to L^2$ with the compact map $j^*\colon L^2 \to E$, so that K is also a continuous and compact mapping, as claimed in the proposition. □

VII.5 Existence of a critical point of Φ

In this section we finally establish a critical point x of the function $\Phi\colon E \to \mathbb{R}$ which satisfies $\Phi(x) > 0$ by means of a special minimax argument.

Let us briefly recall our strategy. We started with the action functional $\Phi\colon \Omega = C^\infty(S^1, \mathbb{R}^{2n}) \to \mathbb{R}$ of classical mechanics,

$$\Phi(x) = \int_0^1 \frac{1}{2}\langle -J\dot{x}, x\rangle\, dt - \int_0^1 H(x(t))\, dt$$

having the *formal* L^2 gradient

$$\nabla\Phi(x) = -J\dot{x} - \nabla H(x).$$

The functional Φ was extended to a continuously differentiable function $\Phi\colon E \to \mathbb{R}$, defined on the Hilbert space $E = H^{\frac{1}{2}} \supset \Omega$ by

$$\Phi(x) = a(x) - b(x) = \frac{1}{2}\|x^+\|^2 - \frac{1}{2}\|x^-\|^2 - \int_0^1 H(x(t))\, dt$$

where $\|\cdot\|$ denotes the norm of $H^{\frac{1}{2}}$ and where the splitting $x = x^- + x^0 + x^+ \in E = E^- \oplus E^0 \oplus E^+$ is defined by means of the Fourier coefficients. The E gradient of Φ is equal to

$$\nabla\Phi(x) = \nabla a(x) - \nabla b(x) = x^+ - x^- - j^*\nabla H(j(x)).$$

The function $H \in C^\infty(\mathbb{R}^{2n}, \mathbb{R})$ is non-negative and has, in addition, the special properties that H vanishes in a neighborhood of $z = 0$ and is equal to the quadratic form

$$H(z) = (\pi + \varepsilon)q(z), \quad |z| \geq R$$

outside of a large ball.

We should point out that the functional $\Phi\colon E \to \mathbb{R}$ has *many critical points*, but most of them are uninteresting! Examples are the critical points z^* of the Hamiltonian function $H\colon \mathbb{R}^{2n} \to \mathbb{R}$ satisfying $\nabla H(z^*) = 0$. These points are *constant* loops and therefore solve the equation

$$\nabla\Phi(z^*) = \nabla a(z^*) - j^*\nabla H(z^*) = 0.$$

Moreover,

$$\Phi(z^*) = a(z^*) - b(z^*) = 0 - H(z^*) \leq 0.$$

The goal of this section is the proof of the following result.

Theorem VII.27. *There exists a critical point $x \in E$ satisfying*

$$\nabla\Phi(x) = 0 \quad \textit{and} \quad \Phi(x) > 0.$$

By the regularity theorem (Proposition VII.24) the critical point x belongs to $C^\infty(S^1, \mathbb{R}^{2n})$ and solves the Hamiltonian equation

$$\dot{x}(t) = J\nabla H(x(t)).$$

In view of $\Phi(x) > 0$ and by Lemma VII.16 the loop $x(t)$ is a *non-constant* 1-periodic solution, lying in the ellipsoid $E \subset Z(1)$ where $H = \bar{H}$ agrees with the original Hamiltonian function. Therefore, Theorem VII.27 above completes the proof of Theorem VII.15. Consequently, also Theorem VII.10 is herewith proved and shows that the map

$$(M,\omega) \mapsto c_0(M,\omega)$$

is, indeed, a symplectic capacity. Together with the existence proof of the symplectic capacity c_0, also all its consequences in symplectic geometry, in particular, Gromov's non-squeezing theorem (Theorem VII.5), are proved. *We see that all the results in symplectic geometry so far follow from the existence of a special critical point of the old action functional of classical mechanics.*

Before presenting a detailed proof of Theorem VII.27 we shall briefly sketch the argument. We define for $\tau \geq 0$ the bounded subset $\Sigma = \Sigma_\tau \subset E$ by

$$\Sigma_\tau = \big\{x \in E \mid x = x^- + x^0 + se^+ \text{ where } \|x^- + x^0\| \leq \tau \text{ and } 0 \leq s \leq \tau\big\}.$$

Here $e^+ \in E^+$ denotes the special loop

$$e^+(t) := e^{2\pi t J} e_1$$

where $e_1 = (1, 0, \dots, 0) \in \mathbb{R}^{2n}$. In addition, we define for $\alpha > 0$ the bounded subset $\Gamma = \Gamma_\alpha \subset E^+$ by

$$\Gamma_\alpha = \{x \in E^+ \mid \|x\| = \alpha\}.$$

It is the sphere of radius α in E^+ centered at the origin.

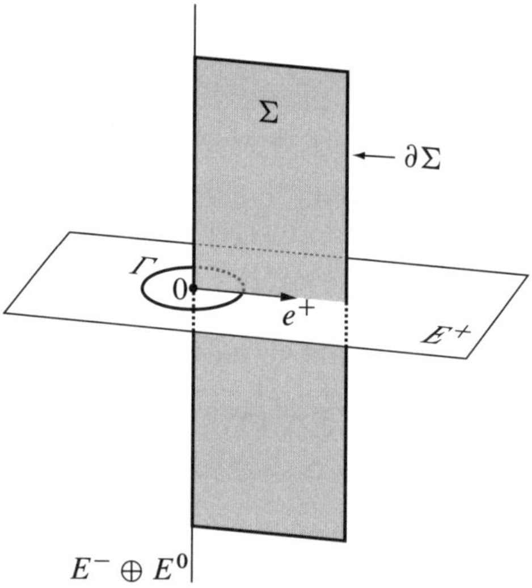

Figure VII.13. The subsets Σ and Γ of E.

Lemma VII.28. *On the boundary $\partial\Sigma_\tau$ of $\Sigma_\tau \subset E^- \oplus E^0 \oplus \mathbb{R}e^+$ the function Φ satisfies the estimate*

$$\Phi|_{\partial\Sigma_\tau} \le 0$$

if $\tau > 0$ is sufficiently large. In addition, there exist real numbers $\alpha > 0$ and $\beta > 0$ such that

$$\Phi|_{\Gamma_\alpha} \ge \beta > 0.$$

We postpone the *proof* of the lemma which uses the *pointwise* behavior of the function H.

Continuing with the proof of Theorem VII.27 we now let the set Σ *move* under the gradient flow ϕ^t of $\dot{x} = -\nabla\Phi(x)$. Since the function $\Phi(\varphi^t(x))$ is decreasing in t for $t \ge 0$, we conclude from Lemma VII.28 not only $\Phi|_{\partial\Sigma} \le 0$, but also

$$\Phi|_{\varphi^t(\partial\Sigma)} \le 0 \quad \text{for all } t \ge 0.$$

It therefore follows from $\Phi|_\Gamma \ge \beta > 0$, where $\Gamma = \Gamma_\alpha$, that

$$\varphi^t(\partial\Sigma) \cap \Gamma = \emptyset \quad \text{for all } t \ge 0.$$

We see that the *deformed frame $\varphi^t(\partial\Sigma)$ never intersects the sphere* Γ for $t \ge 0$. On the other hand, the two sets Γ and Σ do intersect in the point $se^+ \in \Sigma \cap \Gamma$ where $s = \alpha/\|e^+\|$ if τ is sufficiently large. Since the frame $\varphi^t(\partial\Sigma)$ cannot be pulled through the sphere as t increases, one expects the surface $\varphi^t(\Sigma)$, spanned by the frame, to intersect the sphere for every $t \geq 0$. This is intuitively quite clear looking at Figure VII.13 where the sphere is depicted as a circle.

Lemma VII.29 (Topological linking).

$$\varphi^t(\Sigma) \cap \Gamma \neq \emptyset$$

for all $t \geq 0$, *if* $\tau > 0$ *is sufficiently large.*

The proof of this lemma will be based on the Leray–Schauder degree. We postpone it and first complete the proof of Theorem VII.27.

Proof of Theorem VII.27 [Lemmata VII.28 and VII.29, Propositions VII.23 and VII.25, minimax principle]. In order to apply the minimax principle to the function Φ on the Hilbert space E we take as family $\mathcal{F}$ of sets the following one-parameter family

$$\varphi^t(\Sigma), \quad t \geq 0$$

of subsets of E and define the minimax

$$c(\Phi, \mathcal{F}) = \inf_{t \geq 0} \sup_{x \in \varphi^t(\Sigma)} \Phi(x).$$

We verify the assumptions of the minimax principle (Proposition VII.17) and first observe that the family $\mathcal{F}$ is positively invariant under the gradient flow, because of $\varphi^s(\varphi^t(\Sigma)) = \varphi^{s+t}(\Sigma)$. In view of Proposition VII.25 the functional Φ satisfies the P.S.-condition and, due to the global Lipschitz continuity, the gradient equation generates a global flow. It remains to show that $c(\Phi, \mathcal{F})$ is finite. Here, the two lemmata above play a decisive role.

Since $\varphi^t(\Sigma) \cap \Gamma \neq \emptyset$ for all $t \geq 0$ by Lemma VII.29, it follows from Lemma VII.28 that

$$\beta \leq \inf_{x \in \Gamma} \Phi(x) \leq \inf_{x \in \Gamma \cap \varphi^t(\Sigma)} \Phi(x) \leq \sup_{x \in \varphi^t(\Sigma)} \Phi(x) < \infty$$

for every $t \geq 0$. We have used that $\varphi^t(\Sigma)$ is a bounded set and that Φ maps bounded sets onto bounded sets, in view of Proposition VII.23. Hence $c(\Phi, \mathcal{F})$ is a real number so that all the assumptions of the minimax principle are met. The minimax principle now guarantees a critical point $x \in E$ satisfying

$$\nabla\Phi(x) = 0 \quad \text{and} \quad \Phi(x) = c(\Phi, \mathcal{F}) \geq \beta > 0$$

and Theorem VII.27 is proved. □

It remains to prove Lemma VII.28 and Lemma VII.29.

Proof of Lemma VII.28. The lemma follows from the properties of the Hamiltonian function H. We recall that

$$\begin{aligned}\Phi(x) &= a(x) - b(x)\\ &= \frac{1}{2}\|x^+\|^2 - \frac{1}{2}\|x^-\|^2 - b(x)\end{aligned}$$

where $b(x) \geq 0$. Hence, if x satisfies $x^+ = 0$, then $\Phi(x) \leq 0$.

(1) We first verify that

$$\Phi|_{\partial\Sigma_\tau} \leq 0$$

if τ is *sufficiently large*. For this purpose, we use the behavior of the function H at infinity. On the part $\partial\Sigma \cap (E^- \oplus E^0)$ of the boundary where $s = 0$, we have $\Phi \leq 0$, as already observed above.

On the other parts of the boundary $\partial\Sigma_\tau$ we have either $s = \tau$ or $\|x^- + x^0\| = \tau$. By construction of H there exists a constant $\gamma \geq 0$ such that

$$H(z) \geq (\pi + \varepsilon)q(z) - \gamma, \quad \text{for all } z \in \mathbb{R}^{2n}$$

and therefore,

$$\Phi(x) \leq a(x) - (\pi + \varepsilon)\int_0^1 q(x) + \gamma, \quad x \in E.$$

If $x = x^- + x^0 + se^+ \in \Sigma$ we obtain, using the orthogonality of the decomposition that

$$\int_0^1 q(x^- + x^0 + se^+)\,dt = \int_0^1 q(x^-) + \int_0^1 q(x^0) + \int_0^1 q(se^+).$$

Since x^0 is constant and $q(x^-) \geq 0$, one concludes that

$$\Phi(x) \leq a(x) - (\pi+\varepsilon)q(x^0) - (\pi+\varepsilon)\int_0^1 q(se^+) + \gamma, \quad x = x^- + x^0 + se^+ \in \Sigma.$$

Recalling

$$e^+ = e^{2\pi t J}e_1 = \cos(2\pi t)e_1 + \sin(2\pi t)Je_1 = \cos(2\pi t)e_1 - \sin(2\pi t)e_{n+1}$$

and the definition of the quadratic form q one sees that

$$q(se^+) = s^2[\cos(2\pi t)^2 + \sin(2\pi t)^2] = s^2.$$

For the next estimate we note that

$$\|e^+\|_E^2 = 2\pi, \quad \|e^+\|_{L^2} = 1$$

and compute

$$\begin{aligned}\Phi(x) &\leq \frac{1}{2}s^2\|e^+\|^2 - \frac{1}{2}\|x^-\|^2 - (\pi+\varepsilon)q(x^0) - (\pi+\varepsilon)\int_0^1 q(se^+) + \gamma \\ &= -\frac{1}{2}\|x^-\|^2 - \varepsilon s^2\|e^+\|_{L^2}^2 - (\pi+\varepsilon)q(x^0) + \gamma \\ &\leq \gamma - c\|x^- + x^0\|^2 - c\|se^+\|^2\end{aligned}$$

for a constant $c > 0$. In both cases, namely $\|x^- + x^0\| = \tau$ and $s = \tau$, it follows that $\Phi(x) \leq 0$ for sufficiently large τ.

(2) In order to verify the second statement of the lemma,

$$\Phi|_{\Gamma_\alpha} \geq \beta > 0$$

for suitable positive numbers $\alpha, \beta > 0$, we use the behavior of the Hamiltonian function H near $z = 0$. Since $\Gamma \subset E^+$, we have $a(x) = \frac{1}{2}\|x\|^2$ for $x \in \Gamma$. We now make use of the following *Sobolev estimate*, the proof of which can be found in Appendix C in [52].

Lemma VII.30. *The embedding $H^{\frac{1}{2}} \subset L^p(S^1)$ is continuous for every $1 \leq p < \infty$. In other words, there are constants M_p such that*

$$\|u\|_{L^p} \leq M_p\|u\|_{\frac{1}{2}}$$

for all $u \in H^{\frac{1}{2}}$.

Since H vanishes in a neighborhood of $z = 0$ and is quadratic at infinity, it follows that $H(z) \leq c|z|^3$ for all $z \in \mathbb{R}^{2n}$, for a constant c. Therefore, the Sobolev estimate above leads to the estimate

$$\begin{aligned}b(x) &= \int_0^1 H(x(t))\,dt \\ &\leq c\int_0^1 |x(t)|^3\,dt \\ &= c\|x\|_{L^3}^3 \\ &\leq K\|x\|^3,\end{aligned}$$

for all x in E, with a constant $K > 0$. If $x \in E^+$, then

$$\Phi(x) \geq \frac{1}{2}\|x\|^2 - K\|x\|^3 = \|x\|^2\Big(\frac{1}{2} - K\|x\|\Big).$$

Choosing

$$\alpha := \frac{1}{4K},$$

we obtain the desired estimate

$$\Phi(x) \geq \frac{1}{4}\alpha^2 =: \beta > 0 \quad \text{for all } x \in \Gamma_\alpha.$$

The proof of Lemma VII.28 is complete. □

The proof of Lemma VII.29 is of topological nature and based on the Leray–Schauder degree which we shall briefly recall next. We start by recalling the Brouwer mapping degree which is probably known to most our readers. It counts the algebraic number of the solutions of an equation $f(x) = y$ in an open bounded set $\Omega \subset \mathbb{R}^n$ where $f: \bar{\Omega} \to \mathbb{R}^n$ is a continuous mapping and where $y \in \mathbb{R}^n \setminus f(\partial\Omega)$. Thus we consider triples (f, Ω, y) consisting of an open bounded subset $\Omega \subset \mathbb{R}^n$, a continuous map $f: \bar{\Omega} \to \mathbb{R}^n$ and a point $y \in \mathbb{R}^n \setminus f(\partial\Omega)$. Such a triple is called admissible.

Proposition VII.31 (Brouwer). *There exists a unique map d associating with every admissible triple (f, Ω, y) an integer $d(f, \Omega, y) \in \mathbb{Z}$ satisfying the following properties.*

(d1) ***Normalization.*** $d(\mathrm{Id}, \Omega, y) = 1$ *for all* $y \in \Omega$.

(d2) ***Additivity.*** *If Ω_1 and Ω_2 are two disjoint open subsets of Ω and if $y \notin f(\bar{\Omega} \setminus \Omega_1 \cup \Omega_2)$, then $d(f, \Omega, y) = d(f, \Omega_1, y) + d(f, \Omega_2, y)$.*

(d3) ***Homotopy invariance.*** *If $h: \bar{\Omega} \times [0, 1] \to \mathbb{R}^n$ is a continuous map and $y: [0, 1] \to \mathbb{R}^n$ a continuous curve satisfying $y(t) \notin h(\partial\Omega, t)$ for all t, then $d(h(\cdot, t), \Omega, y(t))$ is independent of t.*

(d4) ***Existence principle.*** *If $d(f, \Omega, y) \neq 0$, then there exists a solution of $f(x) = y$ in Ω.*

(d5) ***Continuity.*** *The map $f \mapsto d(f, \Omega, y) \in \mathbb{Z}$ is continuous with respect to the supremum norm. More explicitly, $d(f, \Omega, y) = d(g, \Omega, y)$ for all g satisfying the estimate $\|f - g\|_{C^0(\bar{\Omega})} < \mathrm{dist}(y, f(\partial\Omega))$.*

The map d in the proposition is called a *Brouwer mapping degree*.

Proof. We merely give a rough *sketch* of the proof. We consider an admissible triple (Ω, f, y) and approximate the continuous map f by a smooth map $g \in C^\infty(\Omega, \mathbb{R}^n) \cap C^0(\bar{\Omega}, \mathbb{R}^n)$, so that $\|f - g\|_{C^0(\bar{\Omega})} < r$ where $r := \mathrm{dist}(y, f(\partial\Omega))$ is a positive number, since $f(\partial\Omega)$ is a compact set. If y is a regular value of the function g, so that $dg(x)$ is a linear isomorphism at every preimage $x \in g^{-1}(y)$, then one defines the integer

$$d(f, \Omega, y) := \sum_{x \in f^{-1}(y)} \operatorname{sign} \det dg(x).$$

By the implicit function theorem the set $f^{-1}(y)$ is finite. If y is not a regular value one chooses a regular value y' near y in the same component of $\mathbb{R}^n \setminus f(\partial\Omega)$ making use of Sard's theorem. One proves that the integer $d(f, \Omega, y)$ is independent of the choice of the smooth function g and the regular value y' near y and satisfies the desired properties. This finishes the sketch of our proof. □

For detailed proofs we refer to the monographs [26] by K. Deimling, [5] by H. Amann and [13] by R. Brown. We also recommend the IMPA- lectures [4] by H. Amann. A classic are the marvelous lectures [67] by J. Milnor. A homological construction of the Brouwer mapping degree can be found in Appendix 7 in [52] by H. Hofer and E. Zehnder.

We illustrate the Brouwer degree by a simple example.

Example. We consider the open set $\Omega = (a, b) \subset \mathbb{R}$ whose closure is equal to $\bar{\Omega} = [a, b]$ and a smooth function $f: [a, b] \to \mathbb{R}$. We would like to count the solutions of $f(x) = 0$ and hence consider the triples $(f, \Omega, 0)$. For such triples to be admissible, we have to require that $0 \notin f(\partial\Omega)$, so that $f(a) \neq 0 \neq f(b)$. In addition, we assume that $y = 0$ is a regular value of f, so that

$$d(f, \Omega, 0) = \sum_{x \in f^{-1}(0)} \operatorname{sign} f'(x).$$

In order to compute the index we deform the function f to the affine linear function $f_1: [a, b] \to \mathbb{R}$, determined by $f_1(a) = f(a)$ and $f_1(b) = f(b)$, by means of the admissible homotopy $h(x, t) = tf_1(x) + (1 - t)f(x)$ for t in $[0, 1]$ which is illustrated in Figure VII.14. Using the homotopy invariance of degree 1 obtains

$$d(f, \Omega, 0) = d(f_1, \Omega, 0) = \begin{cases} -1, & f(a) > 0 > f(b), \\ +1, & f(a) < 0 < f(b), \\ 0, & f(a) \cdot f(b) > 0. \end{cases}$$

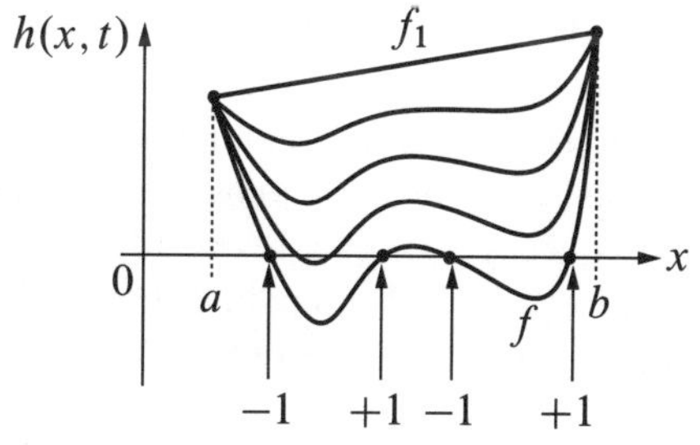

Figure VII.14. Using the homotopy invariance for the calculation of d.

We point out that the index can change during a non-admissible homotopy as the following example illustrates. We choose the functions f_0 and f_1 as illustrated in Figure VII.15 and define the homotopy between them by $h(x, t) =$

$tf_1(x)+(1-t)f_0(x)$. Since there exists a parameter value t_0 satisfying $h(t_0, b) = 0$, the homotopy is not admissible and the invariance of the degree is lost, since $d(f_0, \Omega, 0) = 0$ and $d(f_1, \Omega, 0) = -1$.

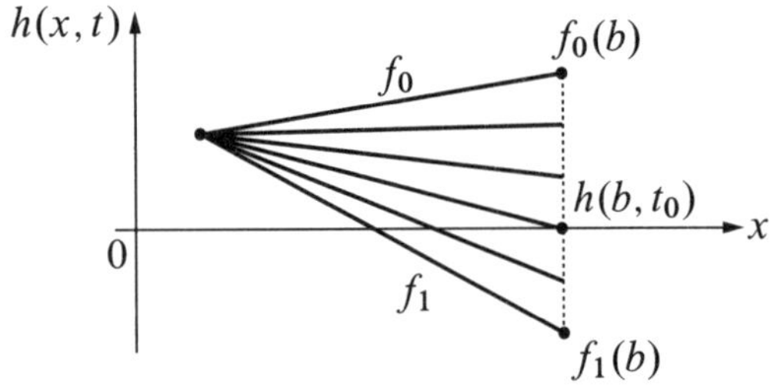

Figure VII.15. An example of a *non*-admissible homotopy.

In order to define the mapping degree in an infinite dimensional normed real vector space F, the class of maps under consideration has to be restricted. We consider triples (f, Ω, y) where $\Omega \subset F$ is an open and bounded subset, $f : \bar{\Omega} \to F$ a continuous map, such that, in addition, the map $f - \mathrm{Id}$ is a compact map, hence mapping bounded sets onto relatively compact sets. Moreover, $y \in F$ satisfies $y \notin f(\partial\Omega)$. These maps f are sometimes called compact vector fields and we abbreviate the set of compact vector fields $f : \bar{\Omega} \to F$ by $\mathrm{KV}(\bar{\Omega}, F)$.

Remark. If Ω is bounded and if $f : \bar{\Omega} \to F$ is a compact vector field, then f is a closed map, that is, if $A \subset \bar{\Omega}$ is a closed set, then its image $f(A)$ is also a closed set.

Proof. We assume that $A \subset \bar{\Omega}$ is a closed set and that the sequence $f(x_n)$ of points in $f(A)$ converges to a point y in F. Since A is a bounded set and $f - \mathrm{Id}$ is a compact map, the image set $(f - \mathrm{Id})(A)$ is relatively compact. Therefore, the sequence $f(x_n) - x_n$ possesses a convergent subsequence and hence the sequence x_n has a convergent subsequence x_j which necessarily converges to an element x in A, because A is a closed set. The continuity of the map f implies that $y = \lim f(x_j) = f(x) \in f(A)$ which proves the remark. □

Theorem VII.32 (Leray–Schauder). *There exists a unique map D associating with every admissible triple (f, Ω, y) in which*

$$\Omega \subset F \textit{ is open and bounded, } f \in \mathrm{KV}(\bar{\Omega}, F) \textit{ and } y \in F \setminus f(\partial\Omega)$$

an integer $D(f, \Omega, y)$ in $\mathbb{Z}$ satisfying the following axioms.

(D1) ***Normalization.*** $D(\mathrm{Id}, \Omega, y) = 1$ *for all* $y \in \Omega$.

(D2) ***Additivity.*** *If Ω_1 and Ω_2 are two disjoint open subsets of Ω and if $y \notin f(\bar{\Omega} \setminus \Omega_1 \cup \Omega_2)$, then $D(f, \Omega, y) = D(f, \Omega_1, y) + D(f, \Omega_2, y)$.*

(D3) ***Homotopy invariance.*** *If $h \colon \bar{\Omega} \times [0,1] \to F$ is a continuous map such that the map $(x,t) \mapsto h(x,t) - x$ is compact and if $y \colon [0,1] \to F$ is a continuous curve satisfying $y(t) \notin h(\partial\Omega, t)$ for all t, then the integer $D(h(\cdot,t), \Omega, y(t))$ is independent of t.*

*This map D is called the **Leray–Schauder degree**.*

Postponing the proof of this theorem, we shall first derive some consequences of the axioms.

Proposition VII.33. *The axioms* (D1)–(D3) *of the Leray–Schauder degree imply the following additional properties:*

(D1′) *$D(\mathrm{Id}, \Omega, y) = 1$ for $y \in \Omega$ while $D(\mathrm{Id}, \Omega, y) = 0$ for $y \notin \bar{\Omega}$.*

(D4) *$D(f, \emptyset, y) = 0$ for all $y \in F$.*

(D5) ***Excision property.*** *For an open set $\Omega_1 \subset \Omega$ satisfying $y \notin f(\bar{\Omega} \setminus \Omega_1)$ we have $D(f, \Omega, y) = D(f, \Omega_1, y)$.*

(D6) ***Existence principle.*** *If $D(f, \Omega, y) \neq 0$, then $y \in f(\Omega)$, i.e., there exists a solution of $f(x) = y$ in Ω.*

(D7) ***Continuity.*** *The map $f \mapsto D(f, \Omega, y) \in \mathbb{Z}$ is continuous with respect to the supremum norm. More explicitly, $D(f, \Omega, y) = D(g, \Omega, y)$ for all $g \in KV(\bar{\Omega}, F)$ satisfying the estimate $\|f - g\|_{C^0(\bar{\Omega})} < \operatorname{dist}(y, f(\partial\Omega))$.*

(D8) ***Translation invariance.*** *$D(f, \Omega, y) = D(f - y, \Omega, 0)$ for all $y \in F$.*

(D9) ***Boundary value property.*** *If $f, g \in \mathrm{KV}(\bar{\Omega}, F)$ satisfy $f|_{\partial\Omega} = g|_{\partial\Omega}$, then $D(f, \Omega, y) = D(g, \Omega, y)$.*

Proof. The property (D1′) follows from the axioms (D2) and (D1) by choosing $\Omega_1 = \Omega$ and Ω_2 a small ball around y satisfying $\Omega_1 \cap \Omega_2 = \emptyset$. From the axiom (D2) we conclude the property (D4) by choosing $\Omega_1 = \Omega$ and $\Omega_2 = \emptyset$. As a consequence, the property (D5) now follows from axiom (D2) and from (D4) by choosing $\Omega_2 = \emptyset$. Property (D9) is deduced from axiom (D3) by using the homotopy $h(x,t) = tf(x) + (1-t)g(x)$, and property (D8) follows from axiom (D3) choosing the homotopy $h(x,t) = f(x) - ty$ and the continuous curve $y(t) = (1-t)y$.

In order to prove property (D6) indirectly we assume that $D(f, \Omega, y) \neq 0$ and $y \notin f(\Omega)$. Then $y \notin f(\bar{\Omega})$, since by assumption $y \notin f(\partial\Omega)$. Choosing $\Omega_1 = \Omega_2 = \emptyset$ in axiom (D2) we conclude by property (D4) that $D(f, \Omega, y) = 0$ contradicting our assumption.

Finally, in order to verify property (D7) we recall that f is a closed map, so that $r = \operatorname{dist}(y, f(\partial\Omega))$ is positive, if $y \notin f(\partial\Omega)$. If $g \in \mathrm{KV}(\bar{\Omega}, F)$ satisfies $\|f - g\|_{C^0(\Omega, F)} < r$, then the homotopy $h(x,t) = f(x) + t(g(x) - f(x))$ is admissible and hence, by axiom (D3), $D(f, \Omega, y) = D(g, \Omega, y)$. Thus the map

$D(\cdot, \Omega, y)$ is locally constant in f and hence continuous in the supremum norm. This completes the proof of Proposition VII.33. □

The proof of Theorem VII.32 follows from the properties of the Brouwer mapping degree together with the following two technical lemmata.

Lemma VII.34. *If $B \subset F$ is a closed and bounded subset of the normed space F, then a continuous and compact mapping $g\colon B \to F$ can be approximated in the supremum norm by continuous compact maps whose images are contained in finite dimensional subspaces of F.*

Proof. Let $\varepsilon > 0$. Since $\overline{g(B)}$ is compact, there exist finitely many points $y_1, \dots, y_m$ in the image $g(B)$ such that

$$\overline{g(B)} \subset \bigcup_{i=1}^{m} B_\varepsilon(y_i).$$

Introducing the continuous functions $\varphi_i(y) := \max\{0, \varepsilon - \|y - y_i\|\}$ for $y \in \overline{g(B)}$ we observe that the functions $\psi_i(y) = \varphi_i(y)/\sum_{j=1}^m \varphi_j(y)$ are well defined continuous functions on $\overline{g(B)}$ and we define the map g_1 on B into F by

$$g_1(x) := \sum_{i=1}^{m} \psi_i(g(x)) y_i, \quad x \in B.$$

The map g_1 is continuous and compact and its image $g_1(B)$ is contained in the finite dimensional subspace $\operatorname{span}\{y_1, \dots, y_m\}$ of F. Further, in view of $\sum_{i=1}^m \psi_i(y) = 1$ for y in $\overline{g(B)}$,

$$\sup_{x \in B} \|g(x) - g_1(x)\| \le \varepsilon,$$

as is readily verified. □

Lemma VII.35. *Let $\Omega \subset F_0 = \mathbb{R}^n$ be an open and bounded subset, and let $f\colon \bar{\Omega} \subset F_0 \to F_0$ be a continuous map satisfying*

$$(f - \mathrm{Id})(\bar{\Omega}) \subset F_1,$$

where $F_1 \subset F_0$ is a linear subspace of F_0. If $y \in F_1 \setminus f(\partial\Omega)$, then the Brouwer degree satisfies

$$d(f, \Omega, y) = d(f|_{\overline{\Omega \cap F_1}}, \Omega \cap F_1, y).$$

Proof. If $m = \dim F_1$ we can assume without loss of generality that the subspace F_1 is represented as

$$F_1 = \{x \in F_0 \mid x_{m+1} = \dots = x_n = 0\}.$$

In view of the construction of the Brouwer degree we may assume that the map f is smooth and that y is a regular value of f. The map $g := f - \mathrm{Id}$ has the representation $g(x) = (g_1(x), \dots, g_m(x), 0, \dots, 0)$. Thus

$$\begin{aligned} df(x) &= dg(x) + \mathbb{1}_n \\ &= \left(\begin{array}{c|c} d(g|_{F_1})(x) & * \\ \hline 0 & 0 \end{array}\right) + \mathbb{1}_n \\ &= \left(\begin{array}{c|c} d(f|_{F_1})(x) & * \\ \hline 0 & \mathbb{1}_{n-m} \end{array}\right). \end{aligned}$$

We read off that $\det df(x) = \det d(f|_{F_1})(x)$ and the claim is proved. □

Proof of Theorem VII.32. Again we merely sketch the construction of the Leray–Schauder degree. Considering admissible triples (f, Ω, y) we first observe that it suffices to consider the special case $y = 0$. Indeed, if we have defined the degree $D(f, \Omega, y)$ satisfying the axioms (D1)–(D3) of the theorem and then use the translation invariance of $KV(\bar{\Omega}, F)$, we obtain the general case by setting $D(f, \Omega, y) := D(f - y, \Omega, 0)$. By assumption, the map $g = f - \mathrm{Id}\colon \Omega \subset F \to F$ is continuous and compact and by Lemma VII.34 there exists a continuous and compact map g_1 close to the map g in the supremum norm whose image lies in a finite dimensional subspace $F_1 \subset F$. Introducing the open set $\Omega_1 := \Omega \cap F_1$ in F_1 we consider the continuous map $f_1 := g_1 + \mathrm{Id}\colon \bar{\Omega}_1 \subset F_1 \to F_1$ and define

$$D(f, \Omega, 0) = d(f_1|_{\bar{\Omega}_1}, \Omega_1, 0)$$

where d is the Brouwer degree.

If we take another approximation g_2 of $f - \mathrm{Id}$ having its image in the finite dimensional subspace F_2 we can similarly define the Brouwer degree $d(f_2|_{\bar{\Omega}_2}, \Omega_2, 0)$. Take now the finite dimensional subspace $F_0 = \mathrm{span}\{F_1, F_2\}$ and set $\Omega_0 = \Omega \cap F_0$. Then it follows from Lemma VII.35 together with the continuity of the Brouwer degree that

$$d(f_1|_{\bar{\Omega}_1}, \Omega_1, 0) = d(f_1|_{\bar{\Omega}_0}, \Omega_0, 0) = d(f_2|_{\bar{\Omega}_0}, \Omega_0, 0) = d(f_2|_{\bar{\Omega}_2}, \Omega_2, 0),$$

so that $D(f, \Omega, 0)$ is well defined. One can demonstrate that the axioms (D1)–(D3) for $D(f, \Omega, 0)$ follow from the properties of the Brouwer degree simply by applying its properties to the finite dimensional approximations. □

We now use the Leray–Schauder degree in order to prove Lemma VII.29.

Proof of Lemma VII.29 [Degree theory]. We shall prove that $\varphi^t(\Sigma_\tau) \cap \Gamma \neq \emptyset$ for all $t \geq 0$, if τ is large enough.

(1) By definition of Γ we look for a point $x \in \Sigma$ satisfying $\varphi^t(x) \in E^+$ and $\|\varphi^t(x)\| = \alpha$. We abbreviate the flow by $\phi^t(x) = x \cdot t$. Then, we have to solve in view of the orthogonal decomposition $E = E^- \oplus E^0 \oplus E^+$, the equation

$$\begin{cases} 0 = (P^- + P^0)(x \cdot t), \\ \alpha = \|x \cdot t\|, \\ x \in \Sigma. \end{cases}$$

Using the representation

$$x \cdot t = e^t x^- + x^0 + e^{-t} x^+ + K(t, x)$$

of the flow from Proposition VII.26, this is equivalent to solving the equation

$$\begin{cases} 0 = e^t x^- + x^0 + (P^- + P^0) K(t, x), \\ 0 = \alpha - \|x \cdot t\|, \\ x \in \Sigma. \end{cases}$$

Since an element x of Σ is by definition of the form $x = x^- + x^0 + se^+$ where $0 \le s \le \tau$, we can rewrite the previous equation as an equation of the form

$$0 = x + B(t, x) \quad \text{for } x \in \Sigma$$

in the normed space $F := E^- \oplus E^0 \oplus \mathbb{R}e^+$, where the map $B\colon \mathbb{R} \times F \to F$ is defined by

$$B(t, x) = (e^{-t} P^- + P^0) K(t, x) + P^+[(\|x \cdot t\| - \alpha)e^+ - x].$$

In view of Proposition VII.26, the map $B\colon \mathbb{R} \times F \to F$ is continuous and compact. We are faced with the following situation. For every fixed $t \ge 0$ we consider the map

$$f_t\colon \Sigma \subset F \to F, \quad f_t(x) = x + B(t, x)$$

and look for a solution $x \in \Sigma$ of the equation

$$f_t(x) = 0,$$

where $f_t(x) - x = B(t, x)$ is a continuous and compact mapping. We know from Lemma VII.28 and the following discussion that $\varphi^t(\partial\Sigma) \cap \Gamma = \emptyset$ for all $t \ge 0$, so that there are *no* solutions $x \in \partial\Sigma$ of $f_t(x) = 0$ on the boundary, that is,

$$0 \notin f_t(\partial\Sigma) \quad \text{for all } t \ge 0.$$

(2) We can, therefore, apply the Leray–Schauder degree in order to show that

$$D(f_t, \mathring{\Sigma}, 0) \ne 0 \quad \text{for all } t \ge 0,$$

where $\overset{\circ}{\Sigma}$ denotes the interior of Σ. Using the homotopy invariance of the degree we conclude

$$D(f_t, \overset{\circ}{\Sigma}, 0) = D(f_0, \overset{\circ}{\Sigma}, 0), \quad t \geq 0.$$

In view of $K(0, x) = 0$, the map $f_0 \colon \Sigma \to F$ has the representation

$$f_0(x) = x + P^+[(\|x\| - \alpha)e^+ - x], \quad x \in \Sigma.$$

We now deform the map f_0 using the homotopy

$$g_\mu(x) = x + P^+[(\mu\|x\| - \alpha)e^+ - \mu x], \quad x \in \Sigma$$

where $\mu \in [0, 1]$, so that

$$g_1(x) = f_0(x),$$
$$g_0(x) = x - \alpha e^+$$

if $\mu = 1$ resp. $\mu = 0$.

Claim. We claim that if τ is sufficiently large, then $g_\mu(x) \neq 0$ for all $x \in \partial\Sigma$ and the homotopy is admissible.

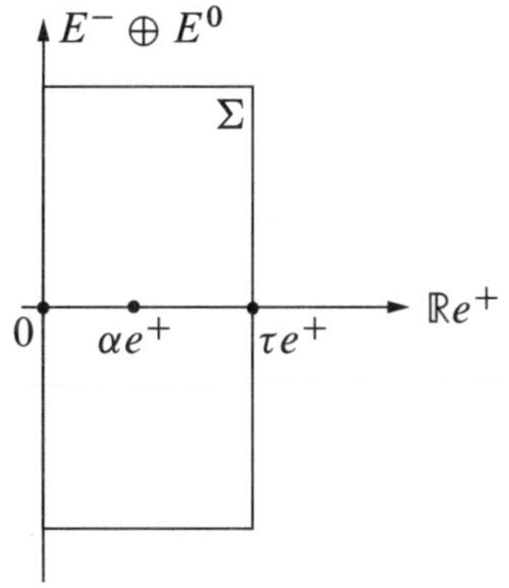

Figure VII.16. Illustrating the proof of Lemma VII.29.

Proof of the claim. If $g_\mu(x) = 0$ for a point $x \in \Sigma$, then, necessarily, $x = se^+$ for some $s \geq 0$ and inserting $x = se^+$ into the equation

$$0 = g_\mu(x) = x + P^+[(\mu\|x\| - \alpha)e^+ - \mu x]$$

we see that s solves the equation

$$s[(1 - \mu) + \mu\|e^+\|] = \alpha.$$

Recalling $\|e^+\| = \sqrt{2\pi} > 1$ and $0 \le \mu \le 1$ it follows that

$$0 < s \le \alpha.$$

If $x = se^+ \in \partial\Sigma$, then $s = 0$ or $s = \tau$. Thus, there exists *no* solution of $g_\mu(x)) = 0$ on the boundary $\partial\Sigma$, if $\tau > \alpha$. Finally, also the compactness condition for the homotopy is satisfied, since the image of the map $g_\mu - \mathrm{Id}$ lies in the subspace $\mathbb{R}e^+$ of dimension 1 and we have verified that the homotopy is indeed admissible, as claimed. □

Consequently, using the homotopy invariance and the translation invariance of the Leray–Schauder degree we obtain

$$\begin{aligned} D(f_0, \mathring{\Sigma}, 0) &= D(g_1, \mathring{\Sigma}, 0) \\ &= D(g_0, \mathring{\Sigma}, 0) \\ &= D(\mathrm{Id} - \alpha e^+, \mathring{\Sigma}, 0) \\ &= D(\mathrm{Id}, \mathring{\Sigma}, \alpha e^+). \end{aligned}$$

From Proposition VII.33 we know that $D(\mathrm{Id}, \mathring{\Sigma}, \alpha e^+) = 1$, if $\alpha e^+ \in \mathring{\Sigma}$. Since the latter is equivalent to $\tau > \alpha$, we have proved that

$$D(f_t, \mathring{\Sigma}, 0) = 1 \quad \text{for all } t \ge 0,$$

if τ is sufficiently large. In view of the existence principle in Proposition VII.33 there exists for every $t \ge 0$ a point x in the interior of Σ which solves the equation $f_t(x) = 0$. Therefore, $\varphi^t(\Sigma) \cap \Gamma \ne \emptyset$ for all $t \ge 0$ and the proof of Lemma VII.29 is complete. □

With Lemma VII.29 the proofs of Theorem VII.27 and hence that of Theorem VII.15 are complete and we have finally established the existence of the Hofer–Zehnder symplectic capacity function c_0. It does not only explain the rigidity phenomena in symplectic geometry described above, but also helps to find global periodic solutions of Hamiltonian vector fields on compact energy surfaces, as the next chapter will demonstrate.

Literature. Chapter VII is based on part of the book [52] by H. Hofer and E. Zehnder which contains many references and remarks about the history of the results. An introduction to symplectic topology is the book [65] by D. McDuff and D. Salamon. As for embedding problems in symplectic geometry we refer to the monograph [99] by F. Schlenk. For recent developments and applications of symplectic capacities as well as for all the references and many open problems we recommend [18] by K. Cieliebak, H. Hofer, J. Latschev and F. Schlenk.

Chapter VIII
Applications of the capacity c_0 in Hamiltonian systems

In Chapter VIII we use the dynamical capacity c_0 to tackle the problem of finding periodic orbits of a Hamiltonian vector field on a prescribed compact energy surface S. We shall establish an abundance of periodic orbits in an open neighborhood U of S if only the Hofer–Zehnder capacity $c_0(U, \omega)$ is finite. However, in order to guarantee periodic orbits on S itself, the class of energy surfaces has to be restricted, for example to the class of contact type hypersurfaces introduced by A. Weinstein. We shall recover, in particular, the pioneering results of A. Weinstein and P. Rabinowitz for convex and starlike hypersurfaces. It turns out that compact energy surfaces of classical mechanical systems of the form kinetic energy plus potential energy always carry a periodic orbit. A classical result due to H. Poincaré shows that, in general, a periodic solution belongs to a smooth family of periodic solutions parametrized by the energy and having similar periods.

VIII.1 Global periodic solutions on prescribed energy surfaces

The flow φ^t of a Hamiltonian vector field X_H on a symplectic manifold (M, ω) leaves the level sets of the smooth Hamiltonian function H on M invariant because $H(\varphi^t(x)) = H(x)$ as long as the flow is defined. Fixing a value of the energy function H which we may assume without loss of generality to be $E = 0$, we shall require that the energy surface

$$S = \{x \in M \mid H(x) = 0\}$$

is compact and regular requiring that $dH(x) \neq 0$ on S. Thus $S \subset M$ is a smooth and compact submanifold of codimension 1 (a compact hypersurface) whose tangent space at the point $x \in S$ is given by $T_x S = \ker dH(x)$. By definition of a Hamiltonian vector field we have $X_H(x) \in T_x S$ so that X_H is a nowhere vanishing vector field on the energy surface S whose flow does exist for all times, since S is compact. Our aim is to find periodic solutions of X_H on the energy surface S.

It is useful to recall from Section VI.3 that the orbits of X_H on S only depend on the hypersurface and the symplectic form ω and not on the choice of the function H which represents S as a regular energy surface. Indeed, if H and F are two Hamiltonian functions representing S as

$$S = \{H = c_1\} = \{F = c_2\}, \quad dH(x) \neq 0,\ dF(x) \neq 0,\ x \in S,$$

then the tangent space $T_xS = \ker dH(x) = \ker dF(x) \subset T_xM$ is of codimension 1 and hence there exists a real number $\rho(x) \neq 0$ satisfying $dF(x) = \rho(x)dH(x)$ for $x \in S$. Consequently,

$$X_F(x) = \rho(x)X_H(x), \quad x \in S,$$

for a smooth nowhere vanishing function $\rho\colon S \to \mathbb{R}$. Of course, this only holds true on the given hypersurface S. The flows φ^t of X_H and ψ^t of X_F on S are related by

$$psi^{s(t,x)}(x) = \varphi^t(x), \quad x \in S,$$

where the function $s\colon \mathbb{R} \times S \to \mathbb{R}$ satisfies

$$s(t,x) = \int_0^t \frac{d\tau}{\rho(\varphi^\tau(x))}, \quad s(0,x) = 0,$$

as has already been mentioned in Chapter VI. We see that the orbits of the Hamiltonian vector fields X_H and X_F are indeed the same, only their parametrizations are different. To formulate this fact geometrically we have introduced the canonical line bundle $\mathfrak{L}_S$ of S whose fibre over the point x is the line

$$\mathfrak{L}_S(x) = \ker(\omega|_{T_xS}) = \{u \in T_xS \mid \omega(u,v) = 0 \text{ for all } v \in T_xS\}$$

in the tangent space T_xS. This line determines the direction of every Hamiltonian vector field whose Hamiltonian function has S as a regular energy surface. Consequently, instead of looking for periodic solutions of X_H on the hypersurface S we can just as well look for closed characteristics of the canonical line bundle $\mathfrak{L}_S$. They agree with the set of unparametrized periodic solutions of every Hamiltonian vector field X_H on S having S as a regular energy surface.

Our trick in the search for periodic solutions on S is *to thicken* the given energy surface $S = \{x \in M \mid H(x) = 0\}$ and to consider a *one-parameter family* of energy surfaces nearby. Since S is compact and regular there is an open and bounded neighborhood U of S which is filled with compact and regular energy surfaces having energy values near $E = 0$,

$$U := \bigcup_{\lambda \in I} S_\lambda$$

where $I = (-\varepsilon, \varepsilon)$ is an open interval around $\lambda = 0$ and where

$$S_\lambda = \{x \in U \mid H(x) = \lambda\}, \quad \lambda \in I$$

is diffeomorphic to the given surface $S = S_0$ which corresponds to the parameter value $\lambda = 0$. Indeed, the following proposition holds true.

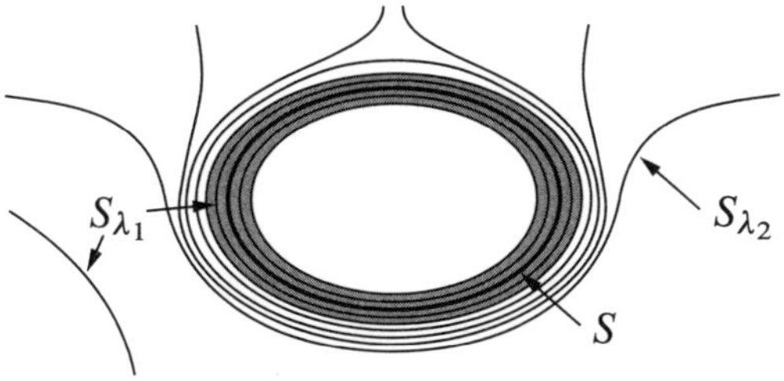

Figure VIII.1. Thickening of the compact energy surface S and the neighborhood U.

Proposition VIII.1. *If $I = (-\varepsilon, \varepsilon)$ is an open interval and $\varepsilon > 0$ sufficiently small, then*

$$U := \bigcup_{\lambda \in I} S_\lambda$$

is the image of a diffeomorphism $\psi\colon S \times I \to U$ satisfying $\psi(S \times \{\lambda\}) = S_\lambda$ where $S_\lambda = \{x \in U \mid H(x) = \lambda\}$.

Proof. We choose a Riemannian metric $\langle \cdot, \cdot \rangle$ on the manifold M and define the gradient $\nabla H(x)$ of the function H at the point $x \in M$ as usual by

$$dH(x)v = \langle \nabla H(x), v \rangle, \quad v \in T_x M.$$

The smooth gradient vector field ∇H on M does not vanish at any point in U so that

$$dH(x)\, \nabla H(x) = \langle \nabla H(x), \nabla H(x) \rangle = |\nabla H(x)|^2 > 0.$$

The vector $\nabla H(x)$ is *orthogonal* to the tangent space $T_x S$, since

$$\langle \nabla H(x), v \rangle = dH(x)v = 0 \quad \text{for all } v \in T_x S = \ker dH(x).$$

Therefore the flow ψ^t of the *modified gradient equation*

$$\dot{x} = \frac{\nabla H(x)}{|\nabla H(x)|^2}, \quad x \in M \text{ near } S$$

is *transversal* to the energy surface S and we claim that

$$H(\psi^t(x)) = t, \quad x \in S,\ |t| < \varepsilon$$

for $\varepsilon > 0$ small, as long as the gradient does not vanish. Indeed,

$$\begin{aligned} \frac{d}{dt} H(\psi^t(x)) &= dH(\psi^t(x))\, \frac{d}{dt}\psi^t(x) \\ &= \left\langle \nabla H(\psi^t(x)), \frac{\nabla H(\psi^t(x))}{|\nabla H(\psi^t(x))|^2} \right\rangle \\ &= 1, \end{aligned}$$

and hence

$$H(\psi^t(x)) = H(\psi^0(x)) + \int_0^t \frac{d}{ds} H(\psi^s(x))\, ds = H(x) + t.$$

If $x \in S$, then $H(x) = 0$ so that the claim is proved. The flow ψ^t defines a diffeomorphism $\psi\colon S \times I \to U$ given by $(x, t) \mapsto \psi^t(x)$ where $I = (-\varepsilon, \varepsilon)$ is an open interval and $\varepsilon > 0$ sufficiently small. The diffeomorphism satisfies $\psi^t(S) = S_t$, and $\psi(0, \cdot)$ is the identity map on S. The proof of Proposition VIII.1 is complete. □

If the capacity c_0 of an open and bounded neighborhood U of S is finite, the following statement guarantees plenty of periodic solutions near S.

Theorem VIII.2 (Hofer–Zehnder). *Let $S = \{x \in M \mid H(x) = 0\}$ be a compact and regular energy surface of the Hamiltonian vector field X_H in the symplectic manifold (M, ω) and let*

$$U := \bigcup_{\lambda \in I} S_\lambda$$

be the bounded open neighborhood of S from the previous proposition where $I = (-\varepsilon, \varepsilon)$ is an open interval. If

$$c_0(U, \omega) < \infty,$$

*then there exists a **dense** subset $\Lambda \subset I$ in the parameter interval such that for every parameter $\lambda \in \Lambda$ the associated energy surface S_λ carries a periodic solution of X_H.*

It should be emphasized that there are no assumptions required for the given compact and regular energy surface S other than that the bounded open neighborhood must have finite capacity c_0. In the special case in which the symplectic manifold is the standard space $(\mathbb{R}^{2n}, \omega_0)$, the assumption $c_0(U, \omega) < \infty$ is *always satisfied*, because the relatively compact neighborhood U is contained in a large ball $B(R)$ of capacity $c_0(B(R)) = \pi R^2 < \infty$ and because the capacity function possesses the monotonicity property. Therefore, every compact regular energy surface $S \subset \mathbb{R}^{2n}$ gives rise to an abundance of periodic solutions of the Hamiltonian vector field whose energies are near the prescribed energy of S.

Proof [Definition of c_0]. The statement is an immediate consequence of the definition of the capacity c_0. If $\lambda_0 \in I$ and $\varepsilon > 0$ are given, we want to show that there exists a parameter $\lambda \in I$ satisfying $|\lambda_0 - \lambda| < \varepsilon$, for which S_λ carries a periodic solution. We can assume without loss of generality that $\lambda_0 = 0$. In order to define a special Hamiltonian function $F \in \mathcal{H}(U, \omega)$ that is constant on every surface S_λ, we choose the auxiliary function $f \in C^{\infty}(\mathbb{R}, \mathbb{R})$, illustrated in Figure VIII.2 and satisfying

$$\begin{aligned} f(s) &= c_0(M,U) + 1, && s \le -\varepsilon,\ s \ge \varepsilon, \\ f(s) &= 0, && -\varepsilon/2 \le s \le \varepsilon/2, \\ f'(s) &< 0, && -\varepsilon < s < -\varepsilon/2, \\ f'(s) &> 0, && \varepsilon/2 < s < \varepsilon. \end{aligned}$$

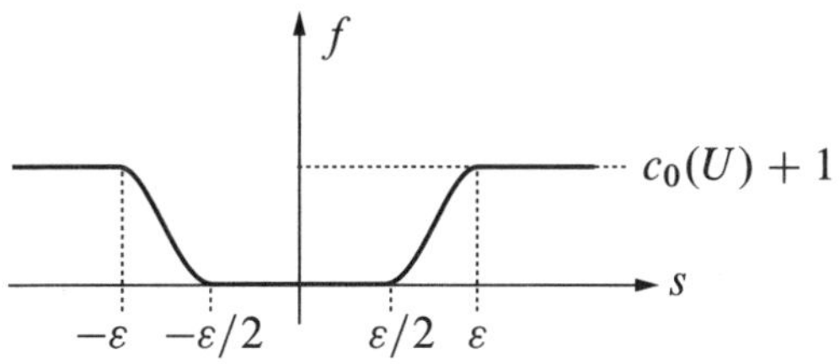

Figure VIII.2. The auxiliary function $f : \mathbb{R} \to \mathbb{R}$.

The smooth function $F : U \to \mathbb{R}$ is now defined by

$$F(x) := f(H(x)),$$

and satisfies $F \in \mathcal{H}(U, \omega)$, since $H(x) = \lambda$ on S_λ. Moreover,

$$m(F) > c_0(U, \omega).$$

Consequently, in view of the definition of the capacity c_0 there exists a non-constant periodic solution $x(t)$ of the Hamiltonian equation

$$\dot{x} = X_F(x)$$

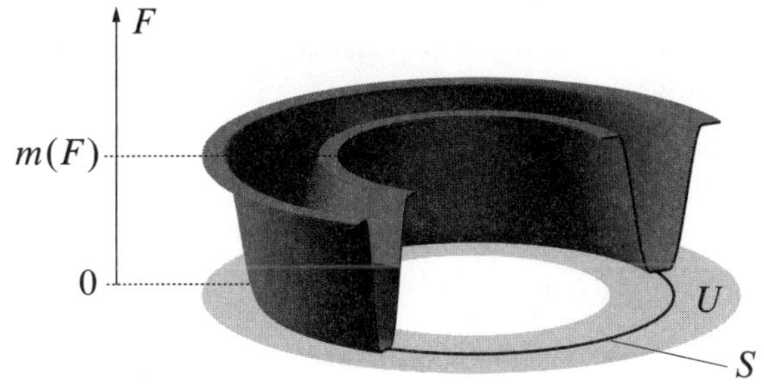

Figure VIII.3. The function $F = f \circ H$ on the neighborhood $U \subset M$ of S.

in U, having period $0 < T \leq 1$. We claim that

$$X_F(x) = f'(H(x))\, X_H(x), \quad x \in U.$$

Indeed, in view of the definition of Hamiltonian vector fields we compute

$$\begin{aligned}\omega(X_F(x), \cdot) &= -dF(x)\\ &= -f'(H(x))\, dH(x)\\ &= \omega\big(f'(H(x))\, X_H(x), \cdot\big),\end{aligned}$$

and the claim follows as ω is nondegenerate. From $dH(X_F) = -\omega(X_H, X_F) = 0$ we deduce that the function H is an integral of the vector field X_F. Hence, H is constant along the periodic solution $x(t)$ of X_F, so that $H(x(t)) =: \lambda$ is independent of t. Thus, also the function $f'(H(x(t))) = f'(\lambda) =: \tau$ is independent of t and since $x(t)$ is a non-constant solution of X_F, the constant satisfies $\tau \neq 0$. In view of the definition of the auxiliary function f, the parameter λ is necessarily contained either in the interval $(-\varepsilon, -\varepsilon/2)$ or in $(\varepsilon/2, \varepsilon)$, in particular $|\lambda| < \varepsilon$. Reparametrizing, the curve

$$y(t) := x(t/\tau)$$

is a solution of X_H, because

$$\begin{aligned}\dot{y}(t) &= \tau^{-1}\dot{x}(t/\tau)\\ &= \tau^{-1} X_F(x(t/\tau))\\ &= \tau^{-1} f'(H(x(t/\tau)))\, X_H(x(t/\tau))\\ &= \tau^{-1}\tau\, X_H(y(t)).\end{aligned}$$

This solution $y(t)$ has period $|\tau| \cdot T > 0$ and, due to $H(y(t)) = H(x(t/\tau)) = \lambda$, lies on the energy surface S_λ. This completes the proof of the theorem. □

If $c_0(U, \omega)$ is finite, there exists in view of Theorem VIII.2 a sequence $\lambda_j \to 0$ such that every energy surface S_{λ_j} carries a periodic solution $x_j(t)$ of period T_j. This raises the question whether these solutions converge to a periodic solution on the given surface $S = S_0$. Under the additional assumption that the *periods T_j are bounded*, the answer is affirmative. Instead of assuming the periods T_j to be bounded, we can equivalently assume that the arc length of the periodic orbits $x_j(t)$ with respect to any Riemannian metric are bounded. Indeed, if T is the minimal period of a solution $x(t)$, then

$$l(x) = \int_0^T |\dot{x}|\, dt = \int_0^T |X_H(x(t))|\, dt$$

is the arc length with respect to the Riemannian metric. Since the neighborhood U of the compact surface S is relatively compact, the Hamiltonian vector field satisfies the estimate

$$c^{-1} \leq |X_H(x)| \leq c, \quad x \in U$$

for a constant $c > 0$, from which we deduce the estimate $c^{-1}T \leq l(x) \leq cT$ for the arc length of the periodic solution $x(t)$ which proves the claim.

Proposition VIII.3. *We consider the family $(S_\lambda)_{\lambda \in I}$ of compact regular energy surfaces of the Hamiltonian vector field X_H in the open neighborhood U as described in Theorem* VIII.2. *We take a converging sequence $\lambda_j \to \lambda^*$ in the interval I having the property that every S_{λ_j} carries a periodic solution $x_j(t)$ of period $T_j > 0$. Then, if these periods are bounded, so that $T_j \leq C$ for all j, there exists a periodic solution on S_{λ^*} of period $0 < T \leq C$.*

Proof [Arzela–Ascoli]. Normalizing the periods to 1, we introduce the 1-periodic orbits

$$y_j(t) := x_j(T_j t), \quad 0 \leq t \leq 1,$$

which solve the equations

$$(*) \quad \dot{y}_j(t) = T_j X_H(y_j(t))$$

and satisfy $H(y_j(t)) = \lambda_j$. Choosing a Riemannian metric and using that the periods are bounded, we see that the derivatives are uniformly bounded,

$$|\dot{y}_j(t)| = |T_j X_H(y_j(t))| \leq C \sup_{x \in U} |X_H(x)| < \infty.$$

Therefore we find, by means of the Arzela–Ascoli theorem, a subsequence such that $T_j \to T$ and $y_j \to y$ with convergence in the C^0-sense. Iteratively using the equations $(*)$, we obtain the convergence $y_j \to y$ in the C^∞-sense. The limit $y(t)$ is a 1-periodic solution of the equation

$$\dot{y}(t) = T X_H(y(t)),$$

which satisfies $H(y(t)) = \lambda^*$, and it remains to prove that $T \neq 0$. Arguing by contradiction we assume that $T = 0$, then $y(t) \equiv y^*$ is constant and since S_{λ^*} is a regular energy surface,

$$X_H(y_k(t)) \to X_H(y^*) =: v \neq 0$$

for all t. We can estimate in local coordinates near the point y^*,

$$T_k^{-1} \langle \dot{y}_k(t), v \rangle = \langle X_H(y_k(t)), v \rangle \geq (1 - \varepsilon)|v|^2$$

for some small and positive ε and for all k large. Integrating over the period one concludes $|v| = 0$, contradicting $v \neq 0$ and proving the proposition. □

Going back to the statement of Theorem VIII.2, we shall show that the set $\Lambda \subset I$ of the parameters whose energy surfaces carry a periodic solution, is not only dense, but has full Lebesgue measure. This result has been proved by H. Hofer and

E. Zehnder in [52] under the additional assumption that the energy surface $S \subset M$ is the boundary of a symplectic manifold. The proof combines the monotonicity of the capacity with the classical result due to Lebesgue which states that a *monotonic* function $f: \mathbb{R} \to \mathbb{R}$ is differentiable in almost every point. That this additional assumption on S is superfluous has recently been observed by L. Macarini and F. Schlenk in [64] .

Theorem VIII.4 (Macarini–Schlenk). *Let $S = \{x \in M \mid H(x) = 0\}$ be a compact and regular energy surface of the Hamiltonian vector field X_H on the symplectic manifold (M, ω) and denote by*

$$U = \bigcup_{\lambda \in I} S_\lambda$$

the bounded open neighborhood of S from Proposition VIII.1 *where $I = (-\varepsilon, \varepsilon)$ is an open interval. If*

$$c_0(U, \omega) < \infty,$$

then there exists a subset $\Lambda \in I$ of full Lebesgue measure $m(\Lambda) = m(I)$ such that for every parameter $\lambda \in \Lambda$ the associated energy surface S_λ carries a periodic orbit of X_H.

Proof. We denote by $\Lambda_n \subset I = (-\varepsilon, \varepsilon)$ the set of parameters λ, for which S_λ carries a periodic solution of period $0 < T \leq n$. Due to Proposition VIII.3 the set Λ_n is closed. So, $\Lambda = \bigcup_{n \geq 1} \Lambda_n$ is a countable union of Lebesgue-measurable sets and, therefore, measurable. In order to show that Λ has full measure we define for $0 < \alpha < \varepsilon$ the open subsets

$$U_\alpha := \bigcup_{|\lambda| < \alpha} S_\lambda \cap U = \{x \in U \mid -\alpha < H(x) < \alpha\}$$

of U.

Lemma VIII.5. *We assume that the increasing function $\alpha \mapsto c_0(U_\alpha) \in \mathbb{R}$ satisfies at the point $0 < \alpha^* < \varepsilon$ the Lipschitz estimate*

$$c_0(U_\alpha) \leq c_0(U_{\alpha^*}) + L(\alpha - \alpha^*)$$

for all $\alpha \geq \alpha^*$ and $|\alpha - \alpha^*|$ sufficiently small and with a constant $L > 0$. Then $\alpha^* \in \Lambda$ or $-\alpha^* \in \Lambda$.

Proof. (1) Fix an $\alpha > \alpha^*$. By definition of the capacity c_0 there exists an admissible Hamiltonian function $G \in \mathcal{H}_a(U_{\alpha^*})$ having the oscillation

$$m(G) > c_0(U_{\alpha^*}) - (\alpha - \alpha^*).$$

We choose a smooth auxiliary function $f : \mathbb{R} \to \mathbb{R}$ illustrated in Figure VIII.4 and satisfying

$$f(s) = \begin{cases} c_0(U_\alpha) + (\alpha - \alpha^*), & s \leq -\alpha \text{ oder } s \geq \alpha, \\ m(G), & s \in [-\alpha^*, \alpha^*], \end{cases}$$

and moreover,

$$\begin{aligned} f'(s) &< 0, s \in (-\alpha, -\alpha^*), \\ f'(s) &> 0, s \in (\alpha^*, \alpha), \\ |f'(s)| &< L', s \in \mathbb{R}. \end{aligned}$$

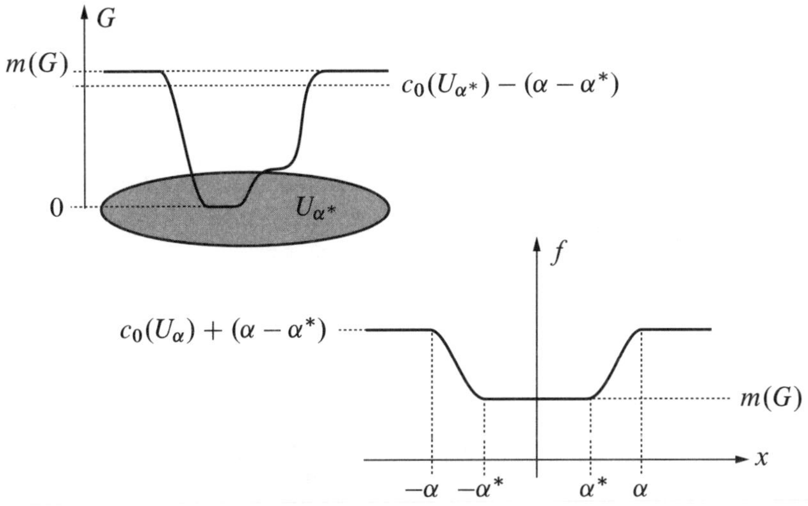

Figure VIII.4. Left: The function $G \in \mathcal{H}_a(U_{\alpha^*})$ with an oscillation $m(G) > c_0(U_{\alpha^*}) - (\alpha - \alpha^*)$. Right: The auxiliary function $f : \mathbb{R} \to \mathbb{R}$.

In view of the Lipschitz assumption on the capacity function $c_0(U_\alpha)$, the auxiliary function f can be chosen in such a way that the constant L' is equal to $L + 3$ independent of the choice of $\alpha > \alpha^*$. Now we define the smooth function $F : M \to \mathbb{R}$ (cf. Figure VIII.5) by

$$F(x) = \begin{cases} f(H(x)), & x \notin U_{\alpha^*} \quad (H(x) \leq -\alpha^* \text{ and } H(x) \geq \alpha^*), \\ G(x), & x \in U_{\alpha^*} \quad (-\alpha^* < H(x) < \alpha^*). \end{cases}$$

Then F is a smooth function belonging to the distinguished set $\mathcal{H}(U_\alpha)$ and has the oscillation $m(F) = c_0(U_\alpha) + (\alpha - \alpha^*) > c_0(U_\alpha)$. Consequently, by the definition

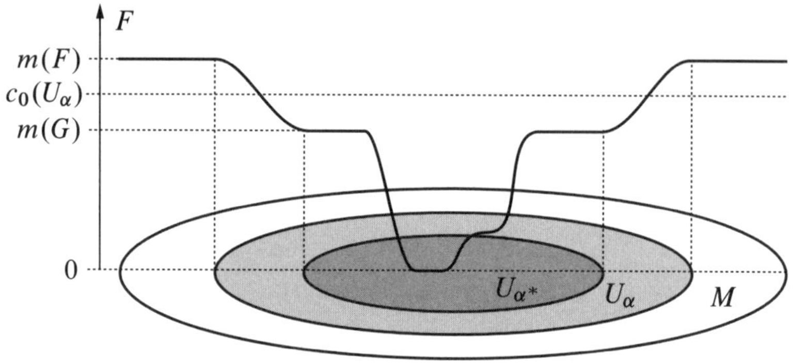

Figure VIII.5. The Hamiltonian function F with oscillation $m(F) > c_0(U_\alpha)$.

of c_0, the Hamiltonian vector field $\dot{x} = X_F(x)$ has a non-constant periodic solution $x(t)$ of period $0 < T \leq 1$ in U_α. Since the function G belonging to the set $\mathcal{H}(U_{\alpha^*})$ is an admissible function on U_{α^*}, the vector field X_G *does not admit* a non-constant periodic solution of period $0 < T \leq 1$ in U_{α^*}. Therefore, $x(t)$ lies in the set $U_\alpha \setminus U_{\alpha^*}$. As in the proof of Theorem VIII.2 one verifies that

$$X_F(x) = f'(H(x))\, X_H(x), \quad \text{if} \quad x \notin U_{\alpha^*},$$

and that the Hamiltonian function H is constant along the periodic solution $x(t)$. In view of the definition of the auxiliary function f, the solution $x(t)$ lies on an energy surface $S_{-\lambda}$ or S_λ for a parameter λ contained in the open interval (α^*, α) and the constant $f'(H(x(t))) =: \tau$ satisfies $0 < \tau < L'$. As in the proof of Theorem VIII.2 the reparametrized curve $y(t) := x(t/\tau)$ is a periodic solution of X_H having period

$$0 < |\tau| \cdot T \leq L'.$$

(2) We next consider a sequence $\alpha_j \searrow \alpha^*$. By the above construction we find for every j a periodic solution $x_j(t)$ of the vector field X_H having period $0 < T_j < L'$ on an energy surface $S_{-\lambda_j}$ or S_{λ_j} for a parameter $\lambda_j \in (\alpha^*, \alpha_j)$. Therefore, we find a subsequence converging to $-\alpha^*$ or to α^*. In view of Proposition VIII.3 there exists on $S_{-\alpha^*}$ or on S_{α^*} a non-constant periodic solution and we have proved that $-\alpha^* \in \Lambda$ or $\alpha^* \in \Lambda$, as claimed in the lemma. □

The function $\alpha \mapsto c_0(U_\alpha) \in \mathbb{R}$ is an increasing function on the interval $I = (-\varepsilon, \varepsilon)$, because c_0 is a capacity and $U_\alpha \subset U_{\alpha'}$ for $\alpha \leq \alpha'$. Therefore, in view of a well-known theorem due to Lebesgue, the function is differentiable almost everywhere and thus satisfies the assumptions of Lemma VIII.5 almost everywhere.

From the lemma we conclude for the Lebesgue measure that

$$(*) \quad m\{0 \le \lambda < \varepsilon \mid -\lambda \in \Lambda \text{ or } \lambda \in \Lambda\} = \varepsilon = \frac{m(I)}{2}.$$

It remains to show that $m(\Lambda) = m(I)$. Arguing indirectly, we first assume that $m(I \setminus \Lambda) > 0$. Then there exists for every small $\delta > 0$ an interval $J \subset I$ satisfying $m(J \cap (I \setminus \Lambda)) \ge (1-\delta)m(J)$. Now, for the interval J we can argue as we did before for the interval I and find analogous to $(*)$ the estimate

$$m(J \cap \Lambda) \ge \frac{m(J)}{2}$$

and choosing $\delta > 0$ sufficiently small we arrive at the contradiction

$$m(J) = m(J \cap (I \setminus \Lambda)) + m(J \cap \Lambda) \ge (1 - \delta + 1/2)m(J) > m(J)$$

which proves Theorem VIII.4. □

We again point out that in the special situation of compact hypersurfaces S in the symplectic standard space $(\mathbb{R}^{2n}, \omega_0)$, the above requirement on the capacity c_0 is always fulfilled. For a direct proof in this case we refer to M. Struwe in [112]

It should be emphasized that not every compact regular energy surface admits a periodic orbit. Examples of hypersurfaces that do not carry periodic orbits are due to V. Ginzburg and M. Herman in the dimensions $2n \ge 6$ and we refer to the survey article [40] by V. Ginzburg. The situation in dimension $2n = 4$ is not yet understood. V. Ginzburg and B. Gürel constructed in [43] a compact regular energy surface of class C^2 in $(\mathbb{R}^4, \omega_0)$ that does not carry a periodic orbit.

In his celebrated non-closing lemma, M. Herman constructs a smooth Hamiltonian function H_0 on a 4-dimensional symplectic manifold possessing a compact regular energy surface S_0 for which in the open neighborhood

$$U = \bigcup_{\lambda \in I} S_\lambda$$

none of the energy surfaces S_λ admits a periodic solution and such that, moreover, the same holds true for all Hamiltonian functions H that are close to H_0 in the C^4-topology. The example of M. Herman can be found in [52].

VIII.2 Hypersurfaces of contact type

We have seen that every compact and regular energy surface S contained in a symplectic manifold (M, ω) gives rise to an abundance of periodic solutions, if only a neighborhood of S has a finite capacity c_0. These solutions are all near but not necessarily on S. To guarantee periodic solutions on the surface S itself one needs additional conditions on the hypersurface S. Prompted by the early existence results on convex and starlike hypersurfaces in $(\mathbb{R}^{2n}, \omega_0)$, A Weinstein formulated in [117] his seminal conjecture that a contact type hypersurface always carries a periodic orbit.

Definition. A compact hypersurface $S \subset (M, \omega)$ is of *contact type*, if, in a neighborhood $U \subset M$ of S, there exists a vector field X possessing the following properties.

(i) X is transversal to S, that is, $X(x) \notin T_x S$ for all $x \in S$.

(ii) $L_X \omega = \omega$ on U.

The condition (ii) is equivalent to the requirement

$$(\varphi^t)^* \omega = e^t \omega \quad \text{on } U$$

for the flow φ^t of the vector field X. Another (equivalent) definition of contact-type will be formulated later on in Proposition VIII.8.

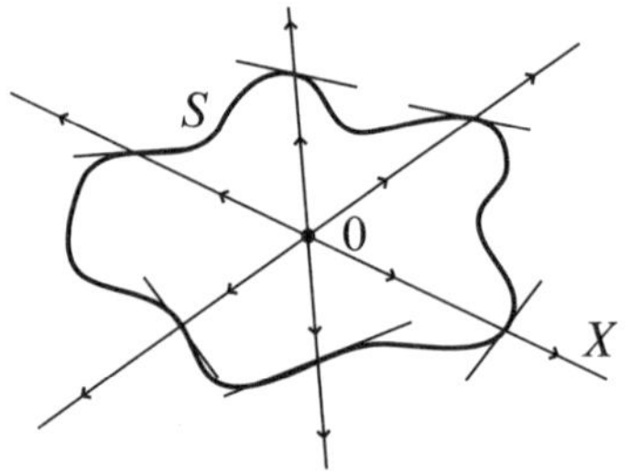

Figure VIII.6. A starlike compact hypersurface $S \subset \mathbb{R}^{2n}$ as an example of a hypersurface of contact type.

Examples. (1) We assume $S \subset (\mathbb{R}^{2n}, \omega_0)$ to be a *starlike compact hypersurface*. This requires the existence of an interior point which we may assume to be $x^* = 0$, having the property that every ray issuing from the point x^* intersects the surface S in exactly one point and does so *transversally*. Then, S is of contact type with respect to the symplectic structure ω_0, since the vector field

$$X(x) := \frac{1}{2} x$$

is transversal to S and its flow φ^t is equal to

$$\varphi^t(x) = e^{t/2} x$$

so that $(\varphi^t)^* \omega_0 = (e^{t/2})^2 \omega_0 = e^t \omega_0$ for all $t \in \mathbb{R}$.

(2) In particular, every *strictly convex hypersurface* $S \subset (\mathbb{R}^{2n}, \omega_0)$ is of contact type. By this we refer to the smooth *boundary* of a strictly convex (bounded) set. An example is the sphere $S^{2n-1} \subset \mathbb{R}^{2n}$.

If S is a hypersurface of contact type in the symplectic manifold (M, ω), then there exists a *distinguished* 1-parameter family of hypersurfaces S_λ, which forms an open neighborhood U of $S = S_0$ by

$$U = \bigcup_{|\lambda|<\varepsilon} S_\lambda.$$

This family is defined by the flow of the vector field X in the following way. Since X is transversal to S, its flow φ^t induces for a small $\varepsilon > 0$ a diffeomorphism $\psi \colon S \times (-\varepsilon, \varepsilon) \to U$ onto a neighborhood U of S defined by

$$(x, \lambda) \mapsto \varphi^\lambda(x) = \psi(x, \lambda),$$

and we set

$$S_\lambda := \varphi^\lambda(S)$$

for $|\lambda| < \varepsilon$. Since S is of contact type, this diffeomorphism is *very special*. If $x \in S$ and if $0 \neq \xi \in \mathcal{L}_S(x)$ is an element of the canonical line through the point x, then it follows for all tangent vectors $\eta \in T_x S$ that

$$\begin{aligned} 0 &= e^t \omega(\xi, \eta) \\ &= (\varphi^t)^* \omega(\xi, \eta) \\ &= \omega\big(d\varphi^t(x)\xi, d\varphi^t(x)\eta\big). \end{aligned}$$

Because $d\varphi^t(x)$ is an isomorphism $T_x S \to T_{\varphi^t(x)} S_t$ between the tangent spaces, it follows that

$$d\varphi^t(x)\xi \in \mathcal{L}_{S_t}(\varphi^t(x))$$

so that

$$d\varphi^t \colon \mathcal{L}_S \to \mathcal{L}_{S_t}$$

induces a (vector bundle-) isomorphism between the canonical line bundles. Thus we obtain a bijection between the characteristics of S and of S_t. In particular, if $\gamma \subset S_t$ is a closed characteristic of the canonical line bundle $\mathcal{L}_{S_t}$, then the closed curve

$$(\varphi^t)^{-1}(\gamma) \subset S$$

is a closed characteristic of $\mathcal{L}_S$. From Theorem VIII.2 we now easily arrive at the following statement.

Theorem VIII.6 (Hofer–Zehnder). *If $S \subset (M, \omega)$ is a compact hypersurface of contact type and if for an open neighborhood $U \subset M$ of S the capacity $c_0(U, \omega)$ is finite, then S possesses a closed characteristic.*

Proof. We take the diffeomorphism $\psi: S \times (-\varepsilon, \varepsilon) \to U$ introduced above and satisfying $S_t = \varphi^t(S) = \psi(S, t)$, and define the Hamiltonian function $H: U \to \mathbb{R}$ by

$$H(\varphi^t(x)) := t, \quad x \in S.$$

This Hamiltonian function has the compact energy surfaces $S_t = \{x \in U \mid H(x) = t\}$ for $t < |\varepsilon|$ which are regular. Indeed, differentiating in t we obtain

$$1 = dH(\varphi^t(x)) \frac{d}{dt}\varphi^t(x) = dH(\varphi^t(x))\, X(\varphi^t(x))$$

so that $dH(\varphi^t(x)) \neq 0$ for all $\varphi^t(x) \in S_t$. In view of Theorem VIII.2 we find a parameter value $|t^*| < \varepsilon$, for which the energy surface S_{t^*} carries a closed characteristic and therefore also $S = S_0$ carries a closed characteristic. □

A compact hypersurface in the symplectic standard space $(\mathbb{R}^{2n}, \omega_0)$ always possesses a bounded neighborhood of finite capacity c_0 and we deduce from Theorem VIII.8 the celebrated solution of A. Weinstein's conjecture due to C. Viterbo [114] in 1987.

Corollary VIII.7 (Viterbo). *Every compact hypersurface $S \subset (\mathbb{R}^{2n}, \omega_0)$ of contact type carries a closed characteristic.*

In particular, convex and starlike hypersurfaces in $(\mathbb{R}^{2n}, \omega_0)$ always carry a periodic orbit and we have confirmed the pioneering results due to A. Weinstein [116] and P. Rabinowitz [90] in 1978, resp. in 1979, which mark the beginning of the investigation of global periodic phenomena of Hamiltonian systems in higher dimensions.

We conclude this section by a useful characterization of hypersurfaces of contact type which is frequently used as a definition of contact type.

Proposition VIII.8. *A compact hypersurface $S \subset (M, \omega)$ is of contact type if and only if S is orientable and if there exists a 1-form α on S satisfying*

(i) $d\alpha = j^*\omega$,

(ii) $\alpha(\xi) \neq 0$ *for all* $0 \neq \xi \in \mathcal{L}_S$,

where $j: S \to M$ denotes the inclusion map and $\mathcal{L}_S$ is the canonical line bundle of S.

Proof. (1) We assume that S is of contact type and let X be the transversal vector field in the neighborhood $U \subset S$ satisfying $L_X\omega = \omega$. Since X is transversal, the hypersurface S is orientable. We introduce the 1-form α on U by

$$\alpha := i_X\omega.$$

Using Cartan's formula and $d\omega = 0$ it follows that

$$\omega = L_X\omega = d(i_X\omega) = d\alpha.$$

We choose a vector $0 \neq \xi \in \mathcal{L}_S(x)$, then $\omega(v,\xi) = 0$ for all $v \in T_xS$. Since ω is nondegenerate, $\omega(v,\xi) \neq 0$ for $v \in T_xM \setminus T_xS$. Therefore, we conclude from the transversality of the vector field X that

$$0 \neq \omega(X(x),\xi) = (i_X\omega)(\xi) = \alpha(\xi), \quad 0 \neq \xi \in \mathcal{L}_S(x),$$

and we have verified the properties (i) and (ii) for the 1-form α.

(2) Conversely, let α be the 1-form satisfying the properties (i) and (ii) in the proposition. We shall extend this form onto a neighborhood U of S by the following lemma.

Lemma VIII.9. *There exists a* 1*-form* τ *on a neighborhood* $U \supset S$*, which satisfies*

$$d\tau = \omega \text{ on } U \quad \text{and} \quad j^*\tau = \alpha \text{ on } S,$$

where $j\colon S \to U$ *is the inclusion mapping.*

Postponing the proof of the lemma, we first complete the proof of Proposition VIII.10. We define the vector field X on U by $i_X\omega = \tau$, and obtain by Cartan's formula,

$$\omega = d\tau = d(i_X\omega) = L_X\omega.$$

In view of property (ii),

$$0 \neq \alpha(\xi) = \tau(\xi) = (i_X\omega)(\xi) = \omega(X(x),\xi), \quad 0 \neq \xi \in \mathcal{L}_S(x),$$

and using the definition of $\mathcal{L}_S(x)$ it follows that $X(x) \notin T_xS$. With the exception of the lemma, the proposition is proved. □

Proof of the lemma. Because S is orientable and of codimension 1, there exists a transversal vector field Y in a neighborhood of S. Similarly to the gradient flow in Proposition VIII.1 the flow of this vector field defines a diffeomorphism $\psi\colon S \times (-\varepsilon,\varepsilon) \to U$ onto an open neighborhood U of S, for a small $\varepsilon > 0$. The diffeomorphism ψ satisfies $\psi(x,0) = x$ for all $x \in S$. Using the projection $\pi\colon S \times (-\varepsilon,\varepsilon) \to S$, given by $\pi(x,t) = x$, we introduce the projection map

$$r = \pi \circ \psi^{-1}\colon U \to S, \quad \psi(x,t) \mapsto x,$$

which on S is the identity map. If we define the 1-form μ on U by $\mu = r^*\alpha$, then $j^*\mu = (r \circ j)^*\alpha = \alpha$. The 2-form $\omega - d\mu$ on U is closed, because ω is closed so that

$$d(\omega - d\mu) = 0.$$

Moreover, it follows from (i) that

$$j^*(\omega - d\mu) = d\alpha - d(j^*\mu) = d\alpha - d\alpha = 0.$$

Claim. We claim that there exists a 1-form ϑ on U, which satisfies $\omega - d\mu = d\vartheta$ and $j^*\vartheta = 0$.

The claim completes the proof of the lemma as follows. We define the 1-form τ on U by $\tau = \mu + \vartheta$ and obtain $d\tau = \omega$ and $j^*\tau = j^*\mu = \alpha$ as required in the lemma and it remains to prove the claim.

Proof of the claim. The claim is a general fact about vector bundles which we apply to the trivial bundle $S \times \mathbb{R} \cong S \times (-\varepsilon, \varepsilon)$ which is diffeomorphic to U. We consider the vector bundle $p \colon E \to N$ over the manifold N and let α be a k-form on E satisfying $d\alpha = 0$ and $j^*\alpha = 0$, where $j : N \to E$ is the inclusion map (the zero section of the bundle). We shall show that there exists a $(k-1)$-form β on E which satisfies $\alpha = d\beta$ and $\beta|_N = j^*\beta = 0$.

Indeed, since the fibers are vector spaces we can define for $t > 0$ the map $\varphi^t \colon E \to E$ by $\varphi^t(x) = tx$. This 1-parameter family of diffeomorphisms defines the time dependent vector field X_t on E by $\frac{d}{dt}\varphi^t(x) = X_t(\varphi^t(x))$ for $t > 0$. In view of $d\alpha = 0$ it follows from Cartan's formula that $L_{X_t}\alpha = d(i_{X_t}\alpha)$ and so,

$$\frac{d}{dt}(\varphi^t)^*\alpha = (\varphi^t)^* L_{X_t}\alpha = d\Big((\varphi^t)^* i_{X_t}\alpha\Big).$$

Using $j^*\alpha = 0$ we obtain $\lim_{s\to 0}(\varphi^s)^*\alpha = p^* j^*\alpha = 0$ and since $\varphi^1 = \mathrm{Id}_E$, we conclude that

$$\begin{aligned}
\alpha &= (\varphi^1)^*\alpha - \lim_{s\to 0}(\varphi^s)^*\alpha \\
&= \lim_{s\to 0}\int_s^1 \frac{d}{dt}(\varphi^t)^*\alpha\, dt \\
&= \int_0^1 d\Big((\varphi^t)^* i_{X_t}\alpha\Big)\, dt \\
&= d\left(\int_0^1 (\varphi^t)^* i_{X_t}\alpha\, dt\right).
\end{aligned}$$

We define β as the integral on the right, then β is a $(k-1)$-form satisfying $d\beta = \alpha$. Moreover, if $x \in N$, then $X_t(x) = 0$ and hence $i_{X_t}\alpha(x) = 0$ and so $\beta(x) = 0$, as we wanted to prove. The proof of Proposition VIII.8 is complete. □

VIII.3 Examples from classical mechanics

We consider the standard symplectic space $(\mathbb{R}^{2n}, \omega_0)$ equipped with the canonical symplectic form ω_0, represented in the coordinates $z = (x, y) \in \mathbb{R}^n \times \mathbb{R}^n$ by

$$\omega_0 = \sum_{j=1}^{n} dy_i \wedge dx_i.$$

The Hamiltonian function $H\colon \mathbb{R}^{2n} \to \mathbb{R}$ defines the associated Hamiltonian vector field

$$X_H(z) = J\nabla H(z) = \begin{pmatrix} \frac{\partial}{\partial y}H(z) \\ -\frac{\partial}{\partial x}H(z) \end{pmatrix},$$

denoted in physics as $\dot{q} = \frac{\partial}{\partial p}H(q,p)$ and $\dot{p} = -\frac{\partial}{\partial q}H(q,p)$, using the coordinates (q,p) instead of (x,y). Problems in physics often lead to Hamiltonian functions of the type

$$H = \text{kinetic energy} + \text{potential energy},$$

and are often of the form

$$H(x,y) = \frac{1}{2}|y|^2 + V(x),$$

with the potential function $V\colon \mathbb{R}^n \to \mathbb{R}$ depending only on the variable x. A somewhat more general classical Hamiltonian function takes the form

$$H(x,y) = \frac{1}{2}\langle A(x)y, y\rangle + V(x),$$

where $A(x)$ is a *positive definite* symmetric matrix, so that the kinetic energy is a Riemannian metric.

We shall show that for such Hamiltonian functions *every* compact regular energy surface $S = \{z \in \mathbb{R}^{2n} \mid H(z) = E\}$ is of contact type. For this purpose we formulate a criterion for energy surfaces in $(\mathbb{R}^{2n}, \omega_0)$ to be of contact type, involving the *Liouville form* λ on $\mathbb{R}^{2n}$ defined by

$$\lambda = \sum_{i=1}^{n} y_j\, dx_j.$$

Then $d\lambda = \sum_{i=1}^{n} dy_j \wedge dx_j = \omega_0$ is the canonical symplectic form.

Theorem VIII.10. *If a smooth function $H\colon \mathbb{R}^{2n} \to \mathbb{R}$ satisfies the inequality*

$$\lambda(X_H(x,y)) > 0 \quad \textit{for all } (x,y) \in \mathbb{R}^{2n} \textit{ with } y \neq 0,$$

then every compact regular energy surface $S = \{z \in \mathbb{R}^{2n} \mid H(z) = E\}$ is of contact type and hence carries a periodic solution.

Proof. We assume the energy surface $S = \{z \in \mathbb{R}^{2n} \mid H(z) = E\}$ to be compact and regular, so that $dH \neq 0$ on S and hence also $X_H \neq 0$ on S and denote by $j\colon S \to \mathbb{R}^{2n}$ the inclusion map. We make use of Proposition VIII.8 and distinguish between two cases.

(1) We first assume that S *does not contain* a point $z = (x, 0)$ where $x \in \mathbb{R}^n$. Then, by assumption, $\lambda(X_H) > 0$ on S. If $\alpha := j^*\lambda$ is the Liouville form induced

on S, then (i) $d\alpha = j^*d\lambda = j^*\omega_0$ and (ii) $\alpha(X_H) = \lambda(X_H) > 0$, since X_H is a vector field on S. Because $0 \neq X_H$ lies in the canonical line bundle $\mathcal{L}_S$ which is 1-dimensional, it follows that $\alpha(\xi) \neq 0$ for all $0 \neq \xi \in \mathcal{L}_S$. Consequently, the energy surface S is of contact type in view of Proposition VIII.8.

(2) If S contains a point $z = (x, 0)$ where $x \in \mathbb{R}^n$, then we modify the Liouville form λ near S and define for $\varepsilon > 0$ the 1-form

$$\alpha := \lambda - \varepsilon dF$$

where the function $F\colon \mathbb{R}^{2n} \to \mathbb{R}$ is defined as

$$F(x, y) := \left\langle \frac{\partial}{\partial x} H(x, 0), y \right\rangle.$$

Clearly, $d\alpha = d\lambda = \omega_0$ and the property (i), namely $d(j^*\alpha) = j^*\omega_0$ on S, follows. In order to apply Proposition VIII.8 to the 1-form $j^*\alpha$, we still have to show that

$$\alpha(X_H) > 0 \quad \text{on } S$$

for a suitable $\varepsilon > 0$, so that the 1-form α meets all the requirements in the definition of contact type. Using the definition of λ our assumption becomes

$$\lambda(X_H) = \left\langle \frac{\partial}{\partial y} H(x, y), y \right\rangle > 0 \quad \text{if } y \neq 0.$$

On the other hand, $\lambda(X_H) = 0$ if $y = 0$. Consequently, for every fixed x satisfying $(x, 0) \in S$, the function $y \mapsto \langle \frac{\partial}{\partial y} H(x, y), y \rangle$ has in $y = 0$ a *minimum*, and thus $\frac{\partial}{\partial y} H(x, 0) = 0$ for all $(x, 0) \in S$. By assumption $dH = (\frac{\partial}{\partial x} H, \frac{\partial}{\partial y} H) \neq 0$ on S, so that

$$\frac{\partial}{\partial x} H(x, 0) \neq 0 \quad \text{if } (x, 0) \in S.$$

By definition of the 1-form α,

$$\alpha(X_H) = \lambda(X_H) + \varepsilon G$$

where $G = -dF\,(X_H) = \frac{\partial F}{\partial y}\frac{\partial H}{\partial x} - \frac{\partial F}{\partial x}\frac{\partial H}{\partial y}$. Explicitly,

$$G(x, y) = \left\langle \frac{\partial H}{\partial x}(x, y), \frac{\partial H}{\partial x}(x, 0) \right\rangle - \sum_{j,k=1}^{n} \frac{\partial^2 H}{\partial x_k \partial x_j}(x, 0) \frac{\partial H}{\partial y_k}(x, y) y_j.$$

At the points $(x, 0) \in S$, we have $G(x, 0) = \|\frac{\partial H}{\partial x}(x, 0)\|^2 > 0$. Since S is compact, it follows that $G(x, y) > 0$ at all the points $(x, y) \in S$ where $|y| < \delta$ is sufficiently small. Therefore, $\alpha(X_H) > 0$ at the points (x, y) in S where $|y| < \delta$ is small, since $\lambda(X_H) \geq 0$. This holds true for every $\varepsilon > 0$. Because S is compact and $\lambda(X_H) > 0$, if $y \neq 0$, it follows that $\lambda(X_H) \geq c$ for a constant $c > 0$, for all (x, y) in S satisfying $|y| \geq \delta$ for some small and positive δ. Finally, choosing ε small enough we achieve that $\alpha(X_H) > 0$ on S. As in step (1) it now follows that S is of contact type. □

For the Hamiltonian vector field $X_H(z) = \big(\frac{\partial}{\partial y}H(z), -\frac{\partial}{\partial x}H(z)\big)$ on $\mathbb{R}^{2n}$ we obtain

$$\lambda(X_H(x,y)) = \sum_{j=1}^{n} y_j \frac{\partial}{\partial y_j} H(x,y) = \Big\langle \frac{\partial}{\partial y} H(x,y), y \Big\rangle.$$

If the function H is of the special form

$$H(x,y) = \frac{1}{2}\langle A(x)y, y\rangle + V(x),$$

then $\frac{\partial}{\partial y}H(x,y) = A(x)y$. If the matrix $A(x)$ is positive definite, then

$$\lambda(X_H)(x,y) = \langle A(x)y, y\rangle > 0, \quad y \neq 0$$

and we deduce from Theorem VIII.10 the following existence statement for systems of classical mechanics.

Theorem VIII.11. *If the Hamiltonian function H on the symplectic standard space $(\mathbb{R}^{2n}, \omega_0)$ has the form*

$$H(x,y) = \frac{1}{2}\langle A(x)y, y\rangle + V(x)$$

where $A(x)$ is positive definite, then every regular compact energy surface $S = \{z \in \mathbb{R}^{2n} \mid H(z) = E\}$ is of contact type and, therefore, carries a periodic solution of the Hamiltonian vector field X_H.

The *phase space* of a general system of classical mechanics is the cotangent bundle $\pi\colon T^*N \to N$ over an n-dimensional manifold N, called a *configuration space*. As a set it is defined as

$$T^*N = \bigcup_{p \in N} (T_pN)^*$$

where $(T_pN)^*$ is the cotangent space, which is the vector space of linear forms defined on the tangent space T_pN. The set T^*N is equipped with a canonical manifold structure defined as follows. If $(x_1, \dots, x_n)$ are the coordinates around the point x in N, a 1-form $\alpha \in (T_xN)^*$ is represented by $\alpha = \sum_{j=1}^{n} y_j\, dx_j$ having the coordinates $(y_1, \dots, y_n)$ and together $(x_1, \dots, x_n, y_1, \dots, y_n)$ form local coordinates in T^*N. In these coordinates one can define the special 1-form λ on T^*N by

$$(*) \quad \lambda = \sum_{j=1}^{n} y_j\, dx_j$$

which so far has only a local meaning. It is remarkable that the above form λ has a global interpretation. In order to define a 1-form λ on the manifold $M = T^*N$ we have to define at every point $P \in M$ a linear form $\lambda_P \in (T_P M)^*$ which assigns to every vector $V \in T_P M$ a real number $\lambda_P(V)$. For the special 1-form this is done as follows. If $V \in T_P M$ then the point $P \in M = T^*N$ is itself a linear form on the tangent space $T_p N$, where $p = \pi(P)$ is the base point of P. Therefore, we can evaluate the form P at the vector $v := d\pi_P V \in T_p N$ where the projection $\pi \colon T^*N \to N$ assigns to each $P \in T^*N$ its base point $p = \pi(P)$ in the manifold N. Hence, $d\pi$ maps the tangent space $T_P M$ onto the tangent space $T_p N$ at the point p and we can define

$$(**) \quad \lambda_P(V) := P(d\pi_P V).$$

This canonically defined 1-form is called the *Liouville form* on the manifold $M = T^*N$. It is sometimes called the *tautological form* on T^*N, since it is defined in terms of itself. In the local coordinates introduced above the form $(**)$ agrees with the form $(*)$. Locally we find

$$d\lambda = \sum_{j=1}^{n} dy_j \wedge dx_j,$$

from which we conclude that $\omega = d\lambda$ is a closed and nondegenerate 2-form on T^*N, hence a symplectic form. It is called the *canonical symplectic structure* on the cotangent bundle T^*N.

The Hamiltonian function of a classical mechanical system on the symplectic manifold $(T^*N, d\lambda)$ is again the sum of the kinetic and the potential energies. The kinetic energy is given by a Riemannian metric on the configuration space N which defines a scalar product $\langle \cdot, \cdot \rangle_p$ in every tangent space $T_p N$. The Riemannian metric induces a linear isomorphism $T_p N \to (T_p N)^*$ at every point p defined by the map $v \mapsto \langle \cdot, v \rangle_p$ and thus an isomorphism $\gamma \colon TN \to T^*N$ between the tangent bundle and the cotangent bundle of the manifold N. The induced Riemannian metric on T^*N is the push-forward $\langle \cdot, \cdot \rangle \circ \gamma^{-1}$ of the metric under the isomorphism γ,

$$\langle \xi, \eta \rangle_p^* := \left\langle \gamma^{-1}(\xi), \gamma^{-1}(\eta) \right\rangle_p .$$

It defines the kinetic energy on the manifold $M = T^*N$ by

$$T(\xi) = \langle \xi, \xi \rangle^*.$$

The potential energy of the mechanical system is presented by a function V defined on the configuration space N and, together with the kinetic energy, the Hamiltonian system on the phase space T^*N is defined by the Hamiltonian function

$$H(\xi) = T(\xi) + V(\pi(\xi)).$$

If $\xi(t)$ is a solution of the Hamiltonian vector field X_H on T^*N, the projection $x(t) = \pi(\xi(t))$ describes the mechanical movement in the configuration space.

Unfortunately little is known about the special capacity c_0 on cotangent bundles except in the special case of the torus where c_0 is finite on open and bounded sets. However, classical systems of the above form can be investigated by different methods. We only mention that S. V. Bolotin [11] showed already in 1978 by means of methods from Riemannian geometry that for a system whose Hamiltonian function $H: T^*N \to \mathbb{R}$ is of the form $H(\xi) = T(\xi) + V(\pi(\xi))$, every compact regular energy surface carries a periodic solution of X_H.

VIII.4 Poincaré's continuation method

In general, a periodic solution of a Hamiltonian vector field is not isolated. It belongs to a 1-parameter family of periodic solutions parametrized by the energy and forming an embedded cylinder as illustrated in Figure VIII.7. This was already known to H. Poincaré who designed perturbation methods for periodic orbits based on the implicit function theorem. The method is local in nature and requires non-degeneracy conditions on the Floquet multiplier of the given unperturbed reference solution which are defined as follows.

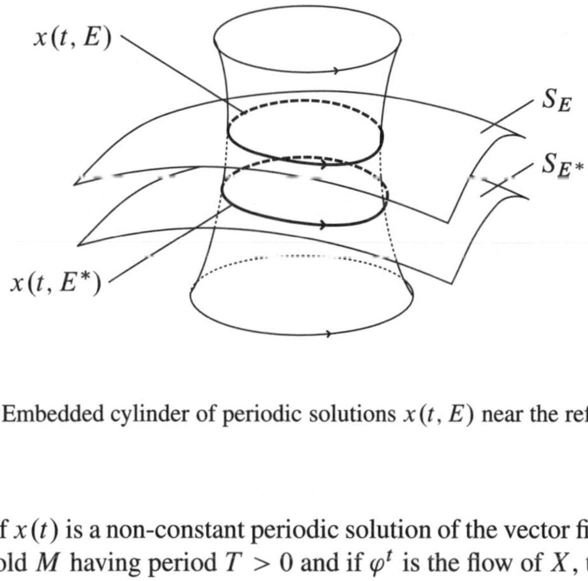

Figure VIII.7. Embedded cylinder of periodic solutions $x(t, E)$ near the reference solution $x(t, E^*)$.

Definition. If $x(t)$ is a non-constant periodic solution of the vector field $\dot{x} = X(x)$ on the manifold M having period $T > 0$ and if φ^t is the flow of X, then

$$\varphi^T(x(0)) = x(T) = x(0)$$

and hence the linearized map at the point $p = x(0)$ is an endomorphism

$$d\varphi^T(p)\colon T_pM \to T_pM.$$

The eigenvalues of the linear map $d\varphi^T(p)$ are called *Floquet multipliers* of the periodic solution $x(t)$ for the period T.

The significance is the following. Looking for periodic solutions near $x(t)$, one is tempted to solve the fixed point problem $\varphi^T(x) = x$ resp. $\varphi^T(x) - x = 0$ for x near p using the implicit function theorem. This leads to an investigation of the linearized map at the reference solution,

$$d\varphi^T(p) - \mathbb{1}.$$

This map is *singular*, if there exists a Floquet multiplier equal to 1 which for time independent vector fields is always the case!

Indeed, differentiating the flow relation $\varphi^t \circ \varphi^s(p) = \varphi^s \circ \varphi^t(p)$ in s at $s = 0$ one obtains

$$d\varphi^t(p)\,X(p) = X(\varphi^t(p))$$

for all t. If $t = T$ is the period and hence $\varphi^T(p) = p$, one finds

$$d\varphi^T(p)\,X(p) = X(p) \in T_pM.$$

We have verified that 1 is a Floquet multiplier. If the vector field X is the Hamiltonian vector field X_H on the symplectic manifold (M, ω), then the linearized map

$$d\varphi^T(p)\colon \big(T_pM, \omega(p)\big) \to \big(T_pM, \omega(p)\big)$$

is symplectic and we conclude from Proposition V.5 that the multiplicity of the Floquet multiplier 1 is at least 2. In the Hamiltonian case there always exist at least two Floquet multipliers equal to 1!

Theorem VIII.12 (H. Poincaré). *We consider the periodic solution $x(t, E^*)$ of the Hamiltonian vector field X_H on the symplectic manifold M having the energy $H(x(t)) = E^*$ and the period $T^* > 0$. If it possesses **precisely** two Floquet multipliers equal to 1, then there exists a smooth 1-parameter family $x(t, E)$ of solutions of periods $T(E)$ near T^* on the energy surfaces $S_E = \{x \in M \mid H(x) = E\}$ for E sufficiently close to E^*. Under these conditions the solutions $x(t, E)$ are unique and the periods depend smoothly on E. If $E \to E^*$, then, in particular, $T(E) \to T(E^*) = T^*$.*

Postponing the proof we observe that the reference solution $x(t, E^*)$ is not isolated in M as a periodic solution. The periodic solutions $x(t, E)$ correspond to different values of the energy E and it turns out in the proof that, on the fixed energy surface E, the periodic solution is isolated among those periodic solutions having periods close to T. Geometrically, the periodic solutions of the theorem fill out an embedded cylinder in M as illustrated in Figure VIII.8.

It is useful to recall the construction of a transversal section map belonging to a non-constant periodic solution $x(t)$ of a vector field X on a manifold M having

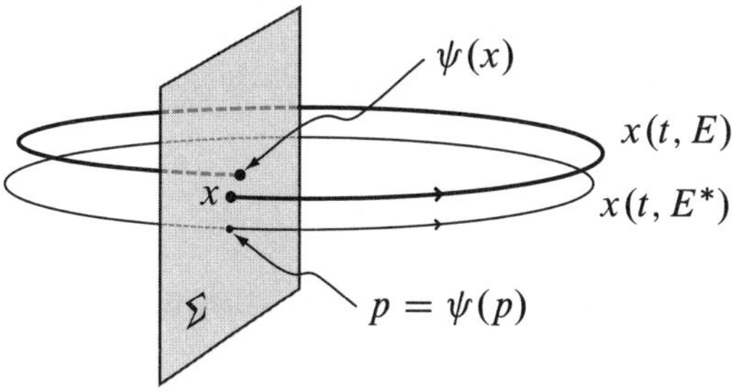

Figure VIII.8. The transversal map ψ on Σ near p.

period $T > 0$. We intersect the periodic solution $x(t)$ in the point $p = x(0)$ with a transversal manifold $\Sigma \subset M$ of codimension 1, so that

$$T_x M = T_x \Sigma \oplus \operatorname{span}\{X(x)\}$$

for all $x \in \Sigma$ near p. Such a manifold Σ is called a *local transversal section* of $x(t)$ in the point p.

In view of the assumption, the flow φ^t of X satisfies at time $t = T$,

$$\varphi^T(p) = p \in \Sigma.$$

In a neighborhood $U \subset \Sigma$ of p the flow φ^t defines a smooth map

$$\psi : U \subset \Sigma \to \Sigma,$$

in the following way. Given a point $x \in U \subset \Sigma$ near $p = x(0)$ we follow its solution $\varphi^t(x)$, until, for the first time, it meets Σ again at a time $\tau(x) > 0$, and we define

$$\psi(x) = \varphi^{\tau(x)}(x) \in \Sigma, \quad x \in U \subset \Sigma.$$

Since the vector field X is transversal to the section Σ, the map $x \mapsto \tau(x) \in \mathbb{R}$ is a smooth map, defined on Σ near the intersection point p and uniquely determined by the conditions

$$\begin{cases} \tau(p) = T, \\ \varphi^{\tau(x)}(x) \in \Sigma, \quad x \in \Sigma \cap U. \end{cases}$$

The map $\psi : \Sigma \to \Sigma$ which is defined near the fixed point $p = \psi(p)$, is called a *transversal map* or *Poincaré map* of the periodic solution $x(t)$. The eigenvalues of its derivative

$$d\psi(p) \colon T_p\Sigma \subset T_pM \to T_p\Sigma \subset T_pM$$

are related to the Floquet multipliers as follows.

Proposition VIII.13. *The derivative $d\varphi^T(p)$ has 1 as an eigenvalue associated with the eigenvector $X(p)$. The other eigenvalues are the eigenvalues of $d\psi(p)$. The characteristic polynomials are related by*

$$\det(\lambda - d\varphi^T(p)) = (\lambda - 1)\det(\lambda - d\psi(p)).$$

Proof. The first statement we have already verified at the beginning of this section. If $\xi \in T_p\Sigma$ is a tangent vector, then

$$\begin{aligned} d\psi(p)\xi &= d\varphi^T(p)\xi + \frac{d}{dt}\Big|_{t=T}\varphi^t(p)\,d\tau(p)\xi \\ &= d\varphi^T(p)\xi + l(\xi)\,X(p), \end{aligned}$$

where the linear form $d\tau(p) \in (T_p\Sigma)^*$ has been abbreviated by l so that $d\varphi^T(p)\xi = -l(\xi)X(p) + d\psi(p)\xi$. With respect to the splitting

$$T_xM = \operatorname{span}\{X(x)\} \oplus T_x\Sigma,$$

we represent $v \in T_pM$ by $v = (\alpha X(p), \xi)$ for an $\alpha \in \mathbb{R}$ and a tangent vector ξ in $T_p\Sigma$. Using $d\varphi^T(p)X(p) = X(p)$ the linear map $d\varphi^T$ has the representation

$$d\varphi^T(p) = \left(\begin{array}{c|c} 1 & -l \\ \hline 0 & \\ \vdots & d\psi(p) \\ 0 & \end{array}\right)$$

with respect to the above splitting from which the proposition follows immediately. □

In view of the construction of the local diffeomorphism ψ of the section Σ, the *fixed points* of ψ are the initial conditions of all periodic solutions near the reference solution $x(t)$ which have periods near T^* and we shall solve the fixed point equation $\psi(x) = x$ near p in suitable coordinates.

Proof of Theorem VIII.12. If $x(t)$ is the non-constant reference solution of period $T^* > 0$ having energy E^*, then $dH(x) \neq 0$ for all $x \in M$ near the orbit $x(t)$. It is convenient to introduce local coordinates $x = (x_1, \dots, x_{2n}) \in \mathbb{R}^{2n}$ locally around $p = x(0)$ having the following properties.

(i) $H(x_1, \dots, x_{2n}) = x_1$,
(ii) $p = (E^*, 0, \dots, 0)$,
(iii) $\Sigma = \{x \mid x_{2n} = 0\}$ defines a local transversal section of $x(t)$.

To find these coordinates, we first choose the coordinates $y = (y_1, \dots, y_{2n}) \in \mathbb{R}^{2n}$

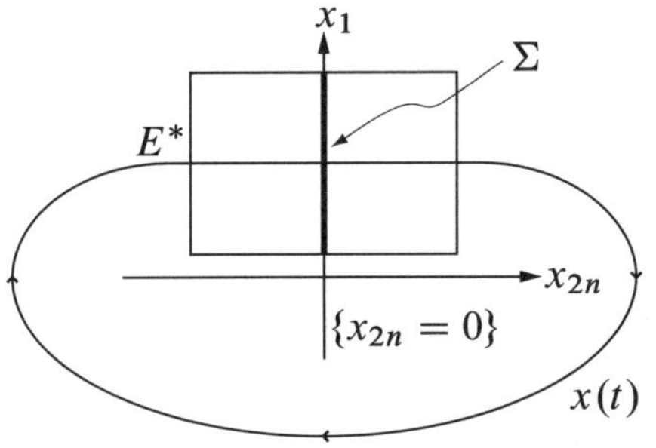

Figure VIII.9. The local situation in the coordinates $x = (x_1, \ldots, x_{2n})$.

in which $p = 0 \in \mathbb{R}^{2n}$ and $\frac{\partial H}{\partial y_1}(0) \neq 0$ using that $dH \neq 0$. In these coordinates we introduce the function $F \colon \mathbb{R}^{2n} \times \mathbb{R} \to \mathbb{R}$ defined by

$$F(y, E) = H(y) - E.$$

It satisfies $F(0, E^*) = H(0) - E^* = 0$ and $\frac{\partial}{\partial y_1} F(0, E) = \frac{\partial H}{\partial y_1}(0) \neq 0$. The implicit function theorem guarantees a unique function

$$y_1 = f(E, y_2, \ldots, y_{2n}) = f(E, y''), \quad y'' := (y_2, \ldots, y_{2n})$$

solving the equations

$$\begin{cases} F\big(f(E, y''), y'', E\big) = H\big(f(E, y''), y''\big) - E = 0, \\ f(E^*, 0) = 0. \end{cases}$$

Near $y = 0$ we introduce the new coordinates $x = (x_1, x_2, \ldots, x_{2n}) = (x_1, x'')$ by means of the local diffeomorphism

$$y = (y_1, y'') = \varphi(x) = (f(x_1, x''), x'')$$

where $x_1 = E$ and $x'' = y''$. In the new coordinates, the function H has the desired representation

$$\widehat{H}(x) = H\big(f(x_1, x''), x''\big) = H\big(f(E, y''), y''\big) = E = x_1.$$

For the function $\widehat{H}$ we shall again write H. From $H(x) = x_1$ we deduce

$$0 = dH(X_H) = dx_1(X_H) = (X_H)_1,$$

where $(X_H)_1$ is the first component of the vector field. Since $X_H \neq 0$ near $p = (E^*, 0)$, we can assume without loss of generality that

$$(X_H)_{2n} \neq 0.$$

Therefore, the set $\{x \mid x_{2n} = 0\}$ represents a local transversal section of $x(t)$ near $(E^*, 0)$ and the coordinates $(x_1, \dots, x_{2n})$ have the desired properties (i), (ii) and (iii).

Recall that $H(\varphi^t(x)) = H(x)$ for all $x \in M$. Then the transversal map

$$\psi \colon \Sigma \to \Sigma, \quad \Sigma = \{x \mid x_{2n} = 0\}$$

has, in the local coordinates

$$x = (x', x'') \in \Sigma, \quad x' = x_1, \ x'' = (x_2, \dots, x_{2n-1})$$

and considering the energy conservation

$$x' = x_1 = H(x) = H(\varphi^{\tau(x)}(x)) = H(\psi(x)) = (\psi(x))_1 =: \psi'(x),$$

the representation

$$\psi \colon \begin{pmatrix} x' \\ x'' \end{pmatrix} \mapsto \begin{pmatrix} \psi'(x', x'') \\ \psi''(x', x'') \end{pmatrix} = \begin{pmatrix} x' \\ \psi''(x', x'') \end{pmatrix}.$$

We are looking for fixed points of the map ψ near the fixed point $(E^*, 0) = \psi(E^*, 0)$, that is, for points (x', x'') satisfying

$$\begin{cases} x' = \psi'(x', x''), \\ x'' = \psi''(x', x''). \end{cases}$$

The first equation has already been solved by our choice of the coordinates. The second equation,

$$\psi''(x', x'') - x'' = 0,$$

possesses the solution $(x', x'') = (E^*, 0)$, so that

$$\psi''(E^*, 0) - 0 = 0.$$

In view of $\psi'(x', x'') = x'$, the linearized map at the point $x^* := (E^*, 0)$ has the form

$$d\psi(x^*) = \left(\begin{array}{c|c} 1 & 0 \\ \hline * & \\ \vdots & \frac{\partial}{\partial x''}\psi''(x^*) \\ * & \end{array}\right).$$

By assumption, the linear map $d\varphi^T(x^*)$ has precisely two eigenvalues equal to 1. From Proposition VIII.13 it therefore follows that 1 is not an eigenvalue of the matrix $\frac{\partial}{\partial x''}\psi''(x^*)$, so that the linear map

$$\frac{\partial}{\partial x''}\psi''(x^*) - \mathbb{1}_{2n-2}$$

is an isomorphism. The implicit function theorem, therefore, guarantees a unique smooth function $x'' = a(x')$ satisfying $a(E^*) = 0$ and

$$\psi''(x', a(x')) - a(x') = 0,$$

for $x' = E$ near E^*. We have constructed a smooth family $(E, a(E)) \in \Sigma$ of fixed points of the section map $\psi \colon \Sigma \to \Sigma$ for energies E close to E^*. These fixed points are the initial conditions of the desired periodic solutions having energies E near E^* and periods $T = \tau(x)$ near T^*. In addition, these solutions are the only such solutions near the reference solution. The proof of Theorem VIII.14 is complete. □

We see that typically, in (time independent) Hamiltonian systems the periodic solutions appear in 1-parameter families, parametrized by the energies and having similar periods. On the other hand, on *a fixed* energy surface the periodic solutions are *isolated* among the periodic solutions having similar periods.

This applies not only to Hamiltonian vector fields. The same statements hold true for every time independent vector field X, possessing a *regular integral* F. All one needs to do is to replace in the proof the Hamiltonian function H by the integral F.

VIII.5 Transversal sections on energy surfaces

As in the previous section, we consider a periodic solution $x(t)$ of the Hamiltonian vector field $\dot{x} = X_H(x)$ on the symplectic manifold (M, ω) having the energy $E = H(x(t))$, and we assume that the energy surface

$$S = \{x \in M \mid H(x) = E\}$$

is regular and hence a submanifold of dimension $2n - 1$. If $\Sigma \subset M$ is a transversal section of the periodic orbit $x(t)$, it follows from the energy conservation that the section map $\psi = \varphi^\tau \colon \Sigma \to \Sigma$ leaves the energy surface invariant,

$$\psi \colon \Sigma \cap S \to \Sigma \cap S.$$

We can, therefore, introduce the transversal sections Σ' on the energy surface S and the associated restricted section map by

$$\Sigma' := \Sigma \cap S \quad \text{and} \quad \psi \colon \Sigma' \to \Sigma'.$$

Theorem VIII.14. *The transversal section $\Sigma' = \Sigma \cap S$ of the periodic solution $x(t)$ in the energy surface S is a symplectic submanifold of (M, ω) of dimension $2n - 2$ equipped with the symplectic form $\omega' = \omega|_{\Sigma'} = j^*\omega$, where $j \colon \Sigma' \to M$ is the inclusion map. The transversal section map $\psi \colon \Sigma' \to \Sigma'$ is* ***symplectic*** *and hence satisfies $\psi^*\omega' = \omega'$.*

Proof. (1) We first show that ω' is a symplectic form. The 2-form ω' is, of course, closed, because $dj^*\omega = j^*d\omega = 0$ and we have to show that ω' is nondegenerate. We recall that the tangent space of S has the representation

$$T_xS = \ker dH(x) = \ker \omega(X_H, \cdot) = \{\xi \in T_xM \mid \omega(X_H, \xi) = 0\}.$$

Since Σ' in S is transversal to X_H,

$$T_xS = T_x\Sigma' \oplus \operatorname{span}\{X_H(x)\}.$$

Assuming that the tangent vector $v \in T_x\Sigma'$ satisfies

$$\omega(v, u) = 0 \quad \text{for all } u \in T_x\Sigma',$$

we shall show that $v = 0$. The tangent vector $\xi \in T_xS$ has the representation $\xi = u + \alpha X_H$ for an $\alpha \in \mathbb{R}$ and a tangent vector $u \in T_x\Sigma'$. In view of $\Sigma' \subset S$ and using the definition of a Hamiltonian vector field, we conclude that

$$\omega(v, \xi) = \omega(v, u) + \alpha\omega(v, X_H(x)) = 0 + 0.$$

This holds true for every tangent vector $\xi \in T_xS$, so that v lies in the 1-dimensional subspace $\ker(\omega|_S) = \operatorname{span}\{X_H\}$. Since $v \in T_x\Sigma'$ and $X_H \notin T_x\Sigma'$ we conclude that $v = 0$. Consequently, $\omega' = \omega|_{\Sigma'}$ is nondegenerate.

(2) It remains to verify that $\psi^*\omega' = \omega'$, explicitly,

$$\omega(\xi, \eta) = \omega\big(d\psi(x)\xi, d\psi(x)\eta\big) \quad \text{for all } \xi, \eta \in T_x\Sigma'.$$

We recall that

$$d\psi(x) = \frac{d}{dx}\varphi^{\tau(x)}(x) = X_H(\varphi^\tau(x))d\tau(x) + d\varphi^\tau(x)$$

for $\tau = \tau(x)$. Since $d\varphi^\tau(x)$ is symplectic, it follows, abbreviating $l := d\tau(x)$ and $y := \varphi^\tau(x)$, for all $\xi, \eta \in T_x\Sigma'$,

$$\begin{aligned}\omega(\xi, \eta) &= \omega\big(d\varphi^\tau(x)\xi, d\varphi^\tau(x)\eta\big)\\ &= \omega\big(d\psi(x)\xi - l(\xi)\, X_H(y), d\psi(x)\eta - l(\eta)\, X_H(y)\big)\\ &= \omega\big(d\psi(x)\xi, d\psi(x)\eta\big).\end{aligned}$$

We have used that $\omega(X_H(y), d\psi(x)\xi) = 0$, in view of $d\psi(x)\xi \in T_y\Sigma' \subset T_yS$, and similarly for η. □

Example. If $\dim M = 4$, the energy surface S has dimension 3 and the transversal section Σ' on S has dimension $\dim \Sigma' = 2$. Near $p = x(0) \in \Sigma'$ we choose a Darboux chart for Σ' in which the symplectic structure ω' of Σ' has the form

$$\omega' = dy \wedge dx.$$

In these local coordinates the transversal map $\psi\colon \Sigma' \to \Sigma'$ is an area preserving local diffeomorphism in the plane $\mathbb{R}^2$ near the fixed point $\psi(p) = p = x(0) \in \mathbb{R}^2$.

Instead of studying the solutions of the Hamiltonian vector field X_H on the energy surface S near the periodic reference solution $x(t)$ for all times, we can as well study the map ψ and all its iterates on Σ'.

By the spectral theorem for linear symplectic maps the Floquet multipliers of $x(t)$ are given by

$$\{1, 1, \lambda, \lambda^{-1}\},$$

where λ and λ^{-1} are the eigenvalues of the linearization $d\psi(p)$. Assuming, for example, that λ is real and that $|\lambda| \neq 1$, we conclude that p is a *hyperbolic fixed point* of the map ψ. Therefore, the periodic solution $x(t)$ is *unstable* on S, in view of the hyperbolic phenomena described in Chapters II and III.

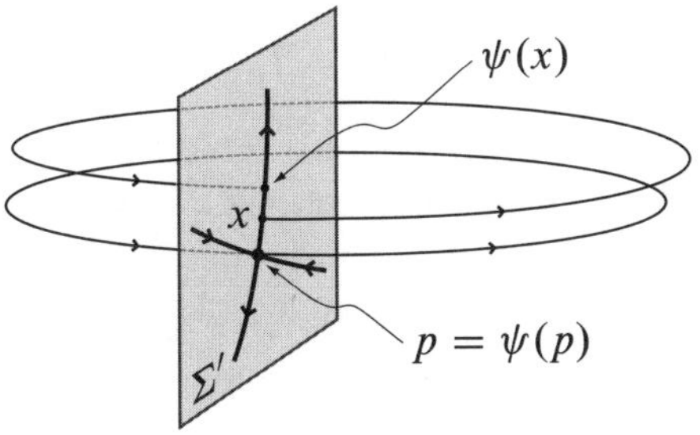

Figure VIII.10. Hyperbolic fixed point p of the section map ψ with invariant manifolds on Σ'. The whole picture is contained in the energy surface S.

We next study the relation between Floquet multipliers and integrals for a vector field X which is not necessarily a Hamiltonian vector field. We look at the periodic solution $x(t)$ having period $T > 0$ of the vector field on the manifold M of dimension $\dim M = n$ and denote by φ^t the flow of X. We assume that the vector field X possesses r smooth integrals

$$G_\alpha\colon M \to \mathbb{R}, \quad \alpha = 1, \dots, r,$$

in a neighborhood of the periodic solution $x(t)$, so that $dG_\alpha(X) = 0$ and hence

$$G_\alpha(\varphi^t(x)) = G_\alpha(x), \quad \alpha = 1, \dots, r$$

for all x near $x(t)$ and all times t as long as the flow is defined. We assume that the integrals are independent, assuming that the 1-forms $dG_\alpha(x) = 0$ are linearly independent. The integral surfaces

$$N = N_c = \{x \in M \mid G(x) = c\}$$

where G is the map defined by $G = (G_1, \dots, G_r)$ and where $c = (c_1, \dots, c_r)$, are invariant under the flow and of dimension

$$\dim N_c = n - r.$$

At the point x in the integral surface N_c the tangent space is equal to

$$T_x N_c = \ker dG(x).$$

In particular, $X(x) \in T_x N_c$. If $\Phi := d\varphi^T(p)$ is the linearized map at the point p belonging to the periodic solution $x(t)$, then

$$dG_\alpha(\Phi\xi) = dG_\alpha(\xi)$$

for all $\xi \in T_p M$ and therefore

$$\Phi(T_p N) = T_p N,$$

N being the integral surface of the periodic orbit $x(t)$. In order to compute the eigenvalues of Φ we choose local coordinates in a neighborhood of $x(0)$ in which the integrals

$$G_\alpha(x) = x_\alpha \quad \text{for all } \alpha = 1, 2, \dots, r,$$

become coordinate functions. If we split the coordinates $x = (x_1, \dots, x_n)$ into $x' = (x_1, \dots, x_r)$ and $x'' = (x_{r+1}, \dots, x_n)$, the map Φ has, with respect to this splitting, the form

$$\Phi = \begin{pmatrix} \mathbb{1}_r & 0 \\ * & \widehat{\Phi} \end{pmatrix}$$

where $\widehat{\Phi} = \Phi|_{T_p N_c}$ is the restriction of Φ to the tangent space $T_p N$. We now proceed as before by considering the flow φ^t of the vector field X restricted to the integral surface N. Clearly,

$$\widehat{\Phi} X(p) = X(p)$$

and choosing a transversal section $\Sigma \subset N$ in the integral surface N, we find with respect to the splitting

$$T_p N_c = \operatorname{span}\{X(p)\} \oplus T_p \Sigma$$

the representation of $\widehat{\Phi}$ as the $(n-r) \times (n-r)$-matrix

$$\widehat{\Phi} = \left(\begin{array}{c|c} 1 & -l \\ \hline 0 & \\ \vdots & d\psi(p) \\ 0 & \end{array}\right).$$

From this matrix we read off the following theorem.

Theorem VIII.15. *We consider a non-constant periodic solution $x(t)$ of the vector field X on the manifold M. If X possesses r independent integrals near $x(t)$, then $r+1$ Floquet multipliers of $x(t)$ are equal to 1. The other Floquet multipliers are the eigenvalues of the derivative $d\psi(x(0))$ of the section map $\psi\colon \Sigma \to \Sigma$ of a transversal section $\Sigma \subset N_c$ of $x(t)$, where N_c is the integral surface on which $x(t)$ lies. The characteristic polynomials are related by* $\det(\lambda - \Phi) = (\lambda - 1)^{r+1} \det(\lambda - d\psi(x(0))$.

Analogous to the proof of Theorem VIII.14, we can introduce new coordinates $x = (x', x'')$ in which the integrals $G(x', x'') = x'$ become coordinate functions and $\{x \mid x_n = 0\} \subset \mathbb{R}^n$ defines a local transversal section. Solving the fixed point equation $x = \psi(x)$ near the reference solution by means of the implicit function theorem, we obtain the following result.

Theorem VIII.16. *We assume that the vector field X on the manifold M possesses r independent integrals $G_1, , G_r$ near a given periodic solution $x(t, c^*)$ of period $T^* > 0$ which lies on the integral surface $N_{c^*} = \{x \in M \mid G(x) = c^*\}$ for the map $G = (G_1, , G_r)$ and for the constants $c^* \in \mathbb{R}^r$.*

*If $x(t)$ has **exactly** $r+1$ Floquet multipliers equal to 1, then there exists a smooth r-parameter family $x(t, c)$ of periodic solutions for c near c^* having periods $T(c)$ near $T^* = T(c^*)$. They are parametrized by the integral values $c = G(x(t, c)) \in \mathbb{R}^r$. In view of these properties, the family is unique. The periods $T(c)$ depend smoothly on the parameter c. In particular, $T(c) \to T^*$ as $c \to c^*$.*

Geometrically the family forms an embedded (generalized) cylinder $S^1 \times B^r$ where $B^r \subset \mathbb{R}^r$ is an open ball.

As a side remark we observe that an additional integral of a Hamiltonian vector field X_H frequently gives rise to two Floquet multipliers equal to 1 and not just one, so that the assumptions of Theorem VIII.16 are not met. In this case we cannot apply the continuation method used above in order to find periodic orbits.

Proposition VIII.17. *Suppose that the Hamiltonian vector field X_{H_1} on the symplectic manifold (M, ω) possesses the r integrals H_α for $1 \le \alpha \le r$, that is,*

$$\{H_1, H_\alpha\} = 0,$$

and assume that the 1-forms dH_α are linearly independent. If, at some point p contained in the periodic solution $x(t)$ of the Hamiltonian vector field X_{H_1}, we have

$$\{H_\alpha, H_\beta\}(p) = 0 \text{ for } 1 \le \alpha \le r \text{ and } 1 \le \beta \le s,$$

for some $s \le r$, then the periodic orbit $x(t)$ has $r+s$ Floquet multipliers equal to 1.

Proof. Introducing locally near the point p the integrals as coordinates, in the notation of Theorem VIII.17 we have

$$\Phi = \begin{pmatrix} \mathbb{1}_r & 0 \\ * & \widehat{\Phi} \end{pmatrix},$$

where $\widehat{\Phi} = \Phi|_{T_pN}$, the manifold N being the integral surface of the periodic solution $x(t)$ whose tangent space at the point p is equal to

$$T_pN = \{\xi \in T_pN \mid \omega(X_{H_\alpha}, \xi) = 0, \text{ for } \alpha = 1, 2, \dots, r\}.$$

It suffices to prove that the matrix $\widehat{\Phi}$ has s eigenvalues equal to 1 belonging to the eigenvectors X_{H_α} for $1 \le \alpha \le s$. Since

$$[X_{H_1}, X_{H_\alpha}] = X_{\{H_1,H_\alpha\}} = 0,$$

the flow φ^t of X_{H_1} and the flow ψ^s of $X_{H_\alpha}(p)$ do commute, so that

$$\varphi^t \circ \psi^s = \psi^s \circ \varphi^t.$$

Differentiating in s at $s = 0$ gives

$$d\varphi^t(p)\, X_{H_\alpha}(p) = X_{H_\alpha}(\varphi^t(p)).$$

Hence, if $t = T$ is the period of the periodic solution $x(t)$,

$$\Phi(X_{H_\alpha}(p)) = X_{H_\alpha}(p)$$

for every $\alpha = 1, 2, \dots, r$. It remains to prove that

$$X_{H_\beta}(p) \in T_pN = \bigcap_{\alpha=1}^{r} \ker dH_\alpha(p)$$

for all $1 \le \beta \le s$. But this is precisely our second assumption, because at the point p,

$$dH_\alpha(X_{H_\beta}) = \omega(X_{H_\beta}, X_{H_\alpha}) = \{H_\beta, H_\alpha\} = 0$$

for all $1 \le \alpha \le r$ and $1 \le \beta \le s$. The proof of Proposition VIII.17 is complete. □

From Proposition VIII.17 we conclude in particular that all the $2n$ Floquet multipliers of a periodic solution of an integrable Hamiltonian system are equal to 1. This situation is, of course, not stable under a perturbation.

Literature. Chapter VIII is, like Chapter VII, also partly based on the book [52] by H. Hofer and E. Zehnder and makes use of the notes on dynamical systems [74] by J. Moser and E. Zehnder. For a recent account of the problem of periodic orbits of Hamiltonian systems having prescribed energies, we refer to [41] and [42] by V. Ginzburg.

Bibliography

The numbers at the end of each item refer to the pages on which the respective work is cited.

[1] A. Abbondandolo, *Morse theory for Hamiltonian systems*. Chapman and Hall/CRC, Boca Raton, Fl., 2001. 181

[2] R. Abraham and J. Marsden, *Foundations of mechanics*. 2nd edition. The Benjamin Cummings Publishing Company, Inc., Reading, Mass., 1978. 238

[3] S. Alpern and V. S. Prasad, *Typical dynamics of volume preserving homeomorphisms*. Cambridge Tracts in Math. 139, Cambridge University Press, Cambridge, 2000.

[4] H. Amann, *Lectures on some fixed point theorems*. IMPA monografias de matemátika. 299

[5] H. Amann, *Ordinary differential equations*. De Gruyter Stud. Math. 13, Walter de Gruter, Berlin, 1990. 80, 127, 130, 148, 299

[6] K. G. Anderson, Poincaré's discovery of homoclinic points. *Arch. History Exact Sci.* 48 (2) (1994), 113–147. 80

[7] D. V. Anosov, Geodesic flows on closed Riemann manifolds with negative curvature. *Trudy Mat. Inst. Steklov* 90, 1967; English translation in *Proc. Steklov Inst. Math.* 90, Amer. Math. Soc., Providence, RI, 1969. 126

[8] V. I. Arnold and A. Avez, *Ergodic problems of classical mechanics*. Adv. Book Classics, Addison-Wesley, Redwood City, Cal., 1989. 237, 238

[9] A. Banyaga and B. Hurtubise, *Lectures on Morse homology*. Kluwer Academic Publishers, Dordrecht, 2004. 180

[10] M. Berger and B. Gostiaux, *Differential geometry: manifolds, curves, and surfaces*. Grad. Texts in Math. 115, Springer-Verlag, New York, 1988.

[11] S. V. Bolotin, Libration motions of natural dynamical systems. *Moscow Univ. Math. Bull.* 33 (1978), 49–53. 327

[12] R. Bott, Lectures on Morse theory, old and new. *Bull. Amer. Math. Soc.* (*N.S.*) 7 (1982), 331–358. 181

[13] R. F. Brown, *A topological introduction to nonlinear analysis*. Birkhäuser, Boston, Mass., 1993. 299

[14] K. C. Chang, *Infinite dimensional Morse theory and multiple solution problems*. Progr. Nonlinear Differential Equations Appl. 6, Birkhäuser, Boston, Mass., 1993. 181

[15] M. Chaperon, Stable manifolds and the Perron-Irwin method. In *Dynamical systems*, ed. by A. Fathi and J.-C. Yoccoz, Cambridge University Press, 2006, Cambridge, 89–124. 80

[16] M. Chaperon, Invariant manifold theory via generating maps. *C. R. Acad. Sci. Paris, Ser. I* 346 (2008), 1175–1180. 80

[17] A. Chenciner, Le dynamique au voisinage d'un point fixe elliptique conservatif: de Poincaré et Birkhoff à Aubry et Mather. Séminaire Bourbaki, *Astérisque* 121/122 (1984), 147–170. 238

[18] K. Cieliebak, H. Hofer, J. Latschev and F. Schlenk, Quantitative symplectic geometry. In *Dynamics, ergodic theory, and geometry*, Math. Sci. Res. Inst. Publ. 54, Cambridge University Press, Cambridge, 2007, 1–44. 306

[19] C. Conley, *Isolated invariant sets and the Morse index.* CBMS Reg. Conf. Ser. in Math. 38, Amer. Math. Soc., Providence, RI, 1978. 181

[20] C. Conley and E. Zehnder, Morse-type index theory for flows and periodic solutions for Hamiltonian equations. *Comm. Pure Appl. Math.* 37 (1984), 207–253. 181

[21] L. Conlon, *Differentiable manifolds.* 2nd edition, Birkhäuser Adv. Texts, Birkhäuser, Boston, Mass., 2001. 164, 180, 206, 229

[22] O. Cornea, G. Lupton, J. Oprea, and D. Tanré, *Lusternik–Schnirelman category.* Math. Surveys Monogr. 103, Amer. Math. Soc., Providence, RI, 2003. 160, 161

[23] I. P. Cornfeld, S. V. Fomin, and Ya. G. Sinai, *Ergodic theory.* Grundlehren Math. Wiss. 245, Springer-Verlag, Berlin, 1982. 45

[24] R. H. Cushman and L. M. Bates, *Global aspects of classical integrable systems.* Birkhäuser, Basel, 1997. 194

[25] B. Dacorogna and J. Moser, On a partial differential equation involving the Jacobian determinant. *Ann. Inst. H. Poincaré Anal. Non Linéaire* 7 (1) (1990) 1–26. 240

[26] K. Deimling, *Nonlinear functional analysis.* Springer-Verlag, Berlin, 1985. 159, 160, 161, 299

[27] R. L. Devaney, *An introduction to chaotic dynamical systems.* 2nd edition, Stud. Nonlinearity, Westview Press, Boulder, CO, 2003.

[28] A. Dold, *Lectures on algebraic topology.* 2nd edition, Grundlehren Math. Wiss. 200, Springer-Verlag, Berlin, 1980.

[29] J. Duistermaat, Stable manifolds. Preprint No. 40, Department of Mathematics, University of Utrecht, 1976. 80

[30] J. Duistermaat, *Fourier integral operators.* Progr. Math. 130, Birkhäuser Verlag, Boston, Mass., 1996. 238

[31] H. Dym and H. P. McKean, *Fourier series and integrals.* 2nd print., Academic Press, New York, 1974. 35

[32] R. W. Easton, *Geometric methods for discrete dynamical systems.* Oxford Engrg. Sci. Ser. 50, Oxford University Press, New York, 1998.

[33] I. Ekeland and H. Hofer, Symplectic topology and Hamiltonian dynamics. *Math. Z.* 200 (1989), 355–378. 252

[34] I. Ekeland and H. Hofer, Symplectic topology and Hamiltonian dynamics II. *Math. Z.* 203 (1990), 553–567. 252

[35] B. Fiedler (ed.), *Handbook of dynamical systems.* Vol. 2, North-Holland, Amsterdam, 2002. 45

[36] J. Franks and M. Misiurewicz, Topological methods in dynamics. In *Handbook of dynamical systems*, Vol. 1A, ed. B. Hasselblatt and A. Katok, North-Holland, Amsterdam, 2002, 547–599. 181

[37] H. Fürstenberg, *Recurrence in ergodic theory and combinatorial number theory.* M.B. Porter lectures. Princeton University Press, Princeton, NJ, 1981. 45

[38] A. M. Garsia, *Topics in almost everywhere convergence.* Lectures in Adv. Math. 4, Markham Publishing, Chicago, Ill., 1970.

[39] C. Genecand, Transversal homoclinic orbits near elliptic fixed points of area-preserving diffeomorphisms of the plane. In *Dynamics reported*, Dynam. Report. Expositions Dynam. Systems (N.S.) 2, Springer-Verlag, Berlin, 1993, 1–30. 238

[40] V. L. Ginzburg, Hamiltonian dynamical systems without periodic orbits. In *Northern California symplectic geometry seminar*, ed. by Ya. Eliashberg et al., Amer. Math. Soc. Transl. Ser. 196, Amer. Math. Soc., Providence, RI, 1999, 35–48. 317

[41] V. L. Ginzburg, The Hamiltonian Seifert conjecture: examples and open problems. In *Proceedings of the third European Congress of Mathematics* (Barcelona 2000), Vol. II, Progr. Math. 202, Birkhäuser, Basel, 2001, 547–555. 338

[42] V. L. Ginzburg, The Weinstein conjecture and theorems of nearby and almost existence. In *The breath of symplectic and Poisson geometry*, Festschrift in Honor of Alan Weinstein, ed. by J. E. Marsden and T. S. Ratiu. Birkhäuser, Boston, Mass., 2005, 139–172. 338

[43] V. L. Ginzburg and B. Z. Gürel, A C^2-smooth counterexample to the Hamiltonian Seifert conjecture in $\mathbb{R}^4$. *Ann. of Math.* (2) 158 (2003), 953–976. 317

[44] V. L. Ginzburg and B. Z. Gürel, Relative Hofer-Zehnder capacity and periodic orbits in twisted cotangent bundles. *Duke Math. J.* 123 (1) (2004), 1–47.

[45] C. Godbillon, *Géométrie différentielle et mécanique analytique.* Collection Méthodes, Hermann, Paris, 1969. 238

[46] L. Grafakos, *Classical and modern Fourier analysis.* Prentice Hall, Upper Saddle River, NJ, 2004. 35

[47] J. Hadamard, Sur l'iteration et les solutions asymptotiques des équations differentielles. *Bull. Soc. Math. France* 29 (1901), 224–228. 80

[48] B. Hasselblatt, Hyperbolic dynamical systems. Chapter 5 in *Handbook of dynamical systems*, Vol. 1 A, ed. by B. Hasselblatt and A. Katok, North-Holland, Amsterdam, 2002. 126

[49] B. Hasselblatt and A. Katok, *Introduction to the modern theory of dynamical systems.* Encyclopedia Math. Appl. 54, Cambridge University Press, Cambridge, 1995. 45, 120, 123, 126

[50] B. Hasselblatt and A. Katok (eds.), *Handbook of dynamical systems*. Vol. 1A, North-Holland, Amsterdam, 2002. 45

[51] H. Hofer and E. Zehnder, A new capacity for symplectic manifolds. In *Analysis, et cetera*, ed. by P. Rabinowitz and E. Zehnder, Academic Press, Boston, Mass., 1990, 405–427 260

[52] H. Hofer and E. Zehnder, *Symplectic invariants and Hamiltonian dynamics.* Birkhäuser Adv. Texts, Birkhäuser, Basel, 1994. vii, 244, 255, 262, 297, 299, 306, 314, 317, 338

[53] M. C. Irwin, *Smooth dynamical systems.* Academic Press, 1980. 80

[54] M. Y. Jiang, Hofer–Zehnder symplectic capacity for two dimensional manifolds. *Proc. Roy. Soc. Edinburgh Sect. A* 123 (1993), 945–950. 262

[55] J. Jost, *Riemannian geometry and geometric analysis.* 3rd edition, Universitext, Springer-Verlag, Berlin, 2002. 177, 180

[56] R. Jost, Poisson brackets. An unpedagogical lecture. *Rev. Mod. Phys.* 36 (1964), 572–579. 214

[57] R. Jost, Winkel- und Wirkungsvariable für allgemeine mechanische Systeme. *Helvetica Physica Acta* 41 (1968), 965–968. 227

[58] A. Katok, Dynamical systems with hyperbolic structure. Three papers on dynamical systems. *Amer. Math. Soc. Transl.* 116 (1981), 43–95. 126

[59] U. Kirchgraber and D. Stoffer, Chaotic behaviour in simple dynamical systems. *SIAM Review* 32 (1990), 424-452. 115

[60] U. Krengel, *Ergodic theorems.* De Gruyter Stud. Math. 6, Walter de Gruyter, Berlin, 1985. 45

[61] R. de la Llave, A tutorial on KAM theory. In *Smooth ergodic theory and its applications*, ed. by A. Katok et al., Proc. Sympos. Pure Math. 69, Amer. Math. Soc., Providence, RI, 2001, 175–292. 238

[62] M. Levi and J. Moser, A Lagrangian proof of the invariant curve theorem for twist mappings. In *Smooth ergodic theory and its applications*, ed. by A. Katok et al., Proc. Sympos. Pure Math. 69, Amer. Math. Soc., Providence, RI, 2001, 733–746 237

[63] R. Mañé, *Ergodic theory and differentiable dynamics.* Ergeb. Math. Grenzgeb. 8, Springer-Verlag, Berlin, 1987. 45

[64] L. Macarini and F. Schlenk, A refinement of the Hofer–Zehnder theorem on the existence of closed characteristics near a hypersurface. *Bull. London Math. Soc.* 37 (2) (2005), 297–300. 314

[65] D. McDuff and D. Salamon, *Introduction to symplectic topology.* Second edition, Oxford Math. Monogr., The Clarendon Press, Oxford University Press, New York, NY, 1998. 238, 306

[66] K. R. Meyer and G. R. Hall, *Introduction to Hamiltonian dynamical systems and the N-body problem.* Applied Math. Sci. 90, Springer-Verlag, New York, NY, 1992. 238

[67] J. Milnor, *Topology from the differential viewpoint.* The University Press of Virginia, Charlottesville, Va., 1965. 299

[68] J. Milnor, *Morse theory.* 5th print., Ann. of Math. Stud. 51, Princeton University Press, Princeton, NJ, 1973. 180

[69] K. Mischaikow and M. Mrozek, The Conley index. In *Handbook of dynamical systems*, Vol. 2, ed. by B. Fiedler, North-Holland, Amsterdam, 2002, 393–460. 181

[70] J. Moser, On invariant curves of area-preserving mappings of an annulus. *Nachr. Akad. Wiss. Göttingen Math.-Phys. Kl. II* 1962 (19562), 1–20. 237

[71] J. Moser, On a theorem of Anosov. *J. Differential Equations* 5 (1969), 411–440. 126

[72] J. Moser, Dynamical systems – past and present. In *Proc. Internat. Congress of Mathematicians* (Berlin, 1998), Vol. I, *Doc. Math.*, 1998, 381-402. 235

[73] J. Moser, *Stable and random motions in dynamical systems: with special emphasis on celestial mechanics.* Ann. of Math. Stud. 77, Princeton University Press, Princeton, NJ, 1973. 238

[74] J. Moser and E. Zehnder, *Notes on dynamical systems.* Courant Lecture Notes in Mathematics, Courant Institute, New York, NY, 2005. vii, 238, 338

[75] J. C. Oxtoby and S. M. Ulam, Measure preserving homeomorphisms and metrical transitivity. *Ann. of Math.* 42 (4) (1941) 874–920.

[76] R. S. Palais, Morse theory on Hilbert manifolds. *Topology* 2 (1963), 299–340. 181

[77] R. S. Palais, Lusternik–Schnirelman theory on Banach manifolds. *Topology* 5 (1966), 115–132. 161, 181

[78] R. S. Palais, Critical point theory and the minimax principle. In *Global analysis*, Proc. Sympos. Pure Math. 15, Amer. Math. Soc., Providence, RI, 1970, 185–212. 181

[79] R. S. Palais and Chun-lian Terng, *Critical point theory and submanifold geometry.* Lecture Notes in Math. 1353, Springer-Verlag, Berlin, 1988. 180

[80] K. J. Palmer, Exponential dichotomies, the shadowing lemma and transversal homoclinic points. In *Dynamics reported*, Vol. 1, Dynam. Report. Ser. Dynam. Systems Appl. 1, B. G. Teubner and John Wiley and Sons, Stuttgart/Chichester, 1988, 265–306. 126

[81] J. Palis, Jr., and W. de Melo, *Geometric theory of dynamical systems: an introduction.* Springer-Verlag, New York, NY, 1982.

[82] O. Perron, Über die Stabilität und asymptotisches Verhalten der Integrale von Differentialgleichungssystemen. *Math. Z.* 29 (1928), 129–160. 80

[83] Ya. B. Pesin, General theory of smooth hyperbolic dynamical systems. Chapter 7 in *Dynamical systems* II, Encyclopedia Math. Sci. 2, Springer-Verlag, Berlin, 1989. 126

[84] Ya. B. Pesin, *Lectures on partial hyperbolicity and stable ergodicity.* Zürich Lect. Adv. Math., European Math. Soc. Publishing House, Zürich, 2004. 126

[85] S. Y. Pilyugin, *Introduction to structurally stable systems of differential equations.* Birkhäuser, Basel, 1992. 126

[86] J. Pöschel, A lecture on the classical KAM theorem. In *Smooth ergodic theory and its applications*, Proc. Sympos. Pure Math. 69, Amer. Math. Soc., Providence, RI, 2001, 707–732. 238

[87] H. Poincaré, Sur le problème des trois corps et les équations de la dynamique. Mémoire couronné du prix de S. M. le roi Oscar III de Suède, *Acta Math.* 13 (1890), 1–270. 80

[88] M. Pollicott, *Lectures on ergodic theory and Pesin theory on compact manifolds.* London Math. Soc. Lecture Note Ser. 180, Cambridge University Press, Cambridge, 1993. 45

[89] P. Rabinowitz, Periodic solutions of Hamiltonian systems. *Comm. Pure Appl. Math.* 31 (1978), 157–184. 270

[90] P. Rabinowitz, Periodic solutions of Hamiltonian systems on a prescribed energy surface. *J. Differential Equations* 33 (1979), 336–352. 270, 320

[91] C. Robinson, *Dynamical systems: stability, symbolic dynamics, and chaos*. 2nd edition, Stud. Adv. Math., CRC Press, Boca Raton, 1999. 45, 114, 123

[92] H. L. Royden, *Real analysis.* 2nd edition, Macmillan Publishing, New York, NY, 1968.

[93] D. Ruelle, *Elements of differentiable dynamics and bifurcation theory.* Academic Press, Boston, Mass., 1989.

[94] K. P. Rybakowski, *The homotopy index and partial differential equations*. Universitext, Springer-Verlag, New York, NY, 1987. 181

[95] D. G. Saari and J. Xia, Off to infinity in finite time. *Notices Amer. Math. Soc.* 42 (5) (1995), 538–546. 3

[96] D. Salamon, Connected simple systems and the Conley index of isolated invariant sets. *Trans. Amer. Math. Soc.* 291 (1985), 1–41. 181

[97] D. Salamon, The Kolmogorov–Arnold–Moser theorem. *Math. Phys. Electron. J.* 10 (2004), Paper 3, 37 pp. 238

[98] D. Salamon and E. Zehnder, KAM theory in configuration space. *Comment. Math. Helv.* 64 (1989), 84–132. 238

[99] F. Schlenk, *Embedding problems in symplectic geometry.* De Gruyter Exp. Math. 40, Walter de Gruyter, Berlin, 2005. 306

[100] M. Schwarz, *Morse homology.* Progr. Math. 111, Birkhäuser Verlag, Basel, 1993. 181

[101] J. T. Schwartz, *Nonlinear functional analysis.* Notes on mathematics and its applications, Gordon and Breach, New York, NY, 1969. 160, 161, 180

[102] L. P. Shil'nikov, Homoclinic trajectories: From Poincaré to the present. In *Mathematical events of the twentieth century*, ed. by A. A. Bolibruch, Yu. S. Osipov and Ya. G. Sinai, Springer-Verlag, Berlin, 2006, 347–370. 80

[103] M. Shub, *Global stability of dynamical systems.* Springer-Verlag, New York, 1987. 123

[104] K. F. Siburg, Symplectic capacities in two dimensions. *Manuscripta Math.* 78 (1993), 149–163. 262

[105] C. L. Siegel, On the integrals of canonical systems. *Ann. of Math.* 42 (3) (1941), 806–822.

[106] C. L. Siegel, Über die Existenz einer Normalform analytischer Hamiltonscher Differentialgleichungen in der Nähe einer Gleichgewichtslösung. *Math. Ann.* 128 (1954), 144–170.

[107] C. L. Siegel and J. Moser, *Lectures on celestial mechanics.* Classics Math., Springer-Verlag, Berlin, 1995. 238

[108] Ya. G. Sinai, *Introduction to ergodic theory.* Math. Notes 18, Princeton University Press, Princeton, NJ, 1977. 45

[109] Ya. G. Sinai, *Topics in ergodic theory.* Princeton Math. Ser. 44, Princeton University Press, Princeton, NJ, 1994 45

[110] S. Smale, Differentiable dynamical systems. *Bull. Amer. Math. Soc.* 73 (1967), 747–817.

[111] K. R. Stromberg, *An introduction to classical real analysis.* Wadsworth and Brooks Cole, Pacific Grove, Cal., 1981. 35

[112] M. Struwe, Existence of periodic solutions of Hamiltonian systems on almost every energy surface. *Bol. Soc. Bras. Mat.* 20 (1990), 49–58. 317

[113] W. Szlenk, *An introduction to the theory of smooth dynamical systems.* PWN - Polish Scientific Publishers/Wiley, Warszawa/Chichester, 1984. 23, 45, 126

[114] C. Viterbo, A proof of Weinstein's conjecture in $\mathbb{R}^{2n}$. *Ann. Inst. H. Poincaré Anal. Non Linéaire* 4 (4) (1987), 337–356. 320

[115] P. Walters, *An introduction to ergodic theory.* Grad. Texts in Math. 79, Springer-Verlag, New York, NY, 2000. 45

[116] A. Weinstein, Periodic orbits for convex Hamiltonian systems. *Ann. of Math.* 108 (1978), 507–518. 320

[117] A. Weinstein, On the hypothesis of Rabinowitz's periodic orbit theorems. *J. Differential Equations* 33 (1979), 353–358. 318

[118] J. Xia. Homoclinic points for area preserving surface diffeomorphisms. Preprint 2006, arXiv:math/0606291. 80

[119] J. C. Yoccoz, An Introduction to hyperbolic dynamics in real and complex dynamical systems. *Nato ASI Series Ser. C* 464 (1995), 265–291. 80, 126

[120] E. Zehnder, Homoclinic points near elliptic fixed points. *Comm. Pure Appl. Math.* 26 (1973), 131–182. 80, 238

List of Symbols

Index